Stochastic Evolution Systems

Mathematics and Its Applications (*Soviet Series*)

Volume 35

Stochastic Evolution Systems

Linear Theory and Applications to Non-linear Filtering

by

B. L. Rozovskii

Formerly: *Institute of Advanced Studies for Chemistry Managers and Engineers, Moscow, U.S.S.R.*

Now at: *Department of Mathematics, The University of North Carolina at Charlotte, Charlotte, U.S.A.*

KLUWER ACADEMIC PUBLISHERS

DORDRECHT / BOSTON / LONDON

Library of Congress Cataloging in Publication Data

Rozovskiĭ, B. L. (Boris L'vovich)
[Évoli͡ut͡sionnye stokhasticheskie sistemy. English]
Stochastic evolution systems : linear theory and applications to non-linear filtering / by B.L. Rozovskiĭ.
p. cm. -- (Mathematics and its applications (Soviet series))
Translation of: Évoli͡ut͡sionnye stokhasticheskie sistemy.
Includes index.
ISBN 0-7923-0037-8
1. Stochastic partial differential equations. I. Title.
II. Series: Mathematics and its applications (Kluwer Academic Publishers). Soviet series.
QA274.25.R6913 1990
519.2--dc19 88-39995

ISBN 0-7923-0037-8

Published by Kluwer Academic Publishers,
P.O. Box 17, 3300 AA Dordrecht, The Netherlands.

Kluwer Academic Publishers incorporates
the publishing programmes of
D. Reidel, Martinus Nijhoff, Dr W. Junk and MTP Press.

Sold and distributed in the U.S.A. and Canada
by Kluwer Academic Publishers,
101 Philip Drive, Norwell, MA 02061, U.S.A.

In all other countries, sold and distributed
by Kluwer Academic Publishers Group,
P.O. Box 322, 3300 AH Dordrecht, The Netherlands.

Printed on acid-free paper

Translated from the Russian by A. Yarkho

This is the translation of the original work
ЭВОЛЮЦИОННЫЕ СТОХАСТИЧЕСКИЕ СИСТЕМЫ
Линейная теория и приложения к статистике
случайных процессов
Published bu Nauka Publishers, Moscow, © 1983.

TO MY MOTHER

SERIES EDITOR'S PREFACE

'Et moi, ..., si j'avait su comment en revenir, je n'y serais point allé.'

Jules Verne

The series is divergent; therefore we may be able to do something with it.

O. Heaviside

One service mathematics has rendered the human race. It has put common sense back where it belongs, on the topmost shelf next to the dusty canister labelled 'discarded non-sense'.

Eric T. Bell

Mathematics is a tool for thought. A highly necessary tool in a world where both feedback and non-linearities abound. Similarly, all kinds of parts of mathematics serve as tools for other parts and for other sciences.

Applying a simple rewriting rule to the quote on the right above one finds such statements as: 'One service topology has rendered mathematical physics ...'; 'One service logic has rendered computer science ...'; 'One service category theory has rendered mathematics ...'. All arguably true. And all statements obtainable this way form part of the raison d'être of this series.

This series, *Mathematics and Its Applications*, started in 1977. Now that over one hundred volumes have appeared it seems opportune to reexamine its scope. At the time I wrote

> "Growing specialization and diversification have brought a host of monographs and textbooks on increasingly specialized topics. However, the 'tree' of knowledge of mathematics and related fields does not grow only by putting forth new branches. It also happens, quite often in fact, that branches which were thought to be completely disparate are suddenly seen to be related. Further, the kind and level of sophistication of mathematics applied in various sciences has changed drastically in recent years: measure theory is used (non-trivially) in regional and theoretical economics; algebraic geometry interacts with physics; the Minkowsky lemma, coding theory and the structure of water meet one another in packing and covering theory; quantum fields, crystal defects and mathematical programming profit from homotopy theory; Lie algebras are relevant to filtering; and prediction and electrical engineering can use Stein spaces. And in addition to this there are such new emerging subdisciplines as 'experimental mathematics', 'CFD', 'completely integrable systems', 'chaos, synergetics and large-scale order', which are almost impossible to fit into the existing classification schemes. They draw upon widely different sections of mathematics."

By and large, all this still applies today. It is still true that at first sight mathematics seems rather fragmented and that to find, see, and exploit the deeper underlying interrelations more effort is needed and so are books that can help mathematicians and scientists do so. Accordingly MIA will continue to try to make such books available.

If anything, the description I gave in 1977 is now an understatement. To the examples of interaction areas one should add string theory where Riemann surfaces, algebraic geometry, modular functions, knots, quantum field theory, Kac-Moody algebras, monstrous moonshine (and more) all come together. And to the examples of things which can be usefully applied let me add the topic 'finite geometry'; a combination of words which sounds like it might not even exist, let alone be applicable. And yet it is being applied: to statistics via designs, to radar/sonar detection arrays (via finite projective planes), and to bus connections of VLSI chips (via difference sets). There seems to be no part of (so-called pure) mathematics that is not in immediate danger of being applied. And, accordingly, the applied mathematician needs to be aware of much more. Besides analysis and numerics, the traditional workhorses, he may need all kinds of combinatorics, algebra, probability, and so on.

In addition, the applied scientist needs to cope increasingly with the nonlinear world and the extra mathematical sophistication that this requires. For that is where the rewards are. Linear models are honest and a bit sad and depressing: proportional efforts and results. It is in the non-linear world that infinitesimal inputs may result in macroscopic outputs (or vice versa). To appreciate what I am hinting at: if electronics were linear we would have no fun with transistors and computers; we would have no TV; in fact you would not be reading these lines.

There is also no safety in ignoring such outlandish things as nonstandard analysis, superspace and anticommuting integration, p-adic and ultrametric space. All three have applications in both electrical engineering and physics. Once, complex numbers were equally outlandish, but they frequently proved the shortest path between 'real' results. Similarly, the first two topics named have already provided a number of 'wormhole' paths. There is no telling where all this is leading - fortunately.

Thus the original scope of the series, which for various (sound) reasons now comprises five sub-series: white (Japan), yellow (China), red (USSR), blue (Eastern Europe), and green (everything else), still applies. It has been enlarged a bit to include books treating of the tools from one subdiscipline which are used in others. Thus the series still aims at books dealing with:

- a central concept which plays an important rôle in several different mathematical and/or scientific specialization areas;
- new applications of the results and ideas from one area of scientific endeavour into another;
- influences which the results, problems and concepts of one field of enquiry have, and have had, on the development of another.

As the author writes, up to fairly recently an Ito stochastic differential equation was almost invariably an ordinary stochastic differential equation. Since then (the mid seventies) the situation has changed drastically and fields like signal processing, magnetohydrodynamics, relativistic quantum mechanics, population dynamics, and genetics, oceanology, and others gave rise to numerous stochastic partial differential equations, mainly linear evolution equations $du = A(t,u)dt + B(t,u)dw$, where w is Wiener noise and where the drift, $A(t,u)$, and diffusion, $B(t,u)$, are operators, usually unbounded. In addition such diffusion equations have turned out to be important in the theory of (nonlinear) nonstochastic second-order partial differential equations of elliptic or parabolic type; see, for instance, Krylov's book on this topic (also published in this series).

Already in 1983 the basic decision was made to translate the book. Also already at that time it was clear that an additional chapter should be included. Indeed the author absolutely refused to allow a straightforward translation. Writing the additional material took longer than anticipated partly because the author moved in this period from the USSR to the USA and mostly because the author wanted to do a thorough job worthy of his reputation and talents. The result is well worth waiting for and the delay has had the advantage that the developments of the so-called Malliavin calculus (stochastic variational calculus) and its applications could be thoroughly included.

The shortest path between two truths in the real domain passes through the complex domain.

J. Hadamard

La physique ne nous donne pas seulement l'occasion de résoudre des problèmes ... elle nous fait pressentir la solution.

H. Poincaré

Never lend books, for no one ever returns them; the only books I have in my library are books that other folk have lent me.

Anatole France

The function of an expert is not to be more right than other people, but to be wrong for more sophisticated reasons.

David Butler

Bussum, 7 October 1990

Michiel Hazewinkel

PREFACE

TO THE ENGLISH EDITION

The subject of this book is linear stochastic partial differential equations and their applications to the theory of diffusion processes and non-linear filtering.

Until recently, the term "stochastic differential equation" did not need any specifications - in 99 cases out of 100 it was applied to ordinary stochastic differential equations. Their theory started to develop at the beginning of the 1940s, based on Ito's stochastic calculus [50], [51], and now forms one of the most beautiful and fruitful branches of the theory of stochastic processes, [36], [52], [49], [63], [90], [132].

In the middle of the 1970s, however, the situation changed: in various branches of knowledge (primarily, in physics, biology, and control theory) a vast number of models were found that could be described by stochastic evolution partial differential equations. Such models were used, for example to describe a free (boson) field in relativistic quantum mechanics, a hydromagnetic dynamo-process in cosmology, diffraction in random-heterogeneous media in statistical physics, the dynamics of populations for models with a geographical structure in population genetics, etc.

The emergence of this new type of equation was simultaneously stimulated by the inner needs of the theory of differential equations. Such equations were effectively used to study parabolic and elliptic second-order equations in infinite-dimensional spaces.

An especially powerful impetus to the development of the theory of evolution stochastic partial differential equations was given by the problem of non-linear filtering of diffusion processes.

The filtering problem (estimation of the "signal" by observing it when it is mixed with a "noise") is one of classical problem in the statistics of stochastic processes. It also belongs to a rare type of purely engineering problems that have a precise mathematical formulation and allows for a mathematically rigorous solution.

The first remarkable results in connection with filtering of stationary processes were obtained by A.N. Kolmogorov [60] and N. Wiener [146]. After the paper by R. Kalman and R. Bucy [54] was published, the 1960s and 1970s witnessed a rapid development of filtering theory for systems whose dynamics could be described by Ito's stochastic differential equations. The results were first summed up by R. Sh. Liptser and A.N. Shiryayev [90] and G. Kallianpur [53].

One of the key results of the modern non-linear filtering theory states that the solution of the filtering problem for the processes described by Ito's ordinary stochastic equations is equivalent to the solution of an equation commonly called the filtering equation. The filtering equation is a typical example of an evolution stochastic partial differential equation. An equation of this type can be regarded as an "ordinary" Ito equation

$$du(t) = A(t, u(t))dt + B(t, u(t))\, dW(t)$$

for the process u(t) taking values in a function space X. The coefficients A, B (of "drift" and "diffusion") in this equation are operators (unbounded, as a rule), and $\dot{W}(t)$ is a "white noise" taking values in a function space. Such an equation may be regarded as a system (an infinite one, if the space X is infinite) of one-dimensional Ito's equations. Below we shall call equations (systems of equations) of this type stochastic evolution systems.

The theory of stochastic evolution systems is quite a young branch of science but it has nevertheless generated many interesting and important results, much more than it would be reasonable to include in one book. Books [5], [8], [144], [81], [19], [17], [105], [141], etc. contain sections devoted to the theory.

In the present monograph, the author has set himself the following program:

(1) to cover the general theory of linear stochastic evolution systems (LSES) with unbounded drift and diffusion operators:

(2) to contstruct the theory of Ito's second-order parabolic equations;

(3) to investigate, on the basis of the latter, the filtering problem and related issues (interpolation, extrapolation) for processes whose trajectories can be described by Ito's ordinary equations.

The first point of the program is dealt with in Chapters 2 and 3, the second one constitutes the subject of Chapters 4 and 5, while Chapter 6 builds on the third point. Chapter 1 is devoted to examples and auxiliary results.

Since the time the present book was finished (the Russian edition of this book was published in 1983), very important discoveries have been made in the theory of stochastic differential equations: namely the development of the Malliavin's calculus and its elaborations, which provided the basis for the stochastic interpretation of Hörmander's results on the hypoellipticity of elliptic-parabolic second-order equations.

This made it necessary to include Chapter 7 in the English version of the book, devoted to hypoellipticity of second-order stochastic partial differential equations and, in particular, to the filtering equations. Necessary extensions of other parts of the book have been done as well.

Here our brief chapter-to-chapter summary of what is covered ends, but we would like to remark that each chapter has its own introduction describing its contents sufficiently thoroughly.

Throughout the book the author has tried to adhere to the universal language of functional analysis and has given preference to functional-analytical methods of proof, the rationale for this is that the book was written to be understood by researchers of different interests and educational background.

The necessary prerequisite for the reader is a familiarity with functional analysis and the theory of stochastic processes within the framework of standard university courses. No preliminary reading on partial differential equations is needed.

Though the book has a strictly hierarchical structure (each chapter to follow is based on the result of the preceding one), the exposition allows the reader interested only in some of the chapters to begin reading them directly and to use the preceding chapters only for references.

Each section of the book is divided into paragraphs that are enumerated but bear no title. Theorems, lemmas, propositions, definitions, notes, and warnings are ascribed the number of the paragraph they belong to (each paragraph contains no more then one theorem, lemma etc.). The formulas are ascribed indices independently within each section. The formula index includes no less than two numbers: the number of the section and that of the formula proper. When formulas are referred to in a subsequent chapter, a third number is added - that of the chapter. Thus (2.3.14) means Formula 14 from Section 3 of Chapter 2. Paragraphs (and therefore theorems, lemmas etc.) when referred to, are indicated in a similar way. When references are made within a section only the number of the paragraph is indicated. A section-paragraph reference is used within chapters but different sections. For reference to a paragraph from another chapter, the number of the chapter is added. E.g. Theorem 3.2.4 belongs to Paragraph 4, Section 2, and Chapter 3.

The author owes his heart felt graditude to N. V. Krylov for numerous important discussions and valuable suggestions on the subject of this book. His thanks are also due to Mrs. A. Yarkho the translator for the English edition and to R. F. Anderson for thoughtful editorial work on it.

The author is much indebted to all the participants of the seminar on the theory of martingales and control at Steklov Mathematical Institute of the USSR Academy of Sciences and of the seminar on stochastic differential equations with partial derivatives at Moscow State Unviersity who read and discussed various parts of the book.

Author's thanks are due to Benita Boyd for the patience and efficiency with which she did the word processing.

Moscow-
Charlotte, N.C.

Boris Rozovskii

CONTENTS

STANDARD NOTATION*

$\mathbb{E}$	expectation
$\mathbb{E}[\mathfrak{F}]$	conditional expectation with respect to σ-algebra $\mathfrak{F}$
:=	equal by definition
$\wedge$	min
$\vee$	max (see Warning 1.4.7)
$\mathbb{N}$	the set of positive integers
a.s.	almost sure
a.a.	almost all
$\mathbb{R}(f)$	the range of values of f
$\mathbb{D}(f)$	the domain of definition of f
$\mathfrak{L}(\mathbf{X},\mathbf{Y})$	the space of continuous linear operators from $\mathbf{X}$ to $\mathbf{Y}$
$1_{\{A\}}$	the indicator function of a set A
$\mathbb{I}$	an identity operator
l_d	the Lebesgue measure on $\mathbf{R}^d$ $(\ell := l_1)$
$\mathfrak{P}$	the σ-algebra of predictable sets
$\mathfrak{B}(\mathbf{X})$	the Borelian σ-algebra in $\mathbf{X}$

*See also notation on pages 8, 9, 128.

$\mathbb{W}_p^m$, $\mathbb{W}_p^m(r,\mathbb{R}^d)$	
$\mathbb{H}^m$, $\mathbb{H}^m(r,\mathbb{R}^d)$	Sobolev spaces (see §§ 3.4.1-3.4.7)
$\lvert\cdot\rvert_d$	the norm in $\mathbb{R}^d$
$(\cdot,\cdot)_m$	the scalar product in $\mathbb{H}^m$
$\mathfrak{L}_1(\mathbf{X},\mathbf{Y})$	the space of nuclear operators from $\mathbf{X}$ into $\mathbf{Y}$
$\mathfrak{L}_2(\mathbf{X},\mathbf{Y})$	the space of Hilbert-Shmidt operators from $\mathbf{X}$ into $\mathbf{Y}$
$[\cdot,\cdot]$	the canonical bilinear functional
$\mathbf{C}^{m,n}$	the space of real-valued functions which are m-times continuously differentiable in the first argument and n-times continuously differentiable in others
$\square$	the end of a statement or a proof

Chapter 1

EXAMPLES AND AUXILIARY RESULTS

1.0. Introduction.

Examples of linear stochastic evolution systems (LSES) arising in various fields of knowledge are presented in the first of the four sections of this chapter. The following three are reference sections. In these sections we collect a number of auxiliary results which are used systematically in the principal part of the book. In particular, §4 deals with the theory of stochastic differential equations.

1.1. Examples of Stochastic Evolution Systems.

1.0. In this section we give examples of LSES arising in the filtering of diffusion processes, physics, chemistry, and biology.

1.1. *Filtering equation.* Suppose that we observe the sum of "signal" ($x(t)$) and "noise" ($w(t)$). The problem of the estimation of the unobservable signal $x(t)$ or a function of $x(t)$ on the basis of the observed path of $y(s) = x(s) + w(s)$, for $s \geq t$, is usually referred to as a filtering problem.

This model arises in may technical problems. For example, it may be that the process $y(t)$ describes the position of a moving object computed on the basis of radar observations, $w(t)$ plays the role of the observational error, and the signal $x(t)$ represents the true coordinates of the object.

Suppose that the signal $x(t)$ is the solution of the ordinary differential equation $dx(t)/dt = b(x(t))$, $x(0) = x_0$, and $w(t)$ is a Wiener process, then the observed process $y(t)$ is a solution of the Ito equation

$$dy(t) = b(x(t))dt + dw(t), \; y(0) = x_0.$$

The variance of $w(t)$ is equal to t, whereas in applications the scattering of the observed process usually depends on its position. Thus, the process satisfying the equation

$$dy(t) = b(x(t))dt + \sigma_0(y))dw(t), \; y(0) = y_0$$

with variance equal to $\int_{[0,t]} \mathbf{E}\sigma_0^2 \,(y(s))ds$, is a more realistic model for the observed process. In various problems of a practical importance, the evolution equation of the signal process includes random perturbation. In these cases it can be described by the Ito equations

$$dx(t) = b(x(t))dt + \sigma(x(t))d\nu(t), \quad x(0) = x_0,$$

where $\nu(t)$ is a Wiener process and x_0 is a random variable with a known distribution.

Let us specify the filtering problem as follows. Given a function $f(x)$ with $\mathbb{E}[|f(x(t))|^2]$ finite, find the function $\hat{f}(y_t^0)$ that minimizes the functional $\mathbb{E}|f(x(t)) - g(y_t^0)|^2$ in the class of square integrable functions $g(y_t^0)$ on the trajectories of the observed process. As will be shown in Chapter 6, if the Wiener process $w(t)$ and $\nu(t)$ are independent, then

$$\hat{f}(y_t^0) = \int_{\mathbb{R}^1} f(x)\varphi(t,x)dx \Big/ \int_{\mathbb{R}^1} \varphi(t,x)dx.$$

Here $\varphi(t,s)$ is a solution of the LSES

$$d\varphi(t,x) = \left(\frac{1}{2}\frac{\partial^2}{\partial x^2}\left(\sigma^2(x)\ \varphi(t,x)\right) - \frac{\partial}{\partial x}\left(b(x)\ \varphi(t,x)\right)\right)dt +$$

$$+ \ b(x)\ \sigma_0^{-1}\ (y,t))\ \varphi(t,x)dy(t), \quad t > 0, \ x \in \mathbb{R}^1,$$

$$\varphi(0,x) = p(x), \ x \in \mathbb{R}^1,$$

where $p(x)$ is the density of the distribution of the random variable x_0.

As a matter of fact, a more general situation is considered in Chapter 6, namely the processes $x(t)$, $y(t)$ are multidimensional, the noises are possibly correlated, and b and σ depend on t, y.

1.2. *The Krylov equation (backward diffusion equation)*. The Ito equation

$$du(t) = b(u(t))dt + \sigma(u(t))dw(t), \quad t>s, \ u(s) = x,$$

simulates the dynamics of a particle (in the process of diffusion) starting from the point x at the moment s. Thus, the solution of this equation also depends on x, s. In 1975 it was pointed out by N.V. Krylov that the function $u(t,x,s)$ as a function of x, s is a solution of the equation

$$- du(t,x,s) = \left(\frac{1}{2}\sigma^2(x)\frac{\partial^2}{\partial x^2}u(t,x,s) + b(x)\frac{\partial}{\partial x}u(t,x,s)\right)ds +$$

$$+ \sigma(x)\frac{\partial}{\partial x}u(t,x,s) * dw(s), \quad s<t, \ x\in R^2, \ u(t,x,t) = x, \quad x\in \mathbf{R}^1.$$

The symbol $*$ before $dw(s)$ indicates that the stochastic integral is interpreted as the backward Ito integral (see 1.4.12). We study this equation explicitly in Chapter 5.

1.3. *The Helmholtz parabolic equation.* In statistical radiophysics LSES are useful in describing diffraction in a random nonuniform medium ([58], [121]). For example, the following version of the Hemholtz equation is exploited ([57]) to describe the propagation of a monochromatic light in a medium with large-scale nonuniformity:

$$\frac{\partial u(t,x)}{\partial t} = (2k)^{-1}\sqrt{-1}\sum_{i=1}^{2}\frac{\partial^2 u(t,x)}{\partial x^2_i} - 8^{-1}k^2 A_0 u(t,x) +$$

$$+ 2^{-1}k\sqrt{-1}\, u(t,x)\, \dot{W}(t,x), \quad t\in \mathbf{R}_+, \ x\in\mathbf{R}^2, \ u(0,x) = u_0(x), \ x\in\mathbf{R}^2,$$

In this equation axis t is chosen in the direction of original propagation of the light, k is the mean wave number, $W(t,x)$ is a "white noise" with respect t and A_0 is the trace of the correlation operator of $W(t,x)$. The white noise here is to simulate the relative magnitude of permitivity.

1.4. *A continuous branching model with geographical structure.* Models connected with the theory of branching processes are extremely useful in various applications. Let us take a look at one such model.

Suppose that some region consists of a series of subregions and the jth subregion at the moment t contains $p(j,t)$ particles of the same type.* Every particle in this subregion within a small time interval $\Delta(t)$ can die, with probability $\lambda(j)\,\Delta t + 0(\Delta t)$, or give birth to a new particle, with probability $\mu(j)\Delta t + 0(\Delta t)$. Apart from that a migration (diffusion through the boundary of the subregion) is possible. If the diameter of the partition of the region tends to zero, we get in the limit of suitably normalized $p(j,t)$, a random field $u(t,x)$ which stands for a number of particles or some function of this number at the time t at the place x.

This model is successfully used in population biology and chemistry (e.g. [2], [21],

*The function $p(j,t)$ can play the role either of the number or of the fraction of the particles.

[40]). In [2], such a model was used for the description of a chemical reaction. It was shown that the function $u(t,x,s)$, being the relative number of the particles of reagent at the time t at the place x, satisfies the equation

$$\frac{\partial u(t,x)}{\partial t} = \frac{\partial}{\partial x}(\lambda - \mu)(\psi(t,x))\ u(t,x) + D\frac{\partial}{\partial x}\ u(t,x) + G(\psi(t,x))\ \dot{W}(t,x),$$

$$x\in]0,1[,\ \ t\in[0,\infty[,\ u(0,x) = u_0(x),\ x\in]0,1[,\ \frac{\partial u}{\partial x}\Big|_{x=0,1} = 0,$$

where D is the diffusion coefficient, ψ is the mean concentration of reagent,

$$GG^* = -D\frac{\partial}{\partial x}\ \psi(t,x)\ \frac{\partial}{\partial x} + (\lambda + \mu)\ (\psi(t,x))$$

and $W(t,x)$ is a random field which is white noise with respect to t.
Nonlinear stochastic evolution equations obtained in a similar way as limit models in population genetics are of great interest for theoretical biology (see [21], [40]).

1.5. *Equation of the free field.* Let $\mathcal{S}$ be the space of all rapidly decreasing real functions on $\mathbf{R}^{d+1}$. Denote by $\mathcal{S}^*$ the dual space, known as the Schwartz space of tempered distributions on $\mathbf{R}^{d+1}$ (see, for example,) [147]). Let $\mathfrak{S}$ be the σ-algebra generated by the cylinder sets. On the space $(\mathcal{S}^*,\mathfrak{S})$ we can construct a probability measure ν associated with the characteristic functional

$$C_\nu(\eta) = \int_{\mathcal{S}^*} e^{i\eta\omega}\ \nu(d\omega) = exp\Big\{-(\eta,\ \tfrac{1}{2}(-\ \Delta_{t,x} + m^2)^{-1}\eta)_{\mathbf{L}_2(\mathbf{R}^{d+1})}\Big\},$$

$$\eta\in\mathcal{S},\ \Delta_{t,x} = \sum_{i=1}^{d}\ \frac{\partial^2}{\partial x_i^2} + \frac{\partial^2}{\partial t^2},$$

where m is a constant and $\eta\omega$ is the result of the application of the functional $\omega\epsilon\mathcal{S}^*$ to $\eta\epsilon\mathcal{S}$. The free (boson) field is one of the simplest objects of relativistic quantum theory. In Euclidean field theory (see, for example, the monograph of B. Simon [127]), this field is interpreted as a canonical generalized field (i.e. $\xi(\omega) = \omega$ for every $\omega\epsilon\mathcal{S}^*$) on the probability space $(\mathcal{S}^*,\mathfrak{S},\nu)$.

Let μ be the Gaussian measure with the characteristic functional

$$C_\mu(\eta) = exp\left\{-\ \tfrac{1}{2}\|\eta\|^2_{\mathbf{L}_2(\mathbf{R}^{d+1})}\right\}$$

on $(\mathcal{S}^*,\mathfrak{S})$. The canonical generalized field $\dot{\text{W}}$ on the probability space $(\mathcal{S}^*,\mathfrak{S},\mu)$ is usually called the generalized white noise. T. Hida and L. Strait have shown [45] that the free field ξ(t,x) is a stationary solution of the equation

$$\frac{\partial\xi(t,x)}{\partial t} = -\sqrt{(m^2 - \Delta_x)}\ \xi(t,x) + \dot{W}(t,x),$$

where $\Delta_x = \sum_{i=1}^{d} \frac{\partial^2}{\partial x_i^2}$, the equation is treated in the sense of the theory of distributions (generalized functions).

1.2. Measurability and Integrability in Banach Spaces.

2.0. We present some classical results concerning measurability and integrability and introduce a number of special spaces which are used systematically in the principal part of the book.

Warning 0. Throughout this book, if the contrary is not stated, we consider only real spaces. Metric spaces will be taken to be separable wherever it is possible. □

2.1. Let $\mathbf{X}$ be a Banach space. We shall denote by $\mathbf{X}^*$ it conjugate. The value of the functional x^* at $x \in \mathbf{X}$ will be denoted by x^*x and the norm in $\mathbf{X}$ by $\|\cdot\|_{\mathbf{X}}$. If $\mathbf{X}$ is a Hilbert space, the scalar product (corresponding to the norm $\|\cdot\|_{\mathbf{X}}$) will be denoted by $(.,.)_{\mathbf{X}}$. The Borel σ-algebra of $\mathbf{X}$ (i.e. the σ-algebra generated by all closed subsets of $\mathbf{X}$) will be denoted by $\mathfrak{B}(\mathbf{X})$.

2.2. Let (S,Σ) be a measurable space (that means that S is an arbitrary set and Σ is some σ-algebra of its subsets). Suppose that a measure μ is countably additive, σ-finite and positive.

Definition 2. *The completion (see for example* [25] *Ch. III*, 5.17, 5.18) *of* Σ *with respect to* μ *will be called the Lebesgue extention of* Σ *and be denoted by* $\overline{\Sigma}$. □

2.3. *Definition* 3. *We say that the triple* (S,Σ,μ) *is a measure space and the triple* $(S,\overline{\Sigma},\mu)$ *is a complete measure space. If* $\mu(S) = 1$ *each of the spaces will be referred to as a probability space.* □

2.4. *Definition* 4. *The mapping (function)* $f: S \to \mathbf{X}$ *is called* $(\overline{\Sigma},\mu)$*-measurable if*

$$f^{-1}(\mathcal{A}) \in \overline{\Sigma} \quad \forall \mathcal{A} \in \mathfrak{B}(\mathbf{X}). \ \square$$

2.5. *Definition* 5. *The mapping (function)* $f: S \to \mathbf{X}$ *is called* Σ*-measurable if*

$$f^{-1}(\mathcal{A})\in\Sigma \quad \forall \mathcal{A}\in\mathfrak{B}(X)^{*}. \ \square$$

Remark 5. *In definitions* 4 *and* 5, *the metric structure of the space* X *is not important. It could be supposed that* (X, $\mathfrak{B}$(X)) *is an arbitrary measurable space. In the case when* X *is a Banach space, it suffices to require the inclusions* $f^{-1}(\mathcal{A})\in\overline{\Sigma}$ *or* $f^{-1}(\mathcal{A})\in\Sigma$ *for all open sets* $\mathcal{A}$. □

In particular, it follows that if S is a topological space and Σ contains all of the open subsets of S, then every continuous mapping f: $S \to X$ is Borel.

2.6. *Theorem* 6. *The mapping* f: $S \to X$ is $(\overline{\Sigma},\mu)$*-measurable* (Σ*-measurable*) *if and only if there exists a sequence of* $(\overline{\Sigma},\mu)$*-measurable* (Σ*-measurable*) *functions* f_n: $S \to X$ *such that* $f_n \to f$ *strongly* μ*-a.s.* (*for every* $s\in S$). □

For the idea of the proof the reader is referred to [25] (Ch. III, 6.10).

Remark 6. *In fact if f is* $(\overline{\Sigma},\mu)$*-measurable* (Σ*-measurable*) *function there exists a sequence of* $(\overline{\Sigma},\mu)$*-measurable* Σ*-measurable*) *funtions* f_n *that converges to* f *uniformly in the norm of* X. *The construction of this sequence is worth demonstrating here.*

Let $\{x_n\}$, $n\in\mathbb{N}$, *be a dense subset of* X, $B(x_n,1/k)$ *be the open sphere in* X *with the radius equal to* $1/k$ *and the center at the point* x_n. *Set* $B_{n,k} := B(x_n, 1/k)/ \bigcup_{m<n} B(x_m, 1/k)$.

The sequence $f_k(s) := \sum_{i=1}^{\infty} x_n 1_{\{f\in B_{n,k}\}}(s)$ *converges to* $f(s)$ *strongly and uniformly.* □

2.7. *Proposition* 7. *If the mapping* f: $S \to X$ *is* $(\overline{\Sigma},\mu)$*-measurable, then there exists a* Σ*-measurable verson of* f. *That means that* f *can be changed on some set* $S_0\in\overline{\Sigma}$, *where* $\mu(S_0) = 0$, *in such a way that the function obtained will be* Σ*-measurable.* □

Remark 7. *The results of* §2.6 *and* §2.7 *are still valid in the case when* X *is taken to be the extended real line.* □

2.8. *Theorem* 8. (*Pettis*) (*see, for example,* [25], *Ch. III,* 6.11). *The mapping* f: $S \to X$ *is* $(\overline{\Sigma},\mu)$*-measurable* (Σ*-measurable*) *if and only if for every* $x^*\in X^*$ *the mapping* x^*f: $S \to \mathbb{R}^1$ *is* $(\overline{\Sigma},\mu)$*-measurable.* □

---

*In the future, wherever it does not confuse the reader, we will take the liberty of referring to the $(\overline{\Sigma},\mu)$-measurable function as $\overline{\Sigma}$-measurable and call the $\overline{\Sigma}$-measurable function Borel.

From this theorem and Proposition 6 it is easy to obtain the following corollary.

Corollary 8. If the sequence of $(\overline{\Sigma},\mu)$*-measurable* (Σ*-measurable) functions* f_n: $S \to \mathbf{X}$ *converges weakly to* f μ*-a.s. (for every* $s\in S$*), then* f *is* $(\overline{\Sigma},\mu)$*-measurable* (Σ*-measurable) function.* □

2.9. Let $\mathbf{Y}$ be a Banach space continuously imbedded in $\mathbf{X}$. This means that $\mathbf{Y}$ is a subset of $\mathbf{X}$ and there exists a constant N that $||x||_{\mathbf{X}} \leq N||x||_{\mathbf{Y}}$ for every $x\in\mathbf{Y}$.

Proposition 9. (i) If f *is a* $(\overline{\Sigma},\mu)$*-measurable* (Σ*-measurable) mapping of* S *into* $\mathbf{Y}$ *then* f*, as a mapping of* S *into* $\mathbf{X}$*, is also* $(\overline{\Sigma},\mu)$*-measurable* (Σ*-measurable).*

(ii). If f *is a* $(\overline{\Sigma},\mu)$*-measurable* (Σ*-measurable) mapping of* S *into* $\mathbf{X}$ *then* $f^{-1}(\Gamma) \in \overline{\Sigma}$ ($f^{-1}(\Gamma) \in \Sigma$) *for every* Γ *in the Borel* σ*-algebra of* $\mathbf{Y}$. □

Really, let f be the imbedding of $\mathbf{Y}$ in $\mathbf{X}$. Then the mapping f: $S \to \mathbf{X}$ is a which in view of Remark 5 yields assertion (i).

Assertion (ii) follows from the fact that imbedding operator is both continuous and one-to-one. It is well known ([82], §39, IV) that for a one-to-one continuous mapping of a complete separable metric space into a metric space, the image of the first is a Borel subset of the second.

2.10. We define an integral of function f: $S \to \mathbf{X}$. Call f a simple function if it is step-function with a finite number of values. Set $Rf:= \{x_1,x_2,\dots,x_n\}$ and $B:= f^{-1}(x_i)$ for $i=1,2,\dots,n$. Define

$$\int_S f(s)\,d\mu(s) := \sum_{i=1}^{n} x_i\, \mu(B_i).$$

Definition 10. The mapping f: $S \to \mathbf{X}$ *is called* μ*-integrable if there exists a sequence of simple functions* $\{f_n\}$, $n\in\mathbf{N}$, *(from* S *into* $\mathbf{X}$*) which converges (strongly) to* f μ*-a.s. and such that*

$$\lim_{n\to\infty} \int_S ||f(s) - f_n(s)||_x \, d\mu(s) = 0.$$

For an arbitrary set $B\in\overline{\Sigma}$ *the integral of* $f(s)$ *over this set is defined by the equality* $\int_S f(s)\,d\mu(s) := \lim_{n\to\infty} \int_S 1_{\{B\}}(s) f_n(s)\,d\mu(s)$. □

It is easy to check that this definition is a proper one and to verify the integral (it is usually referred to as the Bochner integral) is additive.

2.11. We present (without proofs) some of the main properties of the Bochner integral. An interested reader can find all the details in the book [147], (Ch. V, §5).

I. A $(\overline{\Sigma},\mu)$-measurable mapping f is μ-integrable if and only if the norm $\|f\|$ is μ-integrable.

II. $\|\int_B f(s)d\mu(s)\|_x \le \int_B \|f(s)\|_x d\mu(s), \quad \forall B\in\overline{\Sigma}.$

III. Let T be a bounded linear operator on $\mathbf{X}$ with values in the Banach space $\mathbf{Y}$. If f is μ-integrable mapping from S into $\mathbf{X}$, then Tf is μ-integrable mapping from S into $\mathbf{Y}$ and

$$\int_B Tf(s)d\mu(s) = T\int_B f(s)d\mu(s), \quad \forall B\in\overline{\Sigma},$$

2.12. Let both (S_1, Σ_1, μ_1) and (S_2, Σ_2, μ_2) be measure spaces. Denote by Σ_1 sets of the form $\Gamma_1 \times \Gamma_2$, where $\Gamma_1 \in \Sigma_1$ and $\Gamma_2 \in \Sigma_2$. Let $\mu_1 \times \mu_2$ be a measure on $\Sigma_1 \otimes \Sigma_2$ with the property that

$$\mu_1\times\mu_2 = \mu_1(\Gamma_1)\mu_2(\Gamma_2),\ \forall\Gamma = \Gamma_1\times\Gamma_2,\ \Gamma_1\in\Sigma_1,\ \Gamma_2\in\Sigma_2.$$

It can be shown (e.g. [25], Ch. III, 11.2, 11.6) that the measure $\mu_1\times\mu_2$ is uniquely determined by this equality and is also σ-finite and countably additive.

The measure space $(S_1\times S_2, \Sigma_1\otimes\Sigma_2, \mu_1\times\mu_2)$ will be called the direct product of the measure spaces. The completion of $\Sigma_1\otimes\Sigma_2$ with respect to the measure $\mu_1\times\mu_2$ will be denoted by $\overline{\Sigma_1\otimes\Sigma_2}$.

2.13. We introduce some standard function spaces. The proofs of the properties of the space listed below can be found e.g. in [25], Ch. II, IV.

Let $(S, \overline{\Sigma}, \mu)$ be a complete measure space and $p\in]1, \infty[$. Denote by $\mathbf{L}_p(S; \overline{\Sigma}; \mu; \mathbf{X}^*)$ the space (of classes) of $(\overline{\Sigma},\mu)$-measurable mappings $f\colon S \to \mathbf{X}$ such that

$$\|f\|_{\mathbf{L}_p(S;\overline{\Sigma};\mu;\mathbf{X})} := \Big(\int_S \|f(s)\|^p_{\mathbf{X}}\, d\mu(s)\Big)^{1/p} < \infty. \tag{2.1}$$

Every function belonging to a class of functions from $\mathbf{L}_p(S;\overline{\Sigma};\mu;\mathbf{X})$ is called a representative of the class. The space $\mathbf{L}_p(S;\overline{\Sigma};\mu;\mathbf{X})$ is a Banach space with respect to the norm defined by (2.1). The space conjugate to $\mathbf{L}_p(S;\overline{\Sigma};\mu;\mathbf{X})$ is isometrically isomorphic to $\mathbf{L}_q(S;\overline{\Sigma};\mu;\mathbf{X}^*)$, where $1/p + 1/q = 1$, and in the future we shall identify them.

If $\mathbf{X}$ is a Hilbert space then $\mathbf{L}_2(S, \overline{\Sigma}; \mu; \mathbf{X})$ is also a Hilbert space with respect to the scalar product

$$(f,g)_{\mathbf{L}_2(S;\overline{\Sigma};\mu;\mathbf{X})} := \int (f(s),\ g(s))_{\mathbf{X}}\, d\mu(s).$$

In a similar way we define the space $\mathbf{L}_p(S;\ \overline{\Sigma};\ \mu;\ \mathbf{X})$ for $p=1$ and ∞, but in the case of $p=\infty$ the norm is given up by equality

$$\operatorname*{ess\ sup}_{\mu\quad s\in S} \|f(s)\|_{\mathbf{X}} := \|f\|_{\mathbf{L}_\infty(S;\overline{\Sigma};\mu;\mathbf{X})}.$$

Both of the spaces are Banach ones.

2.14. We list some notations which will be used through what follows without special inferences.

$\mathbf{R}^d$	an Euclidean space of dimension of d
G	an (open) domain in $\mathbf{R}^d$
$\|\cdot\|_d$	the Euclidean norm in $\mathbf{R}^d$
$(\cdot,\cdot)_d$	the scalar product in $\mathbf{R}^d$ related to $\|\cdot\|_d$
l_d	the Lebesgue measure on $\mathbf{R}^d$ (subscript d being ommited if $d=1$).

$$\mathbf{L}_p(G) := \mathbf{L}_p(G,\overline{\mathfrak{B}(G)};\ l_d;\ \mathbf{R}^1),\ p\in[1,\infty]$$

$\mathbf{C}^n(G)$, $n\in\mathbf{N} \cup \{0\} \cup \{\infty\}$ the space of continuous real valued functions on G with continuous derivatives (on G) up to the order of n.

$\mathbf{C}_0^n(G)$ the subset of $\mathbf{C}^n(G)$ of functions with compact support and vanishing on the boundary of G together with their derivatives.

$\mathbf{C}_b^n(G)$, $n\in\mathbf{N} \cup \{0\}$ the space of continuous real valued functions on G, which are bounded together with their derivatives up to the order of n.

$\mathbf{C}^{m,n}(B\times G)$ *the space of continuous real valued functions on* $B\times G$ having continuous derivatives in x up to the order of n and in t up to the order of m.

$\mathbf{C}_b^{m,n}(B\times G)$ the subspace of $\mathbf{C}^{m,n}(B\times G)$ consisting of bounded continuous real functions having bounded continuous derivatives in x up to the order of m.

$\mathbf{C}(B;\mathbf{X})$ the space of strongly continuous mappings from B into $\mathbf{X}$.

In the last three definitions B is taken to be an interval of $\mathbb{R}_+$ and $n, m \in \mathbb{N} \cup \{0\} \cup \{\infty\}$.

1.3. Martingales in $\mathbb{R}^1$.

3.0. In this section we state some basic results of the general theory of stochastic processes and the theory of martingales with continuous parameter, taking values in the set of real numbers.

3.1. Let $(\Omega, \mathcal{F}, \mathbb{P})$ be a complete probability space. It is assumed that the family of σ-algebras $\{\mathcal{F}_t\}$, $t \in \mathbb{R}_+$, is given which has the following properties:

1. $\mathcal{F}_t \subset \mathcal{F} \qquad \forall t \in \mathbb{R}_+$.

2. $\mathcal{F}_s \subset \mathcal{F}_t \qquad \forall s < t,\ s, t \in \mathbb{R}_+$.

3. $\bigcap_{\varepsilon > 0} \mathcal{F}_{t+\epsilon} = \mathcal{F}_t \qquad \forall t \in \mathbb{R}_+$.

4. σ-algebra $\mathcal{F}_0$ contains all $\mathbb{P}$-null sets in $\mathcal{F}$.*

The quadruple $\mathbb{F} := (\Omega, \mathcal{F}, \{\mathcal{F}_t\}, \mathbb{P})$ will be called the standard probability space. It will be fixed throughout this section.

Let $(\mathbf{X}, \mathfrak{B})$ be a measurable space. The space $\mathbf{X}$ will be referred to as the state space.

3.2. *Definition 2.* *An $\mathcal{F}$-measurable mapping $\xi: \Omega \to \mathbf{X}$ is called a random variable (taking values in $\mathbf{X}$).*

A mapping $\xi: \mathbb{R}_+ \times \Omega \to \mathbf{X}$, which is a random variable for every $t \in \mathbb{R}_+$, is

---

*The family $\{\mathcal{F}_t\}$ is said to be increasing if it possesses property 2 and right-continuous if property 3 is fulfilled.

called a stochastic process taking values in $\mathbf{X}$($\mathbf{X}$-process)*. If this mapping is $\mathcal{B}(\mathbb{R}_+)\otimes\mathcal{F}$-measurable then it is called a measurable stochastic process. □

Parameter $t\epsilon\mathbf{R}_+$ usually stands for "time" and parameter ω for "random event" (fluctuation). Of course in the scope of this definition we can also consider the process defined on an interval of $\mathbf{R}_+$.

Warning 2. *All stochastic processes considered in this book will be assumed to be measurable.* □

If the state space $\mathbf{X}$ of a random variable (a stochastic process) is the extended real line $\overline{\mathbb{R}}^1$ and $\mathfrak{X}$ is the Borel σ-algebra on $\overline{\mathbb{R}}^1$, we will say this random variable (stochastic process) is real-valued. The attribute "real-valued" will be omitted frequently.

3.3. *Definition* 3. *Let $x(t)$ and $x'(t)$ be stochastic processes**. If $x(t) = x'(t)$ P-a.s. for every $t\epsilon\mathbf{R}_+$, then $x'(t)$ will be called a version*** of $x(t)$.*

3.4. *Definition* 4. *The function $x(\cdot,\ \omega)$: $\mathbf{R}_+ \to \mathbf{X}$ for a fixed ω is called the sample path or the trajectory of the process $x(t)$ (associated with ω). A stochastic process will be referred to as continuous (left-continuous, right-continuous) if its trajectories have this property for P-a.a. ω.* □

3.5. The stochastic process $x(t,\omega)$ is said to be $\mathcal{F}_t$-adapted if for every fixed $t\epsilon\mathbb{R}_+$ the random variable $x(t,\omega)$ is $\mathcal{F}_t$-measurable. It is said to be progressively measurable if for each $t\epsilon\mathbb{R}_+$, the mapping x: $[0,t] \times \Omega \to \mathbf{X}$ is $\mathcal{B}([0,t]\otimes\mathcal{F}_t$ -measurable. As it is easy to see every progressively measurable process is $\mathcal{F}_t$-adapted.

A set $\mathcal{A}\in\mathbb{R}_+\times\Omega$ is said to be progressively measurable if its indicator is a progressively measurable stochastic process.

3.6. Denote by $\mathcal{P}$ the σ-algebra on $\mathbf{R}_+\times\Omega$ generated by all $\mathcal{F}_t$-adapted, left continuous, real-valued stochastic processes. It is called the σ-algebra of

---

*Sometimes if it does not cause misunderstanding, we shall not mention the space where a stochastic process takes it values.

**The argument ω will be dropped, as a rule.

***In general, if $(\mathbf{X},\mathfrak{X})$ is a measurable space, f: $S_1 \times S_2 \to \mathbf{X}$, and for every $s_2 \in S_2$, $f(\cdot,s_2)$ is $\overline{\Sigma}_1$-measurable mapping of the complete measure space. $(S_1, \overline{\Sigma}_1,\mu_1)$ into $\mathbf{X}$, then every function $\tilde{f}$ possessing analogous properties and such that $f(\cdot,s_2) = \tilde{f}(\cdot,s_2)$ μ_1-a.s., for every $s_2 \in S_2$, is called a version of f (with respect to s_1).

predictable sets.

Defintion 6. *A stochastic process is called predictable if it is* $\mathcal{P}$*-measurable.* □

Warning 6. *If besides* $\{\mathcal{F}_t\}$ *some other family of* σ*-algebras on* $(\Omega,\mathcal{F})$ *is given we shall supply the attribute "predictable" by a reference to the family of* σ*-algebras in question.* □

Example 6. *A process* $B(t)$ *which is of the form*

$$B(t) := B_0(\omega)\, 1_{(0)}(t) + \sum_{i=1}^{\infty} B_i(\omega)\, 1_{\{]t_i,t_{i+1}]\}}(t),$$

where $(0 = t_0 < t_1 < \ldots < t_n < \ldots)$, *and* B_i is $\mathcal{F}_{t_i}$*-measurable random variable, for every* i, *is predictable. Processes of this type will be referred to as simple predictable processes.*

It is easy to show by approximation of left-continuous and $\mathcal{F}_t$*-adapted (and consequently predictable) stochastic processes by simple predictable processes that the following statement holds.*

Proposition 6. *Sets of the form* $]a,b] \times \Gamma$, *where* $a, b \in \mathbf{R}_+$, $\Gamma \in \mathcal{F}_a$, *generates the* σ*-algebra* $\mathcal{P}$. □

The σ-slgebra generated by the sets mentioned in the proposition, where $a, b \in [T_0,T]$, T_0, $T \in \mathbf{R}_+$, will be denoted by $\mathcal{P}_{[T_0,T]}$, and if $T_0 = 0$ *by* $\mathcal{P}_T$.

It is clear that $\mathcal{P}_{[T_0,T]}$ is generated by left-continuous, $\mathcal{F}_t$-adapted processes considered on the interval $[T_0,T]$. A $\mathcal{P}_{[T_0,T]}$-measurable stochastic process on $[T_0,T]$ will also be called predictable.

Remark 6. *It is easy to see that every predictable process is progressively measurable and consequently* $\mathcal{F}_t$*-adapted.*

Sometimes instead of "predictable ($\mathcal{F}_t$*-adapted or progressively measurable) stochastic process" we shall say "predictable* ($\mathcal{F}_t$*-adapted or progressively measurable) function" (of the arguments* t,ω). □

3.7. *Definition* 7. *A random variable* τ *taking values in* $\overline{\mathbf{R}}_+$ *is called a stopping time with respect to an increasing family of* σ*-algebras* $\{\mathcal{Y}_t\}$, $t \in \mathbf{R}_+$, *if the random event* $\{\omega: \tau(\omega) \le t\} \in \mathcal{Y}_t$ *for every* $t \in \mathbf{R}_+$. □

Warning 7. *All stopping times are considered with respect to the family* $\{\mathcal{F}_t\}$ *unless otherwise stated.* □

Denote

$$\mathfrak{D}_A(\omega) := \begin{cases} \inf_t \{t:(t,\omega)\in A\} \text{ if } \{t: (t,\omega)\in A\} \neq \emptyset; \\ \infty, \text{ if } \{t: (t,\omega)\in A\} = \emptyset. \end{cases}$$

The random variable $\mathfrak{D}_A$ will be called début of the set A.

Theorem 7. *(e.g.* [100], *Ch. IV, Th.* 48) *If A is a progressively measurable set then the débute of A is a stopping time.* □

Let x be a progressively measurable $\mathbf{X}$-process, $B\epsilon\mathfrak{B}$, and $A: = \{(t,\omega): x(t,\omega)\epsilon B\}$, then the début of A is a stopping time. This stopping time is called the hitting time of B. The following relation provides the reason for this terminology:

$$\mathfrak{D}_A(\omega) := \begin{cases} \inf_t \{t:(t,\omega)\in A\} \text{ if } \{t: (t,\omega)\in A\} \neq \emptyset; \\ \infty, \text{ if } \{t: (t,\omega)\in A\} = \emptyset. \end{cases}$$

3.8. *Definition* 8. *The real-valued stochastic process $M(t)$ will be called a real-valued supermartingale (submartingale) relative to an increasing family of σ-algebras $\{\mathcal{Y}_t\}$, $t\epsilon\mathbf{R}_+$, if it is $\mathcal{Y}_t$-adapted, $\mathbf{P}$-integrable for every $t\epsilon\mathbf{R}_+$, and $\mathbf{E}[M(t)|\mathcal{Y}_s] \leq M(s)$ (respectively $\mathbf{E}[M(t)|\mathcal{Y}_s] \geq M(s)$) for every s, $t\epsilon\mathbf{R}_+$, $s \leq t$.*
If $M(t)$ is both a submartingale and a supermartingale relative to the family $\{\mathcal{Y}_t\}$ it will be called a martingale relative to this family. □

Warning 8. Throughout this section, unless otherwise stated, we will consider real-valued supermartingales, submartingales, and just martingales relative to the family $\{\mathcal{F}_t\}$. Therefore we will omit reference to the family in question. The adjective "real-valued" will also be dropped as a rule. □

It follows from the definition that if $M(t)$ is a martingale then $\mathbf{E}M(t) = \mathbf{E}M(s)$ for all s and $t\epsilon\mathbf{R}_+$. From the definition it follows also that if a martingale is square-integrable then its increments are uncorrelated, i.e. for every $t_1 \leq t_2 \leq t_3 \leq t_4$,

$$\mathbf{E}[(M(t_4) - M(t_3))(M(t_2) - M(t_1))] = 0.$$

3.9. An important feature of martingales is the invariance of the martingale property with respect to "optional sampling transformation", namely: under fairly general conditions if $M(t)$ is a martingale then for each stopping time τ, $M(t\wedge\tau)$ is also a martingale.

Let $M(t)$ be a right-continuous supermartingale. Suppose that there exists a $\mathbf{P}$-integrable, real-valued random variable Y such that

$$M(t) \geq \mathbf{E}[Y|\mathcal{F}_t] \text{ for every } t\epsilon\mathbf{R}_+ . \tag{3.1}$$

Suppose as well that two stopping times τ and σ, ($\tau \leq \sigma$ **P**-a.s.), are given. Set $M(\tau)1_{(\tau=\infty)} := Y1_{(\tau=\infty)}$, $M(\sigma)1_{(\sigma=\infty)} := Y1_{(\sigma=\infty)}$.

Under the assumptions of (3.1), the following result holds, (see [100], Ch. VI, Theorem 13):

Theorem 9. *The random variables* $M(\tau)$ *and* $M(\sigma)$ *are* **P**-*integrable and subject to the supermartingale property*

$$M(\tau) \geq \mathbb{E}[M(\sigma)|\mathcal{F}_\tau] \qquad (\mathbf{P}\text{-}a.s.) \ \square \tag{3.2}$$

Corollary 9. *If* $M(t)$ *is a martingale and the* $M(t)$ *are uniformly* **P**-*integrable, then inequality sign in* (3.2) *can be replaced by equality.* □

3.10. We recall some inequalities for real-valued submartingales and supermartingales (see e.g. [100] Ch. VI, §1).

Let $I := [a,b]$ be a compact interval of $\mathbb{R}_+$. If $M(t)$ is a right-continuous supermartingale and $\lambda \in \mathbb{R}_+$, then

$$\lambda P(\inf_{t \in I} M(t) \leq -\lambda) \leq E|M(b)|. \tag{3.3}$$

Let $M(t)$ be a positive right-continuous submartingales. Let p and q be numbers such that $p \in]1, \infty[$ and $1/p + 1/q = 1$. Then

$$E(\sup_{t \in I} M(t))^p \leq q^p E|M(b)|^p. \tag{3.4}$$

The inequality (3.4) implies, in particular, the following inequality due to J. Doob.

If $M(t)$ is a right-continuous martingale then

$$E(\sup_{t \in I} |M(t)|)^p \leq q^p E|M(b)|^p. \tag{3.5}$$

3.11. *Definition 11. A right-continuous supermartingale is said to belong to class* (D) *if it is uniformly* **P**-*integrable and the collection of random variables* $\{M(\tau), \tau \in T\}$, *where* T *is the set of all stopping times* τ *is also uniformly* **P**-*integrable.* □

The following statement is a basic one for the differential calculus of martingales (see e.g. [22], Ch. V, i. 50).

Theorem 11. *Let* $M(t)$ *be a supermartingale of class* (D). *There exists a unique (up to a version) predictable real-valued stochastic process* $A(t,\omega)$ *such that:*

(*a*) *trajectories of* A *are increasing right-continuous functions for* **P**-*a.a.* ω;

(*b*) $A(0) = 0$ (**P**-*a.s.*) *and* $\sup_t \mathbb{E}\, A(t) < \infty$;

(*c*) $M(t) + A(t)$ *is a martingale.*

If the supermartingale is continuous then the corresponding process $A(t)$ *is also continuous.* □

A decomposition

$$M(t) = Z(t) - A(t), \tag{3.6}$$

where $Z(t)$ is a martingale, is usually referred to as a Doob-Meyer decomposition. An $\mathfrak{F}_t$-adapted real-valued stochastic process $A(t)$ with property (a) and such that $A(0) = 0$, ($\mathbf{P}$-*a.s.*), and $\mathbf{E}\, A(t) < \infty$ for every $t \in \mathbf{R}_+$ is said to be an increasing process.

Remark 11. In fact, much weaker conditions suffice to ensure the continuity of $A(t)$ *(see e.g.* [22], *Ch. V, i.* 52). □

Let $\mathfrak{M}_2^c(\mathbf{R}_+,\mathbf{R}_1)$ be the set of all continuous real-valued martingales M such that $M(0) = 0$, ($\mathbf{P}$-a.s.), and for each $t \in \mathbf{R}_+$, $\mathbf{E}|M(t)|^2 < \infty$. The following corollary is an easy consequence of Theorem 11 (see e.g. [22], Ch. II, §3).

Corollary 11. Let $M \in \mathfrak{M}_2^c(\mathbf{R}_+, \mathbf{R}^1)$. *There exists a stochastic process* $<M>_t$ *such that:*

(*a*) *the trajectories of* $<M>$ *are continuous for* $\mathbf{P}$-*a.s.* ω;

(*b*) $<M>_0 = 0$ ($\mathbf{P}$-*a.s.*);

(*c*) $|M(t)|^2 - <M>_t$ *is a martingale.*

The process $<M>$ *is the only predictable process (up to a version) with the properties* (*a*), (*b*), (*c*). □

We call the process $<M>_t$ the quadratic variation process or simply variation process of $M(t)$. This term is motivated by the following well-known result (see e.g. [23], Ch. 2, §3).

Proposition 11. *If* $M \in \mathfrak{M}_2^c(\mathbf{R}_+, \mathbf{R}^1)$, $T \in \mathbf{R}_+$, *and* $\{0 = t_0^n < ... < t_{k(n)+1}^c = T\}$ *is a sequence of partitions of the interval* $[0,T]$ *such that the diameter of the partition tends to zero as* n *tends to infinity, then*

$$\lim_{n\to\infty} \left| \sum_{i=0}^{k(n)} | M(t_{i+1}^n) - M(t_i^n)|^2 - <M>_T \right| = 0. \quad \square$$

3.12. *Definition* 12. *A real-valued stochastic process* $M(t)$ *is called a (real-valued) local martingale relative to an increasing family of* σ-*algebras* $\{\mathfrak{Y}_t\}$ *if there exists a sequence of stopping times (relative to the same family)* $\{\tau_n\}$, $n \in \mathbb{N}$ $\tau_n \uparrow \infty$, *(P-a.s.), such that for each n,* $M(t \wedge \tau_n)$ *is also a martingale relative to the family* $\{\mathcal{Y}_t\}$. □

The notion of a local submartingale or supermartingale is defined in an analogous way.

Remark 12. Let τ *be a stopping time relative to a family* $\{\mathcal{Y}_t\}$ *and* $M(t)$ *be a* $\mathcal{Y}$-*adapted stochastic process. Then* $M(t \wedge \tau)$ *is a martigale relative to the family* $\{\mathcal{Y}_{t \wedge \tau}\}$ *if and only if it is a martigale with respect to the family* $\{\mathcal{Y}_t\}$.

Thus in the definition 12 *we can require* $M(t \wedge \tau)$ *to be a martigale with respect to the family* $(\mathcal{Y}_{t \wedge \tau_n})$ *to be a martingale with respect to the family* $\{\mathcal{Y}_{t \wedge \tau_n}\}$. □

Warning 12. *Throughout the section we consider only real-valued local submartingales and supermartingales relative to the family* $\{\mathcal{F}_t\}$. *Hence, the phrase "relative to* $\{\mathcal{F}_t\}$*" will usually be dropped as well as the adjective "real-valued".* □

The set of all continuous local martingales with $M(t) = 0$ a.s. $\mathbb{P}$, will be denoted by $\mathfrak{M}^c_{loc}(\mathbb{R}_+, \mathbb{R}^1)$.

Lemma 12. *Let* $M(t)$ *be a continuous local martingale, then there exists a sequence of stopping times* $\{\sigma_n\}$, $n \in \mathbb{N}$, *such that* $\sup_{t \in \mathbb{R}_+} |M(t \wedge \sigma_n)| < n$ $\forall n \in \mathbb{N}$, $(\mathbb{P}$-*a.s*). □

This result is well known and its proof can be easily supplied by the reader.

Let $M(t)$ and $\{\sigma_n\}$ be the local submartingale and the localizing sequence of stopping times described in the lemma. Then, for each $n \in \mathbb{N}$, $-M(t \wedge \sigma_n\}$ is a supermartingale of class ($\mathfrak{D}$) (moreover a continuous one) and consequently in accordance with Theorem 11 there exists unique up to a version a predictable increasing process $A_n(t)$ such that $A_n(t) - M(t \wedge \sigma_n)$ is a martingale.

Set

$$(3.7) \qquad A(t) := A_n(t), \textit{ for } t < \sigma_n$$

It is easy to see that A(t) is well defined and the following theorem holds.

Theorem 12. *Let* $M(t)$ *be a continuous local submartingale. Then there exists a predictable real-valued stochastic process* $A(t)$ *such that*

(*a*) $A(0) = 0$ ($\mathbb{P}$-*a.s.*);

(*b*) $A(t) \geq A(s)$ ($\mathbb{P}$-*a.s.*) *for every* t, $s \in \mathbb{R}_+$, $t \geq s$;

(*c*) $M(t) - A(t)$ *is a local martingale.*

The process $A(t)$ *is continuous. It is the only (up to a version) predictable, continuous stochastic process with properties* (*a*), (*b*),(*c*). □

Corollary 12. *Let* $M \in \mathfrak{M}^c_{loc}(\mathbf{R}_+, \mathbf{R}^1)$, *then there exists a unique* (*up to a version*) *continuous, increasing process* $<M>_t$ *such that* $|M(t)|^2 - <M>_t \in \mathfrak{M}^c_{loc}(\mathbf{R}_+, \mathbf{R}^1)$. □

Explicit proofs of the theorem and other statements presented below in this section can be found e.g. in [23].

The process $<M>_t$ mentioned above is usually called the quadratic variation process or simply variation process of the local martingale $M(t)$.

3.13. Assume that M and $L \in \mathfrak{M}^c_{loc}(\mathbf{R}_+, \mathbf{R}^1)$. Set $<M,L>_t := 1/4\ (<M+L>_t - <M-L>_t)$. It is clear that the trajectories of the process $<M,L>_t$ are (P-*a.s.*) continuous and or bounded variation.

Theorem 13. If $M,L \in \mathfrak{M}^c_{loc}(\mathbf{R}_+, \mathbf{R}^1)$, *then* $(M(t)L(t) - <M,L>_t)$ *also belongs to* $\mathfrak{M}^c_{loc}(\mathbf{R}_+, \mathbf{R}^1)$. *Moreover, suppose that* $A(t)$ *is a real-valued process such that*

(*a*) *trajectories of* $A(t)$ *are of bounded variation* (P-*a.s.*);
Then, for each $t \in \mathbf{R}_+$, $A(t) = <M,L>_t$, (P-*a.s.*). □

Note that if M and $L \in \mathfrak{M}^c_2\ (\mathbf{R}_+, \mathbf{R}^1)$, then $(M(t)L(t) - <M,L>_t)$ is a continuous martingale.

3.14. In this paragraph we construct a stochastic integral with respect to a martingale $M \in \mathfrak{M}^c_2(\mathbf{R}_+, \mathbf{R}^1)$.

Let $B(s)$ be a simple predictable process, defined as in §1.6 on the interval $[0,t]$, where t is some fixed point of $\mathbf{R}_+$.

In addition we suppose that

$$E \int_{[0,t]} |B(s)|^2\ d <M>_s < \infty. \tag{3.8}$$

Define $t_n := t$ and

$$\int_{[0,t]} B(s) dM(s) := \sum_{i=0}^{n-1} B_i(M(t_{i+1}) - M(t_i)). \tag{3.9}$$

It is clear that

$$E \int_{[0,t]} B(s) dM(s) = 0 \tag{310}$$

and

$$(3.14)\quad E\Big|\int_{[0,t]} B(s)dM(s)\Big|^2 = E\int_{[0,t]} |B(s)|^2 d\langle M\rangle_s,$$

we introduce the measure $\langle M\rangle \circ \mathbf{P}_t$ on the σ-algebra $\mathcal{B}([0,t]) \otimes \mathcal{F}$, defined by the equality

$$\langle M\rangle \circ \mathbb{P}_t(A) := \mathbb{E}\int_{[0,t]} 1_{\{A\}}(s)d\langle M\rangle_s,$$

where $\langle M\rangle_s$ is the quadratic variation process of M.

We call $\mathcal{L}_2(\langle M\rangle \circ \mathbb{P}_t)$ the class of real-valued, predictable stochastic process B on $[0,t]$ such that the inequality (3.8) holds true.

It is clear that $\langle M\rangle \circ \mathbb{P}_t$ is a countably additive positive measure (sometimes it is referred to as a Dolean's measure). It could be shown (see e.g. [27], Ch. V, §3) that the Dolean's measure is completely defined by its restriction to the σ-algebra $\mathcal{P}$ of predictable sets.

Now we carry over the definition of the stochastic integral with properties (3.10), (3.11) to the case of integrands from $\mathfrak{L}_2(\langle M\rangle \circ \mathbb{P}_t)$.

We can and do identify $\mathfrak{L}_2(\langle M\rangle \circ \mathbf{P}_t)$ with $\mathbb{L}_2([0,t] \times \Omega;\ \overline{\mathcal{P}}_t;\ \langle M\rangle \circ \mathbb{P}_t;\ \mathbb{R}^1)$ by identifying each element B of the $\mathfrak{L}_2(\langle M\rangle \circ \mathbf{P}_t)$ with the class of functions from $\mathbb{L}_2([0,t] \times\Omega;\ \overline{\mathcal{P}}_t;\ \langle M\rangle \circ \mathbb{P}_t;\ \mathbb{R}^1)$ that contains B.

Let A be some set from $\mathcal{P}$ and $\mathcal{S}$ be the ring of subsets of $[0,t] \times\Omega$, consisting of finite disjoint unions of sets of the form $[a,b] \times \Gamma$, where $\Gamma\in\mathcal{F}_a$, and of the form $\{0\}\times\Gamma$, where $\Gamma\in\mathcal{F}_0$. It follows from §3.6 that there exists a sequence of sets $[\Gamma_n\}$ belonging to $\mathcal{S}$ such that

$$\lim_{n\to\infty} E\int_{[0,t]} |1_{\{A\}}(s) - 1_{\{\Gamma_n\}}(s)|^2 d\langle M\rangle_s = 0.$$

It therefore follows that the subspace generated by simple predictable processes with property (3.8), call it $\mathcal{M}^2$, is dense in $\mathfrak{L}_2(\langle M\rangle \circ \mathbf{P}_t)$.

Thus the mapping $B \to \int_{[0,t]} B(s)dM(s)$ of $\mathcal{M}^2$ into $\mathbb{L}_2(\Omega;\mathcal{F}_t;\mathbf{P};\mathbb{R}^1)$ given by the equality (3.9) is an isometry and has a uniquely determined continuous extention as an isometry from $\mathcal{L}_2(\langle M\rangle \circ \mathbf{P}_t)$ to $\mathbb{L}_2(\Omega;\mathcal{F}_t;\mathbf{P};\mathbb{R}^1)$.

This isometry defines a stochastic integral with respect to M of a predictable stochastic process B and we denote it by $\int_{[0,t]} B(s)dM(s)$.

Obviously, such an integral possesses properties (3.10), (3,11). As a function of t it is a real-valued stochastic process. It is defined up to equivalence and has a measurable version (see below).

3.15. Let us list the basic properties of the integral constructed in 3.14.

(i) If $B_1, B_2 \in \mathfrak{L}_2(<M>\circ \mathbf{P}_t)$ and $\alpha, \beta \in \mathbf{R}^1$, then

$$\int_{[0,t]} (\alpha B_1(s) + \beta B_2(s))dM(s) =$$

$$= \int_{[0,1]} B_1(s)dM(s) + \beta \int_{[0,t]} B_2(s)dM(s), \ (\mathbf{P}\text{-}a.s.).$$

(ii) For every s, u, $t \in \mathbf{R}_+$ such that $0 \leq s \leq u \leq t$ and $B \in \mathfrak{L}_2(<M>\circ \mathbf{P}_t)$

$$\int_{[0,t]} 1_{\{[s,u]\}}(r)B(r)dM(r) = \int_{[0,u]} B(r)dM(r) - \int_{[0,s]} B(r)dM(r) :=$$

$$= \int_{[s,u]} B(r)dM(r), \ (\mathbf{P}\text{-}a.s.).$$

(iii) If $B \in \mathcal{L}_2(<M>\circ \mathbf{P}_t)$, for every $t \in \mathbf{R}^1$, then the stochastic process $\int_{[0,t]} B(r)dM(r)$ has a version $B \circ M(t)$ which belongs to $\mathfrak{M}^c_{loc}(\mathbf{R}_+, \mathbf{R}_1)$ and

$$<B\circ M>_t = \int_{[0,t]} |B(s)|^2 d<M>_s = 0,$$

(iv) If B and $B^n \in \mathfrak{L}_2\ (<M>\circ \mathbf{P})$ for $n \in \mathbf{N}$ and

$$\lim_{n\to\infty} \mathbf{E} \int_{[o,t]} |B^n(s) - B(s)|^2 \ d<M>_s = 0,$$

then

$$\lim_{n\to\infty} \mathbf{E} \sup_{s\leq t} |B^n \circ M(s) - B\circ M(s)|^2 = 0$$

(v) If $B \in \mathfrak{L}_2(<M>\circ \mathbf{P}_t)$ for every $t \in \mathbf{R}_+$ and $L \in \mathfrak{M}^c_{loc}(\mathbf{R}_+, \mathbf{R}^1)$, then

$$<L,B\circ M>_t = \int_{[0,t]} B(s)d<L,M>_s, \ (\mathbf{P}\text{-}a.s.).$$

Remark 15. *The stochastic integral defined above is easy to extend to the case where M is a local continuous martingale. This will be done in the next chapter in a more general situation.*

3.16. Concluding this section we list some important inequalities for the further use.

Theorem **16.** (*The Kunita-Watanabe inequalities*).

(*a*) *If* B *is a predictable bounded process,* L *and* $M \in \mathfrak{M}^c_{loc}(\mathbb{R}_+, \mathbb{R}_1)$, *and* $T \in \mathbb{R}_+$, *then*

$$\int_{[0,T]} B(s)d<L,M>_s \leq \left(\int_{[0,T]} |B(s)|^2 d<M>_s)^{\frac{1}{2}} <L>^{\frac{1}{2}}_T \right), \ (\mathbb{P}\text{-}a.s.)$$

(*b*) *If* B_1 *and* B_2 *are predictable real-valued processes,* L, $M \in \mathfrak{M}^c_{loc}(\mathbb{R}_+, \mathbb{R}^1)$, *and* $T \in \mathbb{R}_+$, *then*

$$\int_{[0,T]} |B_1(s)B_2|d| <L,M>_s| \leq$$

$$\left(\int_{[0,T]} |B_1(s)|^2 d<L>_s)^{\frac{1}{2}} \left(|B_2(s)|^2 d<M>_s \right)^{\frac{1}{2}}, \ (\mathbb{P}\text{-}a.s.).$$

(*c*) *If in addition to the assumptions of* (*b*) L, $M \in \mathfrak{M}^c_2(\mathbb{R}_+, \mathbb{R}^1)$, *then*

$$\mathbb{E} \int_{[0,T]} |B_1(s)B_2(s)|d|<L,M>_s| \leq$$

$$\left(\mathbb{E} \int_{[0,T]} |B_1(s)|^2 d<L>_s \right)^{\frac{1}{2}} \left(\mathbb{E} \int_{[0,T]} |B_2(s)|^2 d<M>_s \right)^{\frac{1}{2}}. \ \square$$

1.4. Diffusion Processes.

4.0. Diffusion processes are one of the most important and thoroughly investigated classes of stochastic processes. The term "diffsion processes" came into use because this kind of processes is a good model of the physical diffusion phenomena.

The mathematical theory of diffusion processes was developed in the works of N. Wiener [145], A.N. Kolmogorov [59], P. Levy [88], I.I. Gikhman [29], [30] etc.

The predominate modern approach to the theory of diffusion processes is the one connected with the theory of the Ito stochastic differential equations, [51].

In this section we present some of the main definitions and results of the theory of diffusion processes which we will need in our subsequent development.

Warning 0. *The standard probability space* $\mathbf{F} := (\Omega, \mathcal{F}, \{\mathcal{F}_t\}_{t\in\mathbf{R}_+}, \mathbf{P})$ *and numbers* d, $d_1 \in \mathbf{N}$ *and* $T > 0$ *are fixed.*

Throughout what follows we shall denote the i-th coordinate of the vector $x\in\mathbf{R}^d$ *by* x^i. *We will also use the following notation*

$$dx := dx^1,\dots dx^d, \text{ for every } x\in\mathbf{R}^d,\ f_i(x) := \frac{\partial}{\partial x^i} f(x),\ f_{ij} := \frac{\partial^2}{\partial x^i \partial x^j} f(x).$$

Repeated indices in monomials are summed over. For example,

$$\int_{[0,t]} a^{il}(s)dw^l(s) := \sum_{i=1}^{d_1} \int_{[0,t]} a^{il}(s)dw^l(s). \ \square$$

4.1. *Definition* 1. *A stochastic process* $w(t)$ *is called a standard Wiener process (martingale) in* $\mathbf{R}^{d_1}$ *if it possesses the following properties:*

(*i*) *The coordinates* $w^1,\dots,w^{d_1}$ *of* w *are independent.*

(*ii*) $w^l(t) \in \mathfrak{M}_2^c(\mathbf{R}_+, \mathbf{R}^1)$ (*with respect to the family* $\{\mathcal{F}_t\}$) *for every* $l = 1,2,\dots d_1$.

(*iii*) *The quadratic variation process* $\langle w^l \rangle_t = t$, $l=1,2,\dots d_1$.

It is well known that there exists a standard probability space $\mathbf{F}$ *such that a standard Wiener process could be constructed on it.*

Theorem 1. (*Levy*) (*see e.g.* [90], *Ch.* 4, §1, *Theorem* 1). *Let* $w(t)$ *be a standard Wiener process in* $\mathbf{R}^1$. *Then for every* $s,t\in\mathbf{R}_+$ *such that* $s\le t$ *the following assertions hold true*

(*a*) *The increment* $w(t) - w(s)$ *is independent of* $\mathcal{F}_s$.

(*b*) *The random variable* $w(t) - w(s)$ *is a Gaussian random variable with zero mean and the variance equal to* $t-s$. $\square$

Suppose that a standard Wiener process $w(t)\dots := (w^1(t),\dots,w^{d_1}(t))$ is given on a standard probability space $\mathbf{F}$ which is fixed in this section.

It follows from §1.3.14 and §1.3.15 that the stochastic integral $\int_{[0,t]} b(s)dw^l(s)$ is defined for each $l=1,2,\dots,d_1$ and every predictable function b: $\mathbf{R}_+ \times \Omega \to \overline{\mathbf{R}}^1$ such that

$$\mathbf{E} \int_{[0,t]} |b(s)|^2 ds < \infty. \tag{4.1}$$

The integral has a continuous (in t) version which belongs to $\mathfrak{M}^c_{loc}(\mathbf{R}_+,\mathbf{R}^1)$ and possesses the property

$$< \int_{[0,\cdot]} b(s)dw^l(s)>_t = \int_{[0,t]} b^2(s)ds \quad \forall t\in\mathbf{R}_+, \ (\mathbf{P}\text{-}a.s.),$$

as well as the other natural properties of a stochastic integral listed in §1.3.15.

In general, a stochastic integral with respect to a one-dimensional Wiener process can be defined for every predictable process f: $\mathbf{R}_+ \times \Omega \to \mathbf{R}^1$ such that

$$\int_{[0,t]} b^2(s)ds < \infty \quad \forall t\in\mathbf{R}_+. \tag{4.2}$$

Such an integral will be constructed below in a much more general situation (see Theorem 2.3.3 and Remark 2.3.3).

A stochastic integral with respect to a Wiener process for a predictable function satisfying condition (4.2) has properties similar to properties (i), (ii) of §1.3.15 as well as some other usual properties (see e.g. §2.3.3). Two of them are presented below.

(a) The integral $\int_{[0,t]} b(s)dw^l(s)$ has a version belonging to $\mathfrak{M}^c_{loc}(\mathbf{R}_+,\mathbf{R}^1)$* and such that

$$< \int_{[0,\cdot]} b(s)dw^l(s)> = \int_{[0,\cdot t]} b^2(s)ds, \ (\mathbf{P}\text{-}a.s.).$$

(b) Let $\{f_n(t)\}$, $n\in\mathbf{N}$, be a sequence of predictable real-valued processes such that

$$\lim_{n\to\infty} \int_{[0,T]} |b_n(t) - b(t)|^2 \, dt = 0$$

*We shall denote this version in the same way as the integral.

(in probability), then

$$\lim_{n\to\infty} \int_{[0,T]} (b_n(t) - b(t))dw^l(t) = 0$$

(in probability).

Warning 1. *Throughout this section when speaking about a stochastic integral with respect to a one-dimensional Wiener process for a predictable function satisfying condition* (4.2) *we have in mind the version of the integral belonging to* $\mathfrak{M}^c_{loc}(\mathbb{R}_+,\mathbb{R}^1)$. □

4.2. Suppose that $\sigma^{il}(t,\omega)$, $f^i(t,\omega)$ are predictable real-valued stochastic processes on $\mathbb{R}_+$, $\sigma^{il}(\cdot,\omega) \in \mathbb{L}_2([0,T])$ and $f^i(\cdot,\omega) \in \mathbb{L}_1([0,T])$ for $i=1,2,...,d$, $l=1,2,...d_1$, and arbitrary $T \in \mathbb{R}_+$ ($\mathbb{P}$-*a.s.*).

Definition 2. *Let* τ *be a stopping time (relative to* $\{\mathcal{F}_t\}$*) and* $\xi(t)$ *be a continuous predictable stochastic process in* $\mathbb{R}^d$. *We say that* $\xi(t)$ *possesses the stochastic differential*

$$d\xi^i(t) = b^i(t)dt + \sigma^{il}(t)dw^l(t), \quad t<\tau \quad i=1\div d, \tag{4.3}$$

if for each $i=1,2,...,d$ *the following equality holds true*

$$P\Big\{ \sup_{t<\tau} | \xi^i(t) - \xi^i(0) - \int_{[0,t]} b^i(s)ds - \int_{[0,t]} \sigma^{il}(s)dw^l(s)| = 0\Big\} = 1. \ \square$$

One of the fundamental results of Ito's stochastic integro-differential calculus is the following change of variable formula.

Theorem 2. *(The Ito formula) (see e.g.* [90], *Ch.*4, §3). *Let* $f \in C^{1,2}(\mathbb{R}_+,\mathbb{R}^d)$ *and* $\xi(t)$ *be the stochastic process possessing the stochastic differential* (4.3), *then* $f(t, \xi(t))$ *also possesses the stochastic differential*

$$df(t, \xi(t)) = [\frac{\partial}{\partial t} f(t,x)|_{x=\xi(t)} +$$

$$+ \frac{1}{2} \sigma^{il}\sigma^{jl} f_{ij}(t, \xi(t)) + b^i f_i(t, \xi(t))]dt +$$

$$+ \sigma^{il} f_i(t, \xi(t))dw^l(t), \quad t<\tau. \ \square$$

4.3. In this paragraph the numbers T_0, $K \in \mathbf{R}_+$, where $T_0 < T$, are fixed.

Suppose that functions $b^i(t,x,\omega)$, $\sigma^{il}(t,x,\omega)$ for $i=1,2,...d$, $l=1,2,...d_1$, which map $[T_0,T] \times \mathbf{R}^d \times \Omega$ into $\overline{\mathbf{R}}^1$ are given as well as a random variable X_0 in $\mathbf{R}^d$. Suppose also that for every $x \in \mathbf{R}^d$ the functions $b^i(t,x) := b^i(t,x,\omega)$ and $\sigma^{il}(t,x) := \sigma^{il}(t,x,\omega)$ are predictable and the random variable X_0 is $\mathcal{F}_{T_0}$-measurable.

Consider the system

$$X^i(t) = X_0^i + \int_{[T_0,t]} b^i(s, X^i(s))ds + \int_{[T_0,t]} \sigma^{il}(s, X^i(s))dw^l(s), \quad t \in [T_0,T], \quad i=1,\div d. \tag{4.4}$$

Equations of this type were considered for the first time by K. Ito and are traditionally called Ito equations. We shall follow this tradition in general but sometimes we shall also add to this term "ordinary" to distinguish them from the Ito stochastic partial differential equations which are the principal subject of this book.

Definition 3. *A continuous predictable stochastic process* $X(t, X_0, T_0)$ *taking values in* $\mathbf{R}^d$ *and satisfying the equality* (4.4) *for all* $t \in [T_0,T]$, $\omega \in \Omega'$, *where* $\Omega' \subset \Omega$ *and* $\mathbf{P}(\Omega') = 1$, *is called a solution of system* (4.4). □

The parameters X_0, T_0 in the notation of the solution of system (4.4) will be dropped when the "origin" of the process does not matter.

Theorem 3. *(see e.g.* [63], *Ch.II,* §5, 17.10) *Let for all* $t \in [0,T]$, $\omega \in \Omega$, x, $y \in \mathbf{R}^d$, *and some* $K \in \mathbf{R}_+$,

$$\sum_{l=1}^{d_1} |\sigma^{\cdot l}(t,x) - \sigma^{\cdot l}(t,y)|_d^2 \le K|x - y|_d^2, \quad |b(t,x) - b(t,y)|_d \le K|x-y|_d,$$

and

$$E\Big(|X_0|_d^2 + \int_{[T_0,T]} \Big(|b(t,0)|_d^2 + \sum_{l=1}^{d_1} |\sigma^{\cdot l}(t,0)|_d^2\Big)\Big)dt < \infty.$$

Then the system (4.4) *possesses a unique* solution* $X(t,X_0,t_0)$, *and for each* $p \in [1,\infty[$ *there exists a constant* N *depending on* p *and* K *such that*

---

*The second inequality is not necessary for the existence and the uniqueness of the solution.

$$E \sup_{s\in[T_0,T]} |X(s, X_0, T_0) - X_0|_d^{2p} \le$$

$$(4.5) \quad \le N|T-T_0|^{p-1} e^{N(T-T_0)} E \int_{[T_0,T]} \Big(|X_0|^{2p} + |b(s,0)_d^{2p} + \sum_{l=1}^{d_1} \sigma^{\cdot l}(s,0)|_d^{2p}\Big) ds$$

and

$$(4.6) \quad E \sup_{s\in[T_0,T]} |X(s, X_0', T_0) - X(s, X_0'', T_0)|_d^{2p} \le Ne^{N(T-T_0)} E|X_0' - X_0''|_d^{2p},$$

where X_0' *and* X_0'' *are* $\mathcal{F}_{T_0}$*-measurable stochastic variables in* $\mathbf{R}^d$. □

We shall say that the process $X(t, X_0, T_0)$ is a diffusion process. The vector $f:= (f^i)$ and the matrix $\sigma:= (\sigma^{il})$ will be called the drift and the diffusion coefficients.

The reason for this terminology is presented below.

Suppose $X_0(\omega) = x\in\mathbf{R}^d$, $T_0 = s$, and the drift and the diffusion coefficients do not depend on ω. If these coefficients are sufficiently smooth in x, then for every $f \in \mathbf{C}_b^2(\mathbf{R}^d)$ the function

$$(4.7) \quad u(s,x):= \mathbb{E}f(X(t,x,s))$$

Is a solution of the backward Komogorov equation

$$(4.8) \quad -\frac{\partial u(s,x)}{\partial s} = \frac{1}{2}\sigma^{il}\sigma^{jl}(s,x)\, u_{ij}(s,x) + b^i(s,x)\, u_i(s,x), \quad (s,x) \in [0,t[\times \mathbf{R}^d,$$

$$(4.9) \quad u(t,x) = f(x), \quad x \in \mathbf{R}^d.$$

As a rule, physicists call a system of this type a diffusion equation and use it for modeling the similarly called physical phenomenon. Thus we can consider a solution of the Ito equation as a mathematical model for the trajectory of a particle in diffusion.

4.4. The representation (4.5) and the backward Kolmogorov equation are explicitly treated in Ch.5. For this purpose the following result will be helpful.

Theorem 4. *Suppose that the conditions of Theorem* 3 *are satisfied and* $X_0(\omega) = x \in \mathbf{R}^d$. *Suppose also that the drift and the diffusion coefficients in* (4.4) *belong to* $\mathbf{C}_b^{0,m+1}([T_0,T] \times \mathbf{R}^d)$, ($\mathbf{P}$-*a.s.*), *and the bounds for the coefficients and their derivatives do not depend on* ω. *Then there exists a function* $X(t,x,s,\omega)$ *on* $[s,T] \times \mathbf{R}^d \times [T_0,T] \times \Omega$ *belonging to* $\mathbf{C}^{0,m}([s,T] \times \mathbf{R}^d)$ *for every* $s \in [T_0,T[$, ($\mathbf{P}$-*a.s.*), *and being a solution of system* (4.4), *for* $s = T_0$. □

The proof of this statement could be derived from the results of [63], (Ch. II, §5) on the basis of the following criterion continuity of random function (see also [15]).

4.5. *Theorem* 5 ***(Kolmogorov's criterion).*** *Let* G *be a domain in* $\mathbf{R}^d$, *and* $y(x,\omega)$ *a* $\mathcal{B}(G) \otimes \mathcal{F}$-*measurable mapping of* $G \times \Omega$ *into* $\mathbf{R}^m$. *Then if there exist numbers* $\nu > 0$, $\epsilon > 0$, *and* $N \in \mathbf{R}_+$ *such that*

$$E|y(x) - y(x')|_m^\nu \leq N|x - x'|_d^{d+\varepsilon}, \quad \forall x,\ x' \in G,$$

$y(x,\omega)$ *has a* $\mathbf{P}$-*a.s. continuous in* x *version.* □

A proof of the theorem $d = 1$ can be found in [24] (Russian translation) on p.576. The general case is considered in the same way.

We now show how to obtain the existence of a version of $X(t,x,s)$ continuous in t, x, s.

Proposition 5. *Suppose that conditions of Theorem* 3 *are satisfied and*

$$|b(t,0)|_d^2 + \sum_{i=1}^{d_1} |\sigma^{\cdot i}(t,0)|_d^2 \leq K < \infty$$

for all t, *and* w, *and* $X_0(w) \equiv x \in R$. *Then there exists a function* $\overline{X}$: $[s,T] \times \mathbf{R}^d \times [T_0,T] \times \Omega \to \mathbf{R}^d$ *which is a solution of equation* (4.4) *for* $T_0 = s$, *which is continuous on* $[s,T] \times \mathbf{R}^d \times [T_0,T]$, ($\mathbf{P}$-*a.s.*). □

Proof. Let x', $x'' \in \mathbf{R}^d$ and $t' \geq t'' \geq s' \geq s''$ be points of the interval $[T_0,T]$. Also let $X(t,x,s)$ be a solution of system (4.4) for $T_0 = s$, $X_0(w) \equiv x$, and $p \in [1,\infty[$. Then

$$|X(t', x', s') - X(t'', x'', s'')|_d^{2p} \leq 9^p(|X(t', x', s') -$$

$$X(t'', x', s')|_d^{2p} + |X(t'', x', s') - X(t'', x'', s')|_d^{2p} +$$

$$|X(t'', x'', s') - X(t'', x'', s'')|_d^{2p} = 9^p(I_1 + I_2 + I_3).$$

It follows from (4.6) that there exists a constant N depending on p, K, T_0, T_1

such that $I_2 \le N_1 |x' - x''|_d^{2p}$.

In view of the uniqueness of solutions system (4.4) we have $X^i(t'', x'', s'') = X^i(t'', X(s', x'', s''), s')$, ($\mathbf{P}$-*a.s.*), for every i. Thus it follows from (4.6) that there exists a constant N_2 which depends only on p, K, T and T_0 such that

$$\mathsf{E}\, I_3 = \mathsf{E}|X(t'', x'', s') - X(t'', X(s', x'', s''), s')|_d^{2p} \le N_2 |s' - s''|^p.$$

We can show similarly that there exists a constant N_3 depending only on p, K, T_0, T such that $\mathsf{E}\, I_1 \le N_3\, |t' - t''|^p$. Thus, making use of Kolmogorov's criterion, continuity follows.

4.6. The assumptions of Theorem 3 are assumed to hold for this paragraph.

The process $X(t, X_0, T_0)$ generates a measure μ_X on $\mathfrak{B}(\mathfrak{C}_d) := \mathfrak{B}(\mathrm{C}(T_0, T] \times \mathbb{R}^d)$ given by

$$\mu_X(\Gamma) := P(X(\cdot, X_0, T_0) \in \Gamma), \quad \forall \Gamma \in \mathfrak{B}(\mathfrak{C}_d).$$

Let $X(t, \tilde{X}_0, T_0)$ be a solution of system (4.4) where the drift coefficient b is replaced

by $\tilde{b} := (b^i + h^l \sigma^{il})$. It is also assumed that the functions $h^l := h^l(t, x, \omega)$ are uniformly bounded and predictable. The following important result holds true (see e.g. [90], Ch. 7).

Theorem 6 (Girsanov's theorem). (i) The measures μ_X and $\mu_{\tilde{X}}$ are mutually absolutely continuous with

$$\frac{d\mu_{\tilde{X}}}{d\mu_X}(X(\cdot)) = \mathsf{E}\Big(exp\Big\{ \int_{[T_0, T]} h^l(t, X(t, X_0, T_0))\, dw^l(t) -$$

$$- \frac{1}{2} \int_{[T_0, T]} h^l h^l(t, X(t, X_0, T_0))\, dt \Big\} | \mathfrak{F}_T^{T_0}(X) \Big),$$

and

$$\mathsf{E}\, \frac{d\mu_{\tilde{X}}}{d\mu_X}(X(\cdot)) = 1, \qquad (\mathbf{P}\text{-}a.s.)$$

where $X(\cdot)$ is a trajectory of the process $X(t, x, T_0)$ and $\mathfrak{F}_T^{T_0}(x)$ is the σ-algebra generated by this process for $t \in [T_0, T]$.

(ii) If Φ is a Borel mapping of $\mathbf{C}([T_0,T];\mathbf{R}^d)$ into $\mathbf{R}^d$, $\tilde{X}(\cdot)$ a trajectory of $\tilde{X}(t, X_0,T)$ for $t\in[T_0,T]$, and $\Phi(\tilde{X}(\cdot))$ is $\mathbf{P}$-integrable, then $\Phi(X(\cdot))\frac{d\mu_{\tilde{x}}}{d\mu_x}(X(\cdot))$ is also a $\mathbf{P}$-integrable random variable and $\mathbf{E}\Phi(\tilde{X}(\cdot))=\mathbf{E}\left(\Phi(X(\cdot))\frac{d\mu_{\tilde{x}}}{d\mu_x}(X(\cdot))\right)$. □

4.7. Let $w(t)$ be a standard Wiener process in $\mathbf{R}^1$. Denote by $\mathfrak{F}_t^s$ the σ-algebra generated by increments $w(u_1) - w(u_2)$ for $u_1, u_2 \in [s,t]$. Let $\mathcal{Y}_0$ be some sub-σ-algebra of $\mathcal{F}_0$ completed with respect to the measure $\mathbf{P}$, and let us denote $\mathcal{Y}_0 \vee \mathfrak{F}_t^0$ by $\mathcal{Y}_t^0$.

Warning 7. Here and in the sequel if $\mathcal{F}$ and G are two σ-algebras, $\mathcal{F}\vee G$ is to be the smallest σ-algebra which contains them both. □

Theorem 7. Let $b(t)$ be a predictable real valued process satisfying inequality (4.1). Then there exists a predictable function $E_{\mathcal{Y}}b(t)$ equal to $E[b(t)|\mathcal{Y}_t]$, ($<w>\circ\mathbf{P}$-a.s.), on $\Omega \times [0,T]$ and such that

$$E\left(\int_{[0,T]} b(t)\,dw(t)|\mathcal{Y}_T^0\right) = \int_{[0,T]} E_{\mathcal{Y}}b(t)\,dw(t), \quad (\mathbf{P}\text{-}a.s.). \ \square$$

Proof. As it was mentioned in §1.3.14, we can approximate the process b (in $\mathbf{L}_2([0,T] \times \Omega; \mathcal{P}_T; <w>\circ\mathbf{P}_T; \mathbf{R}^1)$) by simple predictable processes $b^n(t)$. Thus it suffices to prove the theorem for the process b of the form

$$b(t) = 1_{\{0\}}(t)\,\hat{b}(t_0) + \sum_{i=0}^{n-1} \hat{b}(t_i)\,1_{\{[t_i,t_{i+l}]\}}(t),$$

where $0 = t_0 < t_1 <\ldots< t_n = T$, and $\hat{b}(t_i)$ are $\mathbf{P}$-integrable, $\mathcal{F}_{t_i}$-measurable random variables.

Clearly,

$$E[\hat{b}(t)|\mathcal{Y}_t^0] = E[\hat{b}(t_0)|\,\mathcal{Y}_0 \vee\mathfrak{F}_t^0]\,1_{\{0\}}(t) +$$

$$\sum_{i=0}^{n-1} E[\hat{b}(t_i)|\,\mathcal{Y}_{t_i}^0 \vee\mathfrak{F}_t^{t_i}]\,1_{\{]t_i,t_{i+1}]\}}(t).$$

Since σ-algebras $\mathfrak{F}_t^{t_i}$ and $\mathcal{F}_{t_i}$ are independent for $t_i < t$, then for such t_i,

$$E[\hat{b}(t_i)|\,\mathcal{Y}_{t_i}^0 \vee\mathfrak{F}_t^{t_i}] = E[\hat{b}(t_i)|\,\mathcal{Y}_{t_i}], \quad \mathbf{P}\text{-}a.s.$$

Therefore, $\mathbb{E}[b(t)|\mathcal{Y}_t^0]$ has a predictable (relative to the family $\{\mathcal{Y}_t^0\}$) version

$$\mathbb{E}_{\mathcal{Y}} f(t) := E[\hat{b}(t_0)|\ \mathcal{Y}_t^0]\ 1_{\{0\}}(t) + \sum_{i=0}^{n-1} E[\hat{b}(t_i)|\mathcal{Y}_{t_i}^0]\ 1_{\{]t_i,t_{i+t}]\}}(t).$$

In view of the $\mathcal{Y}_T^0$-measurability of the increments $w(u_1) - w(u)$ for u_1, $u_2 \in [0,T]$, and the independence of $\hat{b}(t_i)$ with respect for $\mathcal{F}_T^i$, we have, ($\mathbb{P}$-a.s.),

$$\mathbb{E}\Big(\int_{[0,T]} b(t)dw(t)|\ \mathcal{Y}_T^0] = \mathbb{E}\left(\sum_{i=0}^{n-1} \hat{b}(t_i)(w(t_{i+1}) - w(t_i))|\ \mathcal{Y}_T^0\right) =$$

$$\sum_{i=0}^{n-1} \mathbb{E}[\hat{b}(t_i)|\ \mathcal{Y}_{t_i}^0]\ (w(t_{i+1}) - w(t_i)) = \int_{[0,T]} E_{\mathcal{Y}} b(t)dw(t).$$

which proves the theorem. □

4.8. The following statement is a stochastic version of Fubini's theorem (cf. [27]).

Theorem 8. Let $f(t,x,w)$ *be a* $\mathcal{B}([0,T] \times \mathbf{R}^d) \otimes \mathcal{F}$*-measurable, predictable for every* x*, real-valued function belonging to* $\mathbb{L}_2\ ([0,T] \times \mathbf{R}^d) \cap \mathbb{L}_2\ ([0,T];\ \mathcal{B}([0,T]);\ l,\ \mathbb{L}_1(\mathbf{R}^d))$ *($\mathbb{P}$-a.s.). Then there exists a predictable function* $Box(t,\omega)$ *equal to the integral* $\int_{\mathbb{R}^d} f(t,x,\omega)ds$*, ($\mathbb{P}$-a.s.), and such that* $\int_{\mathbf{R}^d} \int_{[0,T]} b(t,x)dw(t)dx = \int_{[0,T]} box(t)dw(t)$. □

4.9. Let $\xi(t)$ be a stochastic process possessing stochastic differential (4.3). Suppose that F is a $\mathcal{P}\otimes\mathcal{B}(\mathbf{R}^d)$-measurable function which maps $\mathbf{R}_+ \times \mathbf{R}^d \times \Omega$ *to* $\mathbf{R}^1$, which belongs to $\mathbf{C}^{0,2}(\mathbf{R}_+,\mathbf{R}^d)$ for $\mathbf{P}$-*a.a.* ω, and possesses, for every $x\in\mathbf{R}$ the stochastic differential

$$dF(t,x,\omega) = J(t,x)dt + H^l(t,x)dw^l(t),\ t<\tau,$$

where $F(t,x)$, $H^l(t,x)$, $l = 1,2,...d_1$, are $\mathcal{P}\otimes\mathcal{B}(\mathbf{R}^d)$-measurable functions mapping $\mathbb{R}_+ \times \mathbf{R}^d \times \Omega$ into $\mathbf{R}^1$ and τ is the same stopping times as in (4.3). We also assume that $F(.,.,\omega) \in \mathbf{C}^{0,0}(\mathbf{R}_+ \times \mathbf{R}^d)$ and $H^l \in \mathbf{C}^{0,1}(\mathbf{R}_+ \times(\mathbf{R}^d)$ for all $l = 1,2,...,d_1$ and $\mathbb{P}$-a.a. ω.

Theorem 9. (The Ito-Ventcel formula). The process $F(t,\xi(t))$ *possesses the stochastic differential*

$$dF(t,\xi(t)) = J(t,\xi(t))dt + H^l(t,\xi(t))dw^l(t) +$$

(4.10) $$+ \Big(b^i F_i(t,\xi(t)) + \tfrac{1}{2}\,\sigma^{il}\sigma^{il} F_{ij}(t,\xi(t)) +$$

$$+ H^l_i\sigma^{il}(t,\xi(t))\Big)dt + \sigma^{il}F_i(t,\xi(t))dw^l(t), \quad t<\tau. \ \square$$

4.10. Before proceeding to the proof of the theorem, we consider one method of smoothing which will be used here and in many other places below.

Set

$$\zeta(x) := \begin{cases} exp\{|x|^2_d\,(|x|^2_d - 1)^{-1}, & for\ |x|^2_d \le 1; \\ 0, & for\ |x|^2_d > 1. \end{cases}$$

Let $x/\varepsilon: = (x^1/\varepsilon,\dots,x^d/\varepsilon)$. It is clear that the function $\zeta(x/\varepsilon)$ has the properties as follows:
(a) $\zeta(x/\varepsilon) \in C_0^\infty(\mathbf{R}^d)$, (b) the support of $\zeta(x/\varepsilon)$ is the sphere of radius ε and center zero, (c) $\zeta(0) = 1$, (d) $0 \le \zeta(x) \le 1$.

Let f be a locally integrable function on $\mathbf{R}^d$. Denote

$$T_\varepsilon f(\chi) := \varepsilon^{-d}\chi \int_{\mathbf{R}^d} \zeta\Big(\frac{y-x}{\varepsilon}\Big)\, f(y)dy,$$

where

$$\chi := \Big(\int_{\mathbf{R}^d} \zeta(x)dx\Big)^{-1}.$$

The function $T_\varepsilon f$ is usually called Sobolev's average. Let us consider some important properties of this average.

Lemma 10. *Let* $f \in C^{0,0}(\mathbf{R}_+ \times \mathbf{R}^d)$, *then* $T_\varepsilon f(t,x)^*$ *belongs to* $C^{0,\infty}(\mathbf{R}_+ \times \mathbf{R}^d)$ *and for every* $N \in \mathbf{R}_+$

(4.11) $$\lim_{\varepsilon\to 0}\ \sup_{|x|_d<N} |f(t,x) - T_\varepsilon f(t,x)| = 0.$$

If $f \in C^{0,1}(\mathbf{R}_+ \times \mathbf{R}^d)$, *then*

$$\frac{\partial}{\partial x^i}\, T_\varepsilon f(t,x) = T_\varepsilon \frac{\partial}{dx^i} f(t,x), \quad \forall\ i=1,2,\dots d. \ \square$$

Proof. Similar statements for functions from $\mathbf{L}_p$ are well known (see Lemma 4.1.3 and the reference in 4.1.3). Thus, we prove only equality (4.11) leaving to the reader everything else.

Obviously for every T, $N\in\mathbf{R}_+$

$$\sup_{\substack{t\le T\\ |x|_d\le N}} |f(t,x) - T_\varepsilon f(t,x)| =$$

$$\sup_{\substack{t\le T\\ |x|_d\le N}} |\chi \int_{|z|_d\le 1} \zeta(z)[f(t,x) - f(t,x+\varepsilon z)]dz| \le$$

$$\le \chi \sup_{\substack{t\le T\\ |x|_d\le N}} \Big(\int_{|z|_d\le 1} |\zeta(z)|^2 dz\Big)^{\frac{1}{2}} \Big(\int_{|z|_d\le 1} |f(t,x) - f(t,x+\varepsilon z)|^2 dz\Big)^{\frac{1}{2}},$$

This inequality clearly implies (4.11). □

4.11. *Proof of Theorem* 5. For the sake of simplicity we shall suppose that $\tau=\infty$. To begin with, we prove the following auxliary result.

Lemma 11. Let the conditions of Theorem 5 be satisfied. Suppose also that $F\in C^{0,1}(\mathbf{R}_+\times\mathbf{R}^d)$, ($\mathbf{P}$-.*a.s.*), and for every $x\in\mathbf{R}^d$ and $i=1,2,...d$ the function $F_i(t,x,w)$ possesses the following stochastic differential.

$$dF_i(t,x) = J_i(t,x)dt + H_i^l(t,x)dw^l(t),\ t\in\mathbf{R}_+.$$

Then the assertion of Theorem 5 holds. □

Proof. We shall prove the lemma in the case of coefficients b^i, σ^{il} independent to t,w. The proof is carrying over to the general case in the same way as in the proof of Ito's formula.

Fix s, $t\in\mathbf{R}_t$, $s\le t$, and let $\{s=t < t_1^n <...< t_{k(n)+1}^n = t\}$ be a sequence of partitions of the interval $[s,t]$ for $n\in\mathbf{N}$. We write $\Delta_n(k):= [t_k^n, t_{k+1}^n]$, $\Delta_n:= \max_n |t_{k+1}^n - t_k^n|$ and $\xi_{(n)}(k):= \xi(t_k^n)$. We denote $\mathbf{P}$-lim, the limit in probability as $\Delta_n\to 0$.

---

*Here and below $T_\varepsilon f(t,x)$ is a notation for $T_\varepsilon f(t,\cdot)(x)$. The averaging is done over $\mathbf{R}^d$ only.

Obviously,

$$F(t,\xi(t)) - F(s,\xi(s)) = \sum_{k=0}^{k(n)} [F(t^n_{k+1},\xi_{(n)}(k+1)) -$$

$$F(t^n_k, \xi_{(n)}(k))] = \sum_{k=0}^{k(n)} [F(t^n_{k+1}, \xi_{(n)}(k+1)) -$$

$$F(t^n_k, \xi_{(n)}(k+1)] + \sum_{k=0}^{k(n)} [F(t^n_k, \xi_{(n)}(k+1)) -$$

$$F(t^n_k, \xi_{(n)}(k))] := I_1^{(n)} + I_2^{(n)}.$$

First consider $I_2^{(n)}$. By Taylor's formula

$$\sum_{k=0}^{k(n)} [F(t^n_k, \xi_{(n)}(k+1)) - F(t^n_k, \xi_{(n)}(k))] =$$

$$= \sum_{k=0}^{k(n)} F_i(t^n_k, \xi_{(n)}(k)) (\xi^i_{(n)}(k+1) - \xi^i_{(n)}(k)) +$$

$$+ \sum_{k=0}^{k(n)} \frac{1}{2} F_{ij}(t^n_k, \xi_{(n)}(k) + \theta(\xi_{(n)}(k+1) -$$

$$- \xi_{(n)}(k))) (\xi^i_{(n)}(k+1) - \xi^i_{(n)}(k)) (\xi^i_{(n)}(k+1) - \xi^l_{(n)}(k)),$$

where $\theta := \theta(w) \in [0,1]$ for every ω.

From the last equality it follows that

$$\mathbf{P}\text{-}lim\ I_2^{(n)} = \int_{[s,t]} F_i(u, \xi(u)) d\xi^i(u) + \frac{1}{2} \int_{[s,t]} F_{ij}(u, \xi(u)) \sigma^{il} \sigma^{jl} du.$$

Consider $I_1^{(n)}$, by Taylor's formula

$$I_1^{(n)} = \sum_{k=0}^{k(n)} [F(t^n_{(k+1)}, \xi_{(n)}(k)) - F(t^n_k, \xi_{(n)}(k))] +$$

$$+ \sum_{k=0}^{k(n)} \Big(\sum_{m=0}^{1} (-1)^{m+1} (F_i(t^n_{k+m}, \xi_{(n)}(k) +$$

$$+ \theta_m(\xi_{(n)}(k+1) - \xi_{(n)}(k))) (\xi^i_{(n)}(k+1) - \xi^i_{(n)}(k))\Big) :=$$

$$= I_{11}^{(n)} + I_{12}^{(n)},$$

where $\theta_m := \theta_m(\omega)$ for $m=0,1$, and $\theta_m(\omega) \in]0,1[$ for every m and w.
It is evident that

$$\mathbf{P}\text{-}lim\ I^n_{11} = \int_{[s,t]} J(u, \xi(u))du + \int_{[s,t]} H^l(u, \xi(u))dw^l.$$

Next for $I_{12}^{(n)}$, it is easy to show that

$$\mathbf{P}\text{-}lim \sum_{k=0}^{k(n)} \Big(\sum_{m=0}^{1} (-1)^{m+1} [F_i(t^n_{k+m}, \xi_{(n)}(k) +$$

$$+ \theta_m(\xi_{(n)}(k+1) - \xi_{(n)}(k))) -$$

$$- F_i(t^n_{k+m}, \xi_{(n)}(k))]\Big) (\xi^i_{(n)}(k+1) - \xi^i_{(n)}(k)) = 0.$$

Making use of the representation of F we have

$$\mathbf{P}\text{-}lim \sum_{k=0}^{k(n)} [F_i(t^n_{k+1}, \xi_{(n)}(k)) - F_i(t^n_k, \xi_{(n)}(k))] \times$$

$$\times (\xi^i_{(n)}(k+1) - \xi^i_{(n)}(k)) =$$

$$\mathbf{P}\text{-}lim \Big(\sum_{k=0}^{k(n)} \int_{\Delta_n(k)} J_i(s, \xi_{(n)}(k)) ds (\xi^i_{(n)}(k+1) - \xi^i_{(n)}(k)) +$$

$$+ \sum_{k=0}^{k(n)} \int_{\Delta_n(k)} H^l_i(s, \xi_{(n)}(k)) dw^l(s) b^i (t^n_{k+1} - t^n_k) +$$

$$+ \sum_{k=0}^{k(n)} \int_{\Delta_n(k)} H^l_i(s, \xi_{(n)}(k)) dw^l(s)\ \sigma^{ij}\ (w^i(t^n_{k+1}) - w^i(t^n_k)) :=$$

$$= I^{(n)}_{121} + I^{(n)}_{122} + I^{(n)}_{123}.$$

It is easy to see that

$$\mathbf{P}\text{-}lim\ I^{(n)}_{121} = \mathbf{P}\text{-}lim\ I^{(n)}_{122} = 0,$$

and

$$\mathbf{P}\text{-}lim\ I^{(n)}_{123} = \int_{[s,t]} \sigma^{il} H^1_i(s, \xi(s)) ds.$$

Collecting the above equalities, we get formula (4.10).

Let us fix $\varepsilon > 0$ and write $F_\varepsilon(t,x) := T_\varepsilon F(t,x)$, $J_\varepsilon(t,x) := T_\varepsilon J(t,x)$ and $H^l_\varepsilon(t,x) = T_\varepsilon H^l(t,x)$. We now show that the function $F_\varepsilon(t,x)$ satisfies the conditions of the lemma.

Note that the stochastic integral $\int_{[0,t]} H^l(s,x) dw^l(s)$ has a continuous (in (t,x)) version. This follows, for example, from the continuity of $F(t,x)$ and $\int_{[0,t]} J(s,x) ds$ and the connection between these functions and the stochastic integral via the stochastic differential of $F(t,x)$.

Thus we can and shall suppose that $\int_{[0,t]} H^l(s,x)dw^l(s)$ is a $\mathbf{P}$-a.s. locally l_d-integrable function for every t. Consequently, the operator T_ε is defined on the stochastic integral.

From this, Fubini's theorem and Theorem 8 it follows that $F_\varepsilon(t,x)$ possesses stochastic differential

$$dF_\varepsilon(t,s) = J_\varepsilon(t,x)dt - H^l_\varepsilon(t,x)dw^l(t),\ t\in\mathbf{R}_+.$$

Making use of Lemma 10 we get that the functions $F_\varepsilon(t,x)$ $H^l_\varepsilon(t,x)$ and $J_\varepsilon(t,x)$ belong to $\mathbf{C}^{0,\infty}(\mathbf{R}_+\times\mathbf{R}^d)$ for $\mathbf{P}$-a.s. w. Theorem 8 and Lemma 10 imply that for every $t\in\mathbf{R}_+$ and $i=1,2,\dots,d$,

$$\Big(\int_{[0,t]} H^l_\varepsilon(s,x)dw^l(s))\Big) = \Big(T_\varepsilon\Big(\int_{[0,t]} H^l(s,\cdot)dw^l(s)\Big)(x)\Big)_i =$$

$$= \varepsilon^{-d}\chi\int_{\mathbf{R}_d}\zeta_i\left(\frac{x-y}{\varepsilon}\right)\int_{[0,t]} H^l(s,y)w^l(s)dy =$$

$$= \varepsilon^{d}\chi\int_{[0,t]}\int_{\mathbf{R}_d}\zeta_i\left(\frac{x-y}{\varepsilon}\right) H^l(s,y)dy\, dw^l(s) =$$

$$= \int_{[0,t]}(H^l_\varepsilon(s,x))_i dw^l(s)\quad (\mathbf{P}\text{-}a.s.)$$

It is also clear that $\mathbf{P}$-a.s.

$$\Big(\int_{[0,t]} J_\varepsilon(s,x)ds\Big)^i = \int_{[0,t]}(J_\varepsilon(s,x))_i ds.$$

Consequently, for every $i=1,2,\dots,d$, $(F_\varepsilon(t,x))_i$ possesses the stochastic differential

$$d(F_\varepsilon(t,x))_i = (J_\varepsilon(t,x))_i dt + (H^l_\varepsilon(t,x))_i dw^l(t),\quad t\in\mathbf{R}_+.$$

Thus the function $F_\varepsilon(t,x)$ meets the conditions of the lemma. Hence, it possesses the stochastic differential of form (4.10) where F_ε, H_ε and J_ε are substituted for F, H, and J, respectively.

Passing to the limit and making use of Lemma 10 and property (*b*) of stochastic integral from 1 we complete the proof. □

4.12. In this paragraph we define a backward stochastic integral with respect to a Wiener process.

Suppose that another family of σ-algebras, call it $\{\mathfrak{F}^t_T\}$ $t\in[0,T]$, is given on $(\Omega,\mathfrak{F})$. We also suppose that for every s, $t\in[0,T]$, $\mathfrak{F}^t_T \subset \mathfrak{F}$ and $\mathfrak{F}^s_T \supset \mathfrak{F}^t_T$ if $s\leq t$. It is assumed that $\bigcap_{\varepsilon>0} \mathfrak{F}^{t-\infty}_T = \mathfrak{F}^t_T$ and the $\sigma-$algebra $\mathfrak{F}^T_T$ contains all P-null sets in $\mathfrak{F}$.

Definition 12. *The function* $f(.,.)$: $[T_0,T] \times \Omega \rightarrow \mathbb{R}^1$ *is said to be backward predictable on the compact interval* $[T_0,T]$ (*or* $\overleftarrow{\mathfrak{P}}_{[T_0,T]}$-*measurable*) *relative to the family* $\{\mathfrak{F}^t_T\}$, *if the function* $f(T\text{-}.,.)$ *is predictable with respect to the family* $\{\mathfrak{F}^{T-t}_T\}$ *for* $t\in[T_0,T]$. □

Denote $w_T(t) := w(T) - w(T\text{-}t)$, where $w(t)$ is the standard $\mathbf{R}^{d_1}-$valued Weiner process which is fixed in this section. It is assumed that $w_T(t)$ is a standard Wiener martingale relative to the family $\{\mathfrak{F}^{T-t}_T\}$, where $t\in[0,t]$.

Remark 12. *The* σ-*algebra generated by the "future" of the Wiener process* $w(t)$ (*i.e. by increments* $w(T) - w(s)$ *for* $s\leq t$ *and completed with respect to the measure* $\mathbf{P}$ *usually stands for* $\mathfrak{F}^t_T$. *Obviously, in this case* $\mathfrak{F}^{T-t}_T$ *coincides with the* $\sigma-$*algebra generated by values of the process* $w_T(s)$, *for* $s\leq t$, *which is completed with respect to the measure* $\mathbf{P}$. *It is clear that the process* $w_T(s)$ *is a Wiener martingale with respect to the family of* $\mathbf{P}-$*completed* σ-*algebras generated by itself.* □

4.13. *Let* f be a backward predictable function on $[T_0,T]$ with respect to the family $\{\mathfrak{F}^t_T\}$. Suppose that for every s, $t\in[T_0,T]$ such that $s\leq t$ and $T\text{-}t\leq T_0$

$$\int_{[T_s,T_t]} |f(r)|^2 dr < \infty \quad (\mathbf{P}\text{-}a.s.)$$

where $T_r := T\text{-}r$.

Definition 13. *A backward stochastic integral with respect to a one-dimensional Wiener process* $w^i(t)$, $i=1,2,...d_1$ *is given by the equality*

$$\int_{[s,t]} f(r) * dw^i(r) := \int_{[T_t,T_s]} f(T_r) dw^i(r). \quad \square$$

Remark 13. *It is easily seen that the definition does not depend on the choice of T. This statement can easily be verified in the case that $f(t)$ is a step-function and carried over to the general case by a passage to the limit.*

The reader can easily prove the following result.

Proposition 13. *If $f(t)$ is a simple backward predictable (relative to the family $\{\mathcal{F}^t_T\}$ function on $[T_0,T]$ i.e. $f(T-t)$ is a simple predictable (relative to the family $\{\mathcal{F}^{T-t}_T\}$) stochastic process on the compact interval $[0,T-T_0]$, then $\mathbf{P}-a.s.$*

$$\int_{[s,t]} f(r) * dw^i(r) = \sum_j f(t_{j+1})\,(w^i(t_{j+1}) - w^i(t_j)),$$

where $\{t_j\}$ are the points of jumps of $f(r)$. □

4.14. Suppose that $\sigma^{il}(t,w)$, $f^i(t,w)$, where $i=1,2,\ldots,d$ and $l=1,2,\ldots,d_1$, are backward predictable (relative to the family $\{\mathcal{F}^t_T\}$), real-valued functions on $[0,T]$. Suppose also that $\sigma^{il}(\cdot,w)\in\mathbb{L}_2((0,T))$ and $f^i(t,\cdot,\omega)\in\mathbb{L}_1([0,T])$ for $l=1,2,\ldots,d_1$, and $i=1,2,\ldots,d$ ($\mathbb{P}$-*a.s.*).

Definition 14. *Let τ be a stopping time with respect to the family $\{\mathcal{F}^t_T\}$* such that $\tau\le T$ $\mathbb{P}$-a.s. and $\xi(t)$ be a continuous backward predictable process on $[0,T]$ taking values in $\mathbb{R}^d$. We shall say that $\xi(t)$ possesses the backward stochastic differential*

$$(4.12)\qquad -d\xi^i(t) = b^i(t)dt + \sigma^{il}(t) * dw^l(t),\ t\in]\tau,T],\ \ i=1,2,\ldots,d,$$

if for every $i=1,2,\ldots,d$ it satifies the equality

$$\mathbb{P}\Big(\sup_{t\in[\tau,T]} |\xi^i(t) - \xi^i(T) - \int_{[t,T]} b^i(s)ds -$$

$$- \int_{[t,T]} \sigma^{il}(s)dw^l(s)| = 0\Big\} = 1.\ \square$$

The following result is an obvious version of Ito's formula.

Proposition 14. *Let f be a function from $\mathbf{C}^{1,2}(\mathbb{R}_+,\mathbb{R}^d)$ and $\xi(t)$ be a*

---

*It means that $T-\tau$ is a stopping time with respect to the family $\{\mathcal{F}^{T-t}_T\}$.

stochastic process possessing the backward stochastic differential (4.12), then f(t, ξ(t)) also possesses a backward stochastic differential of the form

$$- df(t, \xi(t)) = \Big(- \tfrac{\partial}{\partial t} f(t,x)|_{x=\xi(t)} +$$

$$+ \tfrac{1}{2} \sigma^{il} \sigma^{jl} f_{ij}(t, \xi(t)) + b^i f_i(t, \xi(t))\Big)\, dt +$$

$$+ \sigma^{il} f_i(t, \xi(t)) * dw^l(t), \qquad t \in [\tau, T]. \ \square$$

We can consider a backward Ito equation in a completely analogous way as it is done for a "forward" one (see 3). The theory of backward Ito equation is identical to the one developed for forward Ito equations.

Warning 14. *In the future when considering backward Ito's equations we shall use corresponding results for forward ones without special references.* □

4.15. In concluding this section we give one technical result frequently used in the theory of differential equations and, in particular, in this book.

Lemma 15 *(Gronwall-Bellman's lemma, cf.* [6]). *Let* $u(t)$ *and* $v(t)$ *be non-negative* $\mathfrak{B}(\mathbb{R}_+)$*−measurable functions mapping* $\mathbf{R}_+$ *into* $\overline{\mathbb{R}}^1$ *and connected for all* $t \geq T_0 \in \mathbb{R}_+$ *by the inequality,*

$$u(t) \leq N + \int_{[T_0,t]} u(s)v(s)\,ds, \quad N \in \mathbb{R}_+.$$

Then for all $t \geq T_0$,

$$u(t) \leq N \exp\Big\{ \int_{[T_0,t]} v(s)\,ds \Big\}. \ \square$$

Chapter 2

STOCHASTIC INTEGRATION IN A HILBERT SPACE

2.0. Introduction.

The essentials of the integrodifferential calculus of continuous martingales and local martingales in a Hilbert space are the topics of this chapter. In particular, definitions and investigations of martingales, local martingales and a Wiener process in a Hilbert space are presented and the construction of stochastic integrals with respect to these processes are given. We also derive Ito's formula for the square of a norm of a continuous semimartingale.*

2.1. Martingales and Local Martingales.

1.0. In this section we introduce the notions of a martingale and a local martingale taking values in a Hilbert space and consider their main properties. For the sake of simplicity, we consider only continuous martingales and local martingales.

1.1. Let $\mathbf{F} := (\Omega, \mathcal{F}, \{\mathcal{F}_t\}_{t \in \mathbf{R}_+}, \mathbf{P})$ be a standard probability space, $\mathbb{H}$ be a Hilbert space and $\mathbb{H}^*$ be its conjugate space.

We denote by $\|\cdot\|$ the norm in $\mathbf{H}$ and by $(\cdot,\cdot)$ the scalar product consistent with the norm. The value of a functional $h^* \in \mathbb{H}^*$ on $h \in \mathbf{H}$ will be denoted by h^*h.

The integral $\int_\Omega x(\omega)\, d\mathbb{P}(\omega)$ of the $\mathbf{P}$-integrable random variable $x(\omega)$ with values in $\mathbb{H}$ is denoted by $\mathbf{E}x$ and called expectation of x.

1.2. Let $\mathcal{G}$ be some sub-σ-algebra of $\mathcal{F}$ and x be a $\mathbf{P}$-integrable $\mathbf{H}$-valued random variable.

---

*A semimartingale is sum of a martingale and a process of bounded variation.

Definition 2. *The random variable* $\mathbb{E}[x|\mathcal{G}]$ *with values in* $\mathbb{H}$ *is called the conditional expectation of* x *with respect to* $\mathcal{G}$ *if for each* $y \in \mathbb{H}^*$,

$$y\ \mathbb{E}[x|\mathcal{G}] = E[yx|\mathcal{G}] \qquad (\mathbb{P}\text{-}a.s.)\ \Box$$

expectation for $\mathbb{H}$-valued random variables has the same properties as the conditional expectation for a scalar random variable. It is clear that the definition is reasonable. By the first property of Bochner's integral (see §1.2.11) the norm of the x is $\mathbb{P}$-integrable and consequently $\mathbb{E}[yx|\mathcal{G}]$ is defined. Obviously, the random variable $\mathbb{E}[x|\mathcal{G}]$ is $\mathcal{G}$-measurable. It is determined uniquely (up to $\mathbb{P}$-negligible sets) by the last equality. The conditional

Warning 2. *A stochastic process taking values in a space* $\mathbb{H}$ *will usually be referred to as an* $\mathbb{H}$*-processes.* $\Box$

1.3. Let $\{\mathcal{G}_t\}_{t\in\mathbb{R}_+}$ be a increasing family of sub-σ-algebras of $\mathcal{F}$.

Definition 3. *An* $\mathbb{H}$*-process* $M(t)$ *is called a martingale relative to the family* $\{\mathcal{G}_t\}$ *if*

(*i*) $M(t)$ *is* $\mathcal{G}$*-adapted;*

(*ii*) $\mathbb{E}||M(t)|| < \infty$ *for every* $t\in\mathbb{R}_+$;

(*iii*) $\mathbb{E}[M(t)|\mathcal{G}_s] = M(s)$, ($\mathbb{P}$-*a.s.*), *for every* s, $t\in\mathbb{R}_+$ *such that* $s \le t$. $\Box$

1.4. *Definition* 4. *An* $\mathbb{H}$*-process* $M(t)$ *is called a local martingale relative to the family* $\{\mathcal{G}_t\}$, *if there exists a sequence of stopping times* $T_n \uparrow \infty$, ($\mathbb{P}$-*a.s.*), *relative to the same family, such that the* $\mathbb{H}$*-process* $M(t\wedge T_n)$ *is a martingale relative to the family* $\{\mathcal{G}_{t\wedge T_n}\}$ *for every* $n\in N$. $\Box$

Warning 4. *In this section we consider martingales and local martingales relative to the family* $\{\mathcal{F}_t\}$. *Thus we shall omit references to the family of* σ*-algebras with respect to which an* $\mathbb{H}$*-process* $M(t)$ *will be a martingale or a local martingale.* $\Box$

We shall also assume, without loss of generality, that all martingales and local martingles have the property $M(0) = 0$. $\Box$

Lemma 4. *An* $\mathbb{H}$*-process* $M(t)$ *is a martingale (a local martingale) if and only if* $\mathbb{E}||M(t)|| < \infty$, $\forall t\in\mathbb{R}_+$, ($\mathbb{E}||M(t\wedge T_n)|| < \infty$, $\forall t\in\mathbb{R}_+$, *for some sequence of stopping times* $\{T_n\}$ *such that* $T_n \uparrow \infty$ *as* $n \to \infty$) *and for every* $h^*\in\mathbb{H}$, *the process* $h^*M(t)$ *is a martingale (a local martingale for the same localizing sequence* T_n).$\Box$

The lemma is an obvious corollary of Theorem 1.2.8.

1.5. Let us fix a complete orthonormal system (CONS) $\{h_i\}$, $i\in\mathbb{N}$, in $\mathbb{H}$ and denote by $\prod_n$ the orthogonal projection on the sub-space generated by the first n elements of the system $\{h_i\}$ i.e.

$$\textstyle\prod_n x := \sum_{i=1}^{n} (x, h_i)h_i$$

Warning 5. *Throughout this section given* $X\in\mathbb{H}$, *we denote* $X_i := (h_i, X)$. □

Theorem 5. *If* $M(t)$ *is a martingale in* $\mathbb{H}$, *then* $\|M(t)\|$ *is a submartingale. Moreover, if* $\mathbb{E}\|M(t)\|^2 < \infty$, $\forall t\in\mathbb{R}_+$, *then* $\|M(t)\|^2$ *is a submartingale.* □

Proof. As a first step, we prove that $\|\prod_n M(t)\|$ is a submartingale. Let us fix s and $t\in\mathbb{R}_+$ such that $s<t$. Denote by $\mathbf{P}_s$ the regular conditional (relative to $\{\mathcal{F}_s\}$) distribution of the random variables $M_1(t),\ldots,M_n(t)$. It is known that such a distribution exists (see e.g. [35], Ch. 1, §3). Then by Minkowski's inequality,

$$\mathbb{E}[\|\textstyle\prod_n M(t)\| \,|\, \mathcal{F}_s] = \displaystyle\int_{\mathbb{R}^n} \left(\sum_{i=1}^{n} x_i^2 \right)^{\frac{1}{2}} d\mathbf{P}_s(x_1,\ldots,x_n) \geq$$

$$\geq \left(\sum_{i=1}^{n} \left(\int_{\mathbb{R}^n} x_i \, d\mathbf{P}_s(x_1,\ldots,x_n) \right)^2 \right)^{\frac{1}{2}} = \left(\sum_{i=1}^{n} [\mathbb{E} M_i(t) | \mathcal{F}_s]^2 \right)^{\frac{1}{2}} \geq \|\textstyle\prod_n M(s)\|$$

$(\mathbb{P}\text{-}a.s.)$

Letting $n \to \infty$ on both sides of the inequality we get the first statement of the theorem by means of a version of the dominated convergence theorem for conditional expectations (see [24], Ch. 1, §8). The second assertion follows immediately from the first one and the generalized Jensen's inequality (see [24], Ch. 1, §9). □

1.6. *Theorem* 6. *If* $M(t)$ *is a weakly continuous (in* t*) martingale in* $\mathbb{H}$ *it is strongly continuous.* □

Proof. From the weak continuity of $M(t)$ it follows that the process $\prod_n M(t)$ is strongly continuous for every $n\in\mathbb{N}$. It is also obvious that the function $\|M(t) - \prod_n M(t)\|$ is a lower semi-continuous function of t, and consequently for each $T\in\mathbb{R}_+$,

$$\sup_{t\leq T} \|M(t) - \textstyle\prod_n M(t)\| = \displaystyle\sup_{I} \|M(t) - \textstyle\prod_n M(t)\|,$$

where I is the set of all rational points of the interval [0,T]. The process $\|M(t) - \prod_n M(t)\|$ is a submartingale by the proposition established above. Thus, we get from the inequality (1.3.3),

$$\mathbb{P}(\sup_{t\le T} \| M(t) - \textstyle\prod_n M(t)\| \ge \varepsilon) = \mathbb{P}(\sup_{I} \|M(t) - \prod_n M(t)\| \ge \varepsilon) \le$$

$$\le \varepsilon^{-1}\, \mathbb{E}\|M(T) - \textstyle\prod_n M(T)\|.$$

Obviously, the right-hand side of the inequality tends to zero as $n \to \infty$. Hence, for some subsequence $\{n'\}$ from $\mathbb{N}$

$$\lim_{n' \to \infty} \sup_{t\le T} \|M(t) - \textstyle\prod_{n'} M(t)\|^2 = 0 \qquad (\mathbb{P}\text{-}a.s.)$$

which implies that $M(t)$ is a strongly continuous process in $\mathbb{H}$. □

Corollary 6. A local martingale $M(t)$ in $\mathbb{H}$ which is weakly continuous is a strongly continuous $\mathbb{H}$-process. □

Proof. Let $\{T_n\}$ be a localizing sequence for $M(t)$. Then $M(t\wedge T_n)$ is a weakly continuous martingale which implies by the theorem that it is a strongly continuous $\mathbb{H}$-process. On the other hand, it is clear that for every $\varepsilon>0$ and $T\in\mathbb{R}_+$,

$$\mathbb{P}(\sup_{t\le T}\|M(t) - M(t \wedge T_n)\| > \varepsilon) \le \mathbb{P}(T_n < T) \to 0$$

as $n \to \infty$.

Hence, for some subsequence $\{n'\} \subset \mathbb{N}$

$$\lim_{n' \to \infty} \sup_{t\le T} \|M(t) - M(t \wedge T_{n'})\| = 0, \qquad (\mathbb{P}\text{-}a.s.),$$

which proves the statement. □

It is now clear that there is no need to distinguish between weakly and strongly continuous martingales and local martingales in $\mathbb{H}$. For this reason we shall omit the attributes "weakly" or "strongly" when referring to continuous martingales and local martingales

1.7. We denote by $\mathfrak{M}_2^c(\mathbb{R}_+,\mathbb{H})$ the set of all continuous martingales in $\mathbb{H}$ such that $\mathbb{E}\|M(t)\|^2 < \infty$ for every $t\in\mathbb{R}_+$. We shall call processes from this class continuous square integrable martingales in $\mathbb{H}$. The set of all continuous local martingales taking values in $\mathbf{H}$ will be denoted by $\mathfrak{M}^c_{loc}(\mathbf{R}_+,\mathbf{H})$.

Remark. It is easy to show that if $M \in \mathfrak{M}^c_{loc}(\mathbf{R}_+,\mathbf{H})$, *then there exists a localizing sequence* $\{\sigma_n\}$, $n\in\mathbf{N}$, *such that*

$$\|M(t \wedge \sigma_n)\| \leq n, \qquad \forall n\in\mathbf{N},$$

for every $t\in\mathbf{R}_+$ *on one and the same* ω*-set of probability one.* □

From this remark and Theorem 5 it follows that if $M(t) \in \mathfrak{M}^c_{loc}(\mathbf{R}_+,\mathbf{H})$ then $\|M(t)\|$ is a continuous local submartingale in $\mathbf{R}^1$.

Given $m(t) \in \mathfrak{M}^c_{loc}(\mathbf{R}_+,\mathbf{H})$ denote by $<M>_t$ the increasing process such that $\|M(t)\|^2 - <M>_t \in \mathfrak{M}^c_{loc}(\mathbf{R}_+,\mathbf{R}_1)$ (cf. corollary 1.3.12). We shall call this process the quadratic variation of $M(t)$.

For future use we need the following auxiliary result.

Proposition 7. Let $M(t) \in \mathfrak{M}^c_{loc}(\mathbf{R}_+,\mathbf{H})$ *and* τ *be a stopping time, then*

(*i*) *If* $\mathbf{E} \sup_{t\leq\tau} \|M(t)\| < \infty$, *then* $M(t\wedge\tau)$ *is a continuous* $\mathbf{H}$*-martingale on* $\mathbf{R}_+$.

(*ii*) *If* $\mathbf{E}<M>_\tau < \infty$, *then* $M(t\wedge\tau) \in \mathfrak{M}^c_{loc}(\mathbf{R}_+,\mathbf{H})$.*

(*iii*) *If* $\mathbf{E}<M>_\tau^{1/2} < \infty$, *then* $M(t\wedge\tau)$ *is a continuous* $\mathbf{H}$*-martingale on* $\mathbf{R}_+$. □

Proof. We start with a proof of the first part of the statement. Let $\{\sigma_n\}$ be a sequence of stopping times defined in Remark 7. By the Optional Sampling Theorem (see §1.3.9) and Lemma 4, $M(t\wedge\sigma_n\wedge\tau) \in \mathfrak{M}^c_{loc}(\mathbf{R}_+,\mathbf{H})$. On the other hand, in view of the assumption, $\mathbf{E} \sup_{n\in\mathbf{N}} \|M(t\wedge\sigma_n\wedge\tau)\| < \infty$. Hence, the set $\{M(t\wedge\sigma_n\wedge\tau)\}$, $n\in\mathbf{N}$, is uniformly integrable and consequently $\|M(t\wedge\sigma_n\wedge\tau)\| \to \|M(t\wedge\tau)\|$ in $\mathbf{L}_1(\Omega; \mathfrak{F}; \mathbf{P}; \mathbf{R}^1)$.

By an analogous passage to the limit, it follows that for every s, $t\in\mathbf{R}_+$, $s\leq t$,

$$\mathbf{E}[M(t\wedge\tau)|\mathfrak{F}_s] = M(s\wedge\tau) \qquad (\mathbf{P}\text{-}a.s.).$$

To prove the second assertion, observe that by inequality (1.3.4) for every $T\in\mathbf{R}_+$,

---

*Here and below on the set $\{\omega: \tau=\infty\}$ we define $<M>_\tau$ by the equality $<M>_\infty := \lim_{t\to\infty} <M>_t$.

$$\mathbb{E} \sup_{t \le T} \|M(t\wedge\sigma_n\wedge\tau)\|^2 \le 4\mathbb{E}<M>_{T\wedge\sigma_n\wedge\tau}.$$

Considering the limit in both sides of the inequality first as $n \to \infty$ and then as $T \uparrow \infty$ we obtain by Fatou's lemma that

$$\mathbb{E} \sup_{t \le \tau} \|M(t)\|^2 \le 4\mathbb{E}<M>_\tau < \infty.$$

From this and the first part of our proposition it follows that $M(t\wedge\tau) \in \mathfrak{M}^c_{loc}(\mathbb{R}_+,\mathbb{H})$.

We obtain the third assertion in the same way making use of the inequality given below.

Theorem 7. (*Burkholder-Davis' inequality, see* [97], [106]). *Let* $M \in M^c_{loc}(\mathbb{R}_+,\mathbb{H})$ *and* τ *be a stopping time. Then*

$$\mathbb{E} \sup_{t\le\tau} \|M(t)\| \le 3\mathbb{E}<M>_\tau^{1/2}. \;\square$$

Corollary 7. Let $M(t) \in \mathfrak{M}^c_{loc}(\mathbb{R}_+,\mathbb{H})$, τ *be a stopping time, and* a *and* $b \in \mathbb{R}_+$. *Then*

$$\mathbb{P}(\sup_{t\le\tau} \|M(t)\| \ge a) \le 3a^{-1}\, \mathbb{E}(<M>_\tau^{1/2}\wedge b) + \mathbb{P}(<M>_\tau^{1/2} \ge b). \;\square$$

1.8. Let M and $L \in \mathfrak{M}^c_{loc}(\mathbb{R}_+,\mathbb{H})$. Write

$$<M,L>_t := 1/4(<M,L>_t - <M\text{-}L>_t).$$

Similarly to the finite-dimensional case, it is easy to verify that the following result holds.

Theorem 8. Let $M(t)$, $L(t) \in \mathfrak{M}^c_{loc}(\mathbb{R}_+,\mathbb{H})$, *then* $(M(t), L(t)) - <M,L>_t \in \mathfrak{M}^c_{loc}(\mathbb{R}_+,\mathbb{R}^1)$. *Moreover, if* $A(t)$ *is a continuous predictable process such that its full variation over every compact from* $\mathbb{R}_+$ *is bounded,* ($\mathbb{P}$-*a.s.*), *and* $(M(t), L(t)) - A(t)) \in \mathfrak{M}^c_{loc}(\mathbb{R}_+,\mathbb{R}^1)$, *then* $A(t) = <M,L>_t$, ($\mathbb{P}$-*a.s.*), *for every* $t \in \mathbb{R}_+$. $\square$

In the future we will denote by $<M(\cdot),h>_t$ the quadratic variation of the continuous local martingale $(M(t), h)$, where h is an element of $\mathbb{H}$. We will use for $<M(\cdot), h_i>_t$ the shorter notation $<M_i>_t$ when h is the i-th element of the CONS $\{h_i\}$ which was fixed above.

Proposition 8. *If* $h \in \mathbb{H}$ *then for all* $w \in \Omega_h \subset \Omega$, *where* $\mathbf{P}(\Omega_h) = 1$, *and every* $s,\ t \in \mathbb{R}_+$ *such that* $s \leq t$

$$(1.1) \qquad <(M(\cdot),\, h)>_t - <M(\cdot),\, h)>_s \ \leq\ \|h\|^2\,(<M>_t - <M>_s). \ \Box$$

Proof. Without loss of generality we may assume that $h\|h\|^{-2} = h_1$ is the first element of the CONS $\{h_n\}$. Then $\|M(t)\|^2 - \|h\|^2\,(M(t),h)^2 = \sum_{i \geq 2} (M(t),h_i)^2$ is obviously a continuous local submartingale (see §1.7). On the other hand, $B(t) := <M>_t - \|h\|^{-2} <(M(\cdot)),\, h>_t$, is continuous process of bounded variation on compacts such that $\sum_{i \geq 2} (M(t),h_i)^2 - B(t) \in \mathfrak{M}^c_{loc}(\mathbb{R}_+,\mathbb{R}^1)$. From the theorem it follows that B(t) is the only process with such properties and therefore it is the increasing process given in the Doob-Meyer decomposition of $<M>_t - \|h\|^{-2} <M(\cdot)),\, h>_t$. This proves inequality (1.1). $\Box$

Lemma 8. *For all* $w \in \Omega' \subset \Omega$, *where* $P(\Omega') = 1$,

$$(1.2) \qquad <M>_t = \sum_{i=1}^{dim\,\mathbb{H}} <M_i>_t, \quad \forall t \in \mathbb{R}_+. \ \Box$$

Proof. By the same argument used for Proposition 8, we can prove that for every positive integer $n \leq dim\ \mathbb{H}$, $\|M(t)\|^2 - \sum_{i=1}^{n} M_i^2(t)$ is a continous local submartingale and $<M>_t - \sum_{i=1}^{n} <M_i>_t$ is the corresponding increasing process in the Doob-Meyer decomposition. If $\mathbb{H}$ is finite-dimensional the lemma is proved.

If $\mathbb{H}$ has infinite dimensions making use of the fact that $(<M>_t - \sum_{i=1}^{n} <M_i>_t)$ is a continuous increasing process we obtain for every $T \in \mathbb{R}_+$ and $\varepsilon > 0$,

$$\lim_{n \to \infty} \mathbf{P}\Big(\sup_{t \leq T} \Big(<M>_t - \sum_{i=1}^{n} <M_i>_t\Big) > \varepsilon \Big) \leq$$

$$\leq \lim_{n \to \infty} \varepsilon^{-1} \mathbf{E}\Big(<M>_T - \sum_{i=1}^{n} <M_i>_T\Big) =$$

$$= \lim_{n\to\infty} \varepsilon^{-1} \mathbb{E}(\|M(T)\|^2 - \|\Pi_n M(T)\|^2) = 0.$$

Equality (1.2) easily follows from this fact. □

From the lemma and Proposition 1.3.11 we get the following result.

Corollary 8. *If* $M(t) \in \mathfrak{M}_2^c(\mathbb{R}_+, \mathbb{H})$, $T \in \mathbb{R}_+$ *and* $\{0 = t_0^n < t_1^n < \dots < t_{k(n)+1}^n = T\}$ *is a sequence of partitions of the interval* $[0, T]$ *with the diameter converging to zero, then*

$$\lim_{n\to\infty} \mathbb{E}\Big| \sum_{i=0}^{k(n)} \|M(t_{i+1}^n) - M(t_i^n)\|^2 - \langle M\rangle_T\Big| = 0. \quad \square$$

1.9. We state here for the later use in this section, the following simple auxiliary result.

Proposition 9. *If* $M^n(t) \in \mathfrak{M}_2^c(\mathbb{R}_+,\mathbb{H})$, $\forall n \in \mathbb{N}$, *and for every* $T \in \mathbb{R}_+$

$$\lim_{n\to\infty} \mathbb{E} \sup_{t\le T} \|M^n(t) - M^\infty(t)\|^2 = 0.$$

then $M^\infty(t) \in \mathfrak{M}_2^c(\mathbb{R}_+,\mathbb{H})$ *and for every* $T \in \mathbb{R}_+$,

$$\lim_{n\to\infty} \mathbb{E} \sup_{t\le T} |\langle M^n\rangle_t - \langle M^\infty\rangle_t| = 0. \quad \square$$

1.10. For use later in the study of martingales and local martingales in $\mathbb{H}$, we state here some important facts from functional analysis and introduce some additional definitions.

Let $\mathbb{X}$ be a Hilbert space. The set of all continuous linear operators from $\mathbb{H}$ and $\mathbb{X}$ will be denoted by $\mathfrak{L}(\mathbb{H},\mathbb{X})$.

Warning 10. *In the sequel we shall write* $\mathfrak{L}(\mathbb{H})$ *instead of* $\mathfrak{L}(\mathbb{H},\mathbb{H})$. *Analogous contraction will be used for the subspaces of* $\mathfrak{L}(\mathbb{H})$ *below. Namely, we shall write* $\mathfrak{L}_1(\mathbb{H})$ *instead of* $\mathfrak{L}_1(\mathbb{H},\mathbb{H})$ *and* $\mathfrak{L}_2(\mathbb{H})$ *instead of* $\mathfrak{L}(\mathbb{H},\mathbb{H})$.

Let $\mathbb{Y}$ *be a Hilbert space and* $\mathbb{Y}^*$ *be its conjugate space. The mapping* $\mathfrak{J}_{\mathbb{Y}}$: $\mathbb{Y} \to \mathbb{Y}^*$ *given by the formula* $y \in \mathbb{Y} \to (\cdot, y)_{\mathbb{Y}} \in \mathbb{Y}^*$ *is (by the Riesz representation of the continuous linear functional in a Hilbert space) an isometric isomorphism. In the sequel we call it the canonical isomorphism. The operator* $A \in \mathfrak{L}(\mathbb{H})$ *is said to be self-adjoint if* $A = \mathfrak{J}_{\mathbb{H}}^{-1} A^* \mathfrak{J}_{\mathbb{H}}$, *where* A^* *is the adjoint of* A. *In the future for* $A \in \mathfrak{L}(\mathbb{H},\mathbb{Y})$, *we shall use the following disignation*

$$A' := \mathfrak{J}_{\mathbb{H}}^{-1} A^* \mathfrak{J}_{\mathbb{Y}}$$

1.11. *Definition* 11. *An operator* $B \in \mathfrak{L}(\mathbb{H},\mathbb{X})$ *is called a Hilbert-Schmidt operator if*

$$\sum_{i=1}^{\dim \mathbb{H}} \|Bh_i\|_{\mathbb{H}}^2 < \infty. \ \Box$$

We denote by $\mathfrak{L}_2(\mathbb{H},\mathbb{X})$ the space of all Hilbert-Schmidt operators. The following theorem shows that this definition is meaningful and describes some additional properties of Hilbert-Schmidt operators.

Theorem 11. (*i*) *If* $\{h_i'\}$ *is any other CONS in* $\mathbb{H}$ *and* $B \in \mathfrak{L}_2(\mathbb{H},\mathbb{X})$, *then*

$$\sum_{i=1}^{\dim \mathbb{H}} \|Bh_i\|_{\mathbb{X}}^2 = \sum_{i=1}^{\dim \mathbb{H}} \|Bh_i'\|_{\mathbb{X}}^2.$$

(*ii*) *Let* A, $B \in \mathfrak{L}_2(\mathbb{H},\mathbb{X})$. *The formula*

$$|||B||| := \Big(\sum_{i=1}^{\dim \mathbb{H}} \|Bh_i\|_{\mathbb{X}}^2\Big)^{1/2}$$

defines a norm in $\mathfrak{L}_2(\mathbb{H},\mathbb{X})$. *The space* $\mathfrak{L}_2(\mathbb{H},\mathbb{X})$ *is a separable Hilbert space with respect to the scalar product defined by the formula*

$$((B,A)) := \sum_{i=1}^{\dim \mathbb{H}} (Bh_i, Ah_i)_{\mathbb{X}}$$

(*iii*) *The operator* $B \in \mathfrak{L}_2(\mathbb{H},\mathbb{X})$ *is compact.*

(*iv*) *If* B *is a self-adjoint operator on* $\mathfrak{L}_2(\mathbb{H})$, *then* $\|B\|_{\mathfrak{L}_2(\mathbb{H})} = (\sum_i \lambda_i^2)^{1/2}$, *where* $\{\lambda_i\}$ *are non-zero eigen-values taken with due regard to their multiplicity.*

(*v*) *Let* $\mathbb{Y}$ *be a Hilbert space,* $B \in \mathfrak{L}_2(\mathbb{H},\mathbb{X})$, $A \in \mathfrak{L}(\mathbb{X},\mathbb{Y})$ *and* $C \in \mathfrak{L}(\mathbb{Y},\mathbb{H})$, *then* $AB \in \mathfrak{L}_2(\mathbb{H},\mathbb{Y})$, $BC \in \mathfrak{L}_2(\mathbb{Y},\mathbb{X})$ *and also*

$$\|AB\|_{\mathfrak{L}_2(\mathbb{H},\mathbb{Y})} \le \|A\|_{\mathfrak{L}(\mathbb{X},\mathbb{Y})} \cdot \|B\|,$$

$$\|BC\|_{\mathfrak{L}_2(\mathbf{Y},\mathbf{X})} \leq |||B||| \cdot \|C\|_{\mathfrak{L}(\mathbf{Y},\mathbf{H})}, \quad \|B\|_{\mathfrak{L}(\mathbf{H},\mathbf{X})} \leq |||B|||,$$

where $\|\cdot\|_{\mathfrak{L}(\mathbf{X},\mathbf{Y})}$, $\|\cdot\|_{\mathfrak{L}(\mathbf{Y},\mathbf{H})}$ *and* $\|\cdot\|_{\mathfrak{L}(\mathbf{H},\mathbf{X})}$ *are "usual" operator norms.* □

1.12. *Definition* 12. *An operator* $B \in \mathfrak{L}(\mathbf{H})$ *is called a nuclear operator if it may be represented as follows:*

$$B = \sum_{i=1}^{N} A_i C_i$$

where A_i, $C_i \in \mathfrak{L}_2(\mathbf{H})$ *and* $N<\infty$.

The set of all nuclear operators from $\mathfrak{L}(\mathbf{H})$ *will be denoted by* $\mathfrak{L}_1(\mathbf{H})$. □

Theorem 12. 1. *Given* $B \in \mathfrak{L}_1(\mathbf{H})$ *define* trB *(the trace of B) to be the sum*

$$\sum_{i=1}^{dim\mathbf{H}} (Bh_i, h_i).$$

Then $tr\, B < \infty$ *and does not depend on the choice of the CONS.*

2. *If* B *is a compact self-adjoint operator in* $\mathfrak{L}(\mathbf{H})$ *then* $B \in \mathfrak{L}_1(\mathbf{H})$ *if and only if*

$$\sum_j |\lambda_j| < \infty$$

where λ_j *are non-zero eigen-values of* B *taken with due regard to their multiplicity. It is clear that* $trB = \sum_j \lambda_j$. *Also* $\mathfrak{L}_1(\mathbf{H})$ *is a Banach space with respect to the norm*

$$\|\cdot\|_{\mathfrak{L}_1(\mathbf{H})} := \sup_{\substack{B\in\mathfrak{L}(\mathbf{H}) \\ \|B\|\leq 1}} |tr(BA)|$$

3. *If* $B \in \mathfrak{L}_1(\mathbf{H})$ *and* $C \in \mathfrak{L}(\mathbf{H})$, *then* $\|B\|_{\mathfrak{L}_2(\mathbf{H})} \leq \|B\|_{\mathfrak{L}_1(\mathbf{H})}$, *and* BC *and* $CB \in \mathfrak{L}_1(\mathbf{H})$. □

The results concerning Hilbert-Schmidt nuclear operators presented above are classical (see e.g. [28], [126] etc.).

1.13. *Theorem* 13 (*cf.* [99]). *Let* $M(t) \in \mathfrak{M}^c_{loc}(\mathbf{R}_+,\mathbf{H})$. *Then there exists a predictable process* $Q_M(t,\omega)$ *taking values in* $\mathfrak{L}_1(\mathbf{H})$ *with properties:*

1) *for all* $(t,\omega)\in\mathbb{R}_+\times\Omega$, $trQ_M = 1$;

2) *for all* $(t,\omega)\in\mathbf{R}_+\times\Omega$ *and* $h\in\mathbf{H}$, $(Q_M h,h) \geq 0$;

3) *for every* y, $z\in\mathbf{H}$ *and* $(t,\omega)\in\mathbf{R}_+\times\Omega$ $(Q_M y,z) = (y, Q_M z)$ *and*

$$<(M(\cdot), y), (M(\cdot), z)>_t = \int_{[0,t]} (Q_M(s)y, z)d <M>_s. \ \square$$

Proof. Suppose for the moment that $M(t) \in \mathfrak{M}^c_2(\mathbf{R}_+, \mathbf{H})$. From Proposition 8 it follows that the process $<(M(\cdot), y)>_t$ is absolutely continuous with respect to $<M>_t$ on every finite interval of $\mathbf{R}_+$ ($\mathbb{P}$-a.s.). Indeed, denote $y(t) := <M>_t - \|y\|^{-2} <(M(\cdot), y)>_t$. From (1.1) it follows that $y(t)$ is an increasing process ($\mathbb{P}$-a.s.). Hence, for every interval $[a,b]$, where a, $b\in\mathbf{R}_+$, and every bounded non-negative function f on this interval

$$\int_{[a,b]} f(t)d <M>_t = \|y\|^{-2} \int_{[a,b]} f(t)d <(M(\cdot),y)>_t +$$

$$(\mathbb{P}\text{-}a.s.)$$

$$+ \int_{[a,b]} f(t)dy(t) \geq \|y\|^{-2} \int_{[a,b]} f(t)d <(M(\cdot),y)>_t$$

This yields (see also [22], Theorem 33) that there exists a non-negative predictable process $q(t,y,y)$ such that

$$\mathbb{E} \int_{[0,t]} q(s,y,y)d <M>_1 < \infty \qquad \forall t\in\mathbf{R}_+$$

and for all w from a set of probability 1

$$<(M(t),y)>_t =$$

$$= \int_{[c,t]} q(s,,,y)d <M>_1, \qquad \forall t\in\mathbf{R}_+.$$

Thus, taking into account that

$$<(M(\cdot),y), (M(\cdot),z)>_t =$$

$$= \tfrac{1}{4}\,(<(M(\cdot),y+z)>_t - <(M(\cdot),y-z)>_t,$$

we have that there exists a predictable stochastic process $q(t,y,z)$ such that for all ω from a set $\Omega_{y,z}$ of probability 1

$$<(M(\cdot),y),\ (M(\cdot),z)>_t = \int_{[0,t]} q(s,y,z)\,d<M>_s, \quad \forall t \in \mathbb{R}_+ \tag{1.3}$$

and, of course,

$$\mathbb{E} \int_{[0,t]} q(s,y,z)\,d<M>_s < \infty \qquad \forall t \in \mathbb{R}_+$$

From now on we shall denote by $\mathfrak{D}$ a set of finite linear combinations with rational coefficients of elements from CONS $\{h_i\}$. It is clear that $\mathfrak{D}$ is dense in $\mathbb{H}$. From equality (1.3) we get that $<M>\circ\mathbb{P}$-a.s.:

(a) $q(t,\cdot,\cdot)$ is a bilinear symmetric functional on $\mathfrak{D} \times \mathfrak{D}$;

(b) $q(t,y,z) \le \|y\|\,\|z\|, \qquad \forall y,z \in \mathfrak{D}$,

(c) $q(t,y,y) \ge 0 \qquad \forall y \in \mathfrak{D}$.

From (1.2), (1.3) it follows that $<M>\circ\mathbb{P}$-*a.s.*

$$<M>_t = \int_{[0,t]} \sum_{i=1}^{\dim \mathbb{H}} q(s,h_i,h_i)\, d<M>_s,$$

and therefore for $<M>\circ\mathbb{P}$-*a.a.* t,ω the following property holds

(d) $\sum_{i=1}^{\dim \mathbb{H}} q(s,h_i,h_i) = 1$

Denote by S an (s,ω)-set where $q(s,y,z)$ does not satisfy properties (a) to (d). For $(s,\omega) \in (\mathbb{R}_+ \times \Omega)\backslash S$ by a continuity argument we extend q up to a bilinear functional on the whole of $\mathbb{H}\times\mathbb{H}$. Clearly, the resulting functional will satisfy (a) to (d) on all of $\mathbb{H}\times\mathbb{H}$. Thus for every $(s,\omega) \in \mathbb{R}_+ \times \Omega\backslash S$, there exists a non-negative, selfadjoint continuous linear operator $Q_M(s,\omega)$ on $\mathbb{H}$ such that for all y, $z \in \mathbb{H}$

$$(Q_M(s,\omega)y,z) = q(s,y,z)$$

By Theorem 12 $Q_M(s,\omega)$ is a nuclear operator. Let Q^0 be a non-negative, selfadjoint, nuclear operator on $\mathbb{H}$ and $trQ^0 = 1$. We write $Q_M(s,\omega) = Q^0$ for $(s,\omega)\in S$. By definition S is a predictable set (i.e. S$\in\mathcal{P}$). Thus for every y, $z\in\mathbb{H}$ the mapping $(Q_M(\cdot,\cdot)y,z)$: $\mathbb{R}_+\times \Omega \to \mathbb{R}^1$ is predictable. Consequently for every $C\in\mathfrak{L}(\mathbb{H})$, the process $trCQ_M(s,\omega)$ is also a predictable stochastic process.

Thus by Pettis' theorem (§1.2.8) $Q_M(s,\omega)$ is a predictable stochastic process in $\mathfrak{L}_1(\mathbb{H})$.

From (1.3) it follows that there exists a set $\Omega' \subset \Omega$ of probability 1 such that for all $(t,\omega)\in\mathbb{R}_+\times\Omega'$ and y, $z\in\mathfrak{D}$

$$<(M(\cdot),y), (M(\cdot),z)>_t = \int_{[0,t]} (Q_M(s,\omega)y,z)d <M>_s \tag{1.4}$$

since $\mathfrak{D}$ is dense in $\mathbb{H}$ and both parts of equality (1.4) are continuous in probability (with respect to (y,z), uniformly on every compact subset of $\mathbb{R}_+$). for every $(y,z)\in\mathbb{H}$, there exists an ω-set $\Omega_{y,z}$ of probability 1 such that equality (1.4) holds for all $(t,\omega)\in\mathbb{R}_+\times \Omega_{y,z}$.

Let $M(t) \in \mathfrak{M}^c_{loc}(\mathbb{R}_+,\mathbb{H})$ and $\{\sigma_n\}$ be the localizing sequence of stopping times from Lemma 7. Denote by $Q^n_M(s,\omega)$ the operator defined by the above for the martingale $M(t\wedge\sigma_n)$. It is easy to verify that on a set $\{(s, \omega \quad s<t_n(\omega)\}$ $Q^n_M(s,\omega) = Q^{n+1}_M(s,\omega) \quad <M>\circ\mathbf{P}$-$a.s.$

Thus defining $Q_M(s,\omega):= Q^n_M(s,\omega)$ for $s < \sigma_n(\omega)$, we obtain the operator required for the proof of the theorem. □

Warning 13. *In the future we shall sometimes call* Q_M *the correlation operator of the martingale* $M(t)$. □

Definition 13. *Let* Q *be a nuclear symmetric non-negative operator on* $\mathbb{H}$. *A martingale* $W(t)\ \varepsilon\ \mathfrak{M}^c_2(\mathbb{R}_+,\mathbb{H})$ *with correlation operator* $Q_W:= (trQ)^{-1}Q$ *and quadratic variation* $<W>_t:= t\cdot trQ$ *will be called a Wiener process (martingale) with the covariance operator* Q. □

Remark 13. *It is easy to show that a Wiener process* $\mathbf{H}$ *with the covariance operator* Q *can be decomposed as follows:*

$$W(t) = \sum_{i=1}^{dim\mathbb{H}} \sqrt{\lambda_i}\, w^i(t)\varphi_i,$$

where $\{\varphi_i\}$ *is a CONS in* $\mathbb{H}$ *of the eigen-functions of* Q, λ_i *is eigen-value corresponding to* φ_i *and* $\{w^i(t)\}$ *is a set of independent one-dimensional Wiener processes. The series in the decomposition coverges* $\mathbf{P}$-$a.s.$ □

2.2. Stochastic Integral with Respect to Square Integrable Martingale.

2.0. Let $M(t) \in \mathfrak{M}^c_{loc}(\mathbf{R}_+,\mathbf{H})$. This assumption will be in force throughout the section.

Denote by $\mathfrak{L}_2(<M>\circ\mathbf{P}, \mathfrak{L}_2(\mathbf{H},\mathbf{X}))$ the set of mappings $\phi: \Omega \times \mathbf{R}_+ \to \mathfrak{L}_2(\mathbf{H},\mathbf{X})$ such that*

$$(2.1) \qquad \mathbf{E} \int_{[0,t]} |||\Phi(s)|||^2 \, d<M>_s < \infty, \quad \forall t \in \mathbf{R}_+ \quad (\mathbf{P}\text{-}a.s.)$$

In this section we define an integral

$$\int_{[0,t]} B(s)\, dM(s)$$

for $B \in \mathfrak{L}_2(<M>\circ\mathbf{P}, \mathfrak{L}_2(\mathbf{H},\mathbf{X}))$.

We suppose that $dim\mathbf{H} = dim\mathbf{X} = \infty$. The finite-dimensional case is similar to the infinite-dimensional one and will provide the reader an easy exercise.

2.1. Let us fix in $\mathbf{X}$ a CONS $\{e_j\}$. For every $h_i \in \{h_i\}$, $e_j \in \{e_j\}$ (we should remind the reader that $\{h_i\}$ is the fixed CONS in $\mathbf{H}$), the following inequality holds true (see Theorem 1.11 (V)),

$$(2.2) \qquad |(\mathbf{B}(s,\omega)h_i, e_i)_{\mathbf{X}}| \le |||B(s)||| \quad \forall s, \omega.$$

In view of Pettis' thoerem (see §1.2.8), $(B,(s,\omega)h_i, e_j)$ is a predictable process in $\mathbf{X}$. Thus the stochastic integral

$$(2.3) \qquad B\circ M^{ij}(t) := \int_{[0,t]} (B(s)h_i, e_j)_{\mathbf{X}} \, dM_i(s)$$

is defined and belongs to $\mathfrak{M}^c_2(\mathbf{R}_+,\mathbf{R}^1)$.

Indeed, in view of (2.1), (2.2) and Theorem 1.13 we have

$$\mathbf{E} \int_{[0,t]} (B(s)h_i, e_j)^2_{\mathbf{X}} d <M_i>_s = \mathbf{E} \int_{[0,t]} (B(s)\, h_i, e_j)^2_{\mathbf{X}} \, (Q_M(s)h_i, h_i)\, d<M>_s$$

$$\le \mathbf{E} \int_{[0,t]} |||B(s)|||^2 d<M>_s < \infty.$$

*The notation $|||\cdot|||$ is used for norm in $\mathfrak{L}_2(\mathbf{H},\mathbf{X})$ (see §1.1.1).

From this inequality and the results of sections 1.3.14, 1.3.15 it follows that the stochastic integral (2.3) is defined and belongs to $\mathfrak{M}_2^c(\mathbf{R}_+,\mathbf{R}^1)$.

Write

$$B\circ M_{m,n}(t):=\sum_{i=1}^{m}\sum_{j=1}^{n} B\circ M^{ij}(t)e_j.$$

It is clear that $B\circ M_{m,n}(t) \in \mathfrak{M}_2^c(\mathbb{R}_+,\mathbb{H})$. Let use now calculate $<B\circ M_{m,n}>_t$.

In accordance with Lemma 1.8

$$<B\circ M_{m,n}>_t:=\sum_{j=1}^{n}\left\langle\sum_{i=1}^{n} B\circ M^{ij}\right\rangle_t \quad (\mathbf{P}\text{-}a.s.).$$

Thus it is sufficient to calculate $\left\langle \sum_{i=1}^{n} B\circ M^{ij}\right\rangle_t$. From the results of §§ 1.3.14, 1.3.15 it follows that $s,\ t\in\mathbf{R}_+$,

$$I:=\mathbb{E}\left[\left(\sum_{i=1}^{n} B\circ M^{ij}(t)\right)^2-\left(\sum_{i=1}^{n} B\circ M^{ij}(s)\right)^2\middle|\ \mathfrak{F}_s\right]=$$

$$=\mathbb{E}\int_{[s,t]}\sum_{i,k=1}^{n}(D(s)h_i,\ e_j)_{\mathbf{X}}(B(s)h_k,\ e_j)_{\mathbf{X}}d<M_i,M_k>_s.$$

This, due to Theorem 1.13, gives

$$I=\mathbb{E}\Bigg(\int_{[s,t]}\sum_{i,k=1}^{m}(B(u)h_i,\ e_j)_{\mathbf{X}}(B(u)h_k,\ e_j)_{\mathbf{X}}\times$$

$$\times\ (Q_M(u)h_i,\ h_k)d<M>_u|\ \mathfrak{F}_s\Bigg)=$$

$$=\mathbb{E}[\int_{[s,t]}(Q_M(u)\sum_{i=1}^{n}h_i(h_i,\ B'(u)e_j),$$

$$\sum_{k=1}^{n}h_k(h_k,\ B'(u)e_j))d<M>_u|\ \mathfrak{F}_s]=$$

$$= \mathbb{E}\left[\int_{[s,t]} \|Q_M^{1/2}(u)\, \Pi_n B'(u)\, e_j\|^2 \, d\langle M\rangle_u \,\Big|\, \mathcal{F}_s \right]^{*}$$

Note that from Theorems 1.11 to 1.13 it follows that

$$\|Q_M^{1/2}(u)\, \Pi_n B'(u) e_j\|^2 \le \|Q_M^{1/2}(u)\|^2 \, \|B'(u) e_j\|^2 \le$$

$$\le \|Q_M^{1/2}(u)\|^2_{\mathcal{L}_2(\mathbf{H})} \cdot \|B'(u)\|^2_{\mathcal{L}_2(\mathbf{H},\mathbf{X})} =$$

$$= tr\, Q_M(u)\, |||B(u)|||^2 = |||B(u)|||^2.$$

Because of the uniqueness of the [illegible]-Meyer decomposition (§ 1.3.11), we get

$$\Big\langle \sum_{i=1}^{n} B \circ M^{ij} \Big\rangle_t = \int_{[0,t]} \|Q_M^{1/2}(u)\, \Pi_n B'(u)\, e_j\|^2 d\langle M\rangle_u \quad (\mathbf{P}\text{-}a.s.),$$

which implies that for every $t \in \mathbf{R}_+$, $\mathbf{P}$-$a.s.$

$$\langle B \circ M_{m,n} \rangle_t = \sum_{j=1}^{m} \int_{[0,t]} \|Q_M^{1/2}(u)\, \Pi_n B'(u) e_j\| d\langle M\rangle_u. \tag{2.4}$$

Remark 1. *It follows from the derivation of the formula* (2.4), *that the process* $\|Q_M^{1/2}(u)\, \Pi_n B(u) e_j\|$ *is predictable.* □

Let m, n and $m', n' \in \mathbb{N}$ and $m' > m$, $n' > n$ and $T \in \mathbf{R}_+$. From the Doob inequality (1.3.5) we have

---

*See the notation introduced in §§1.5, 1.10, fractional power of a positive definite self-adjoint linear operator can be defined in many ways (see e.g. [42] or §3.2.1).

$$U := \mathbb{E} \sup_{t \le T} \|B \circ M_{m',n'}(t) - B \circ M_{m,n}(t)\|_{\mathbb{X}}^2 \le$$

$$\le 4\mathbb{E}\left\| \sum_{j=1}^{m'} \sum_{i=n+1}^{n'} B \circ M^{ij}(T) e_j + \sum_{j=m+1}^{m'} \sum_{i=1}^{n'} B \circ M^{ij}(T) e_j \right\|_{\mathbb{X}}^2 \le \tag{2.5}$$

$$\le 8 \sum_{j=1}^{m'} \mathbb{E}\left| \sum_{i=n+1}^{n'} B \circ M^{ij}(T) \right|^2 + 8 \sum_{j=m+1}^{m'} \mathbb{E}\left| \sum_{i=1}^{n'} B \circ M^{ij}(T) \right|^2 .$$

In the same way as it was done for the proof of formula (2.4) we can show that

$$\mathbb{E}\left| \sum_{i=n+1}^{n'} B \circ M^{ij}(T) \right|^2 \le \mathbb{E} \int_{[0,t]} \|Q_M^{1/2}(t) \, (\mathbf{I} - \Pi_{n'}) \, B'(t) e_j\|^2 d\langle M\rangle_t . \tag{2.6}$$

where $\mathbf{I}$ is the identity operator, and

$$\mathbb{E}\left| \sum_{i=1}^{n'} B \circ M^{ij}(T) \right|^2 = \mathbb{E} \int_{[0,T]} \|Q_M^{1/2} \, \Pi_{n'} B'(t) e_j\|^2 d\langle M\rangle_t . \tag{2.7}$$

From (2.5) - (2.7) it follows that

$$U \le 8\mathbb{E} \int_{[0,T]} \left(\|\mathbf{I} - \Pi_n\| \sum_{j=1}^{m} \|B'(t)e_j\|^2 + \sum_{j=m+1}^{m'} \|B'(t)e_j\|^2 \right) d\langle M\rangle_t . \tag{2.8}$$

Since $\|B'(t)\|^2_{\mathfrak{L}_2(\mathbb{X},\mathbb{H})} = |||B(t)|||^2$ and $\|B'(t)\|^2_{\mathfrak{L}_2(\mathbb{X},\mathbb{H})} = \sum_{j=1}^{\infty} \|B'(t)e_j\|^2$, it follows from (2.1) and (2.8) that $U \to 0$ as m, n, m' and $n' \to \infty$. Consequently, there exists a stochastic process $B \circ M(t) \in \mathfrak{M}_2^c(\mathbb{R}_+, \mathbb{X})$ such that for every $T \in \mathbb{R}_+$

$$\lim_{\substack{m\to\infty \\ n\to\infty}} \mathbb{E} \sup_{t \le T} \|B \circ M(t) - B \circ M_{m,n}(t)\|^2_{\mathbb{X},\mathbb{H}} = 0. \tag{2.9}$$

We define the stochastic integral by the equality

$$\int_{[0,t]} B(s) dM(s) := B \circ M(t) \qquad (\mathbf{P}\text{-}a.s.).$$

It is clear that this integral is defined uniquely up to stochastic equivalence. From (2.4), (2.9) and Proposition 1.9 it follows easily that

$$<B\circ M>_t = \lim_{\substack{m\to\infty \\ n\to\infty}} \int_{[0,t]} \sum_{j=1}^{m} \|Q_M^{1/2}\, \Pi_n B'(s) e_j\|^2 d<M>_s =$$

(2.10)

$$= \int_{[0,t]} \sum_{j=1}^{\infty} \|Q_M^{1/2} B'(s) e_j\|^2 \; d<M>_s = \int_{[0,t]} |||B(s)\; Q_M^{1/2}(s)|||^2 \; d<M>_s.$$

Note that from Remark 1 and the results of section 1.2.6 it follows that

$|||B(s)\; Q_M^{1/2}(s)|||^2$ is a predictable process. It is also obvious that

$|||B(s)\; Q_M^{1/2}(s)|||^2 \le |||B(s)|||^2.$

2.2. We show next that the definition of the stochastic integral presented above does not depend on the choice of the CONS's in $\mathbf{H}$ and $\mathbf{X}$.

Given $y \in \mathbf{X}$ define the operator B_y by the formula

$$B_y\, h = (B'_y, h), \qquad \forall h \in \mathbf{H}$$

It is clear that $B_y \in \mathfrak{L}_2(<M>\circ\mathbf{P},\ \mathfrak{L}_2(\mathbf{H},\mathbf{R}^1))$. Write $B_y \circ M^{h_i}(t)$ for the integral $\int_{[0,t]} (B'(s)y,\ h_i) dM_i(s)$.

As it was shown in the previous section, for every $T \in \mathbf{R}_+$,

$$\text{(2.11)} \qquad \lim_{n\to\infty} \mathbf{E} \sup_{t\le T} \left| B_y\circ M(t) - \sum_{i=1}^{n} B_y\circ M^{h_i}(t)\right|^2 = 0.$$

We now prove that $B_y\circ M(t)$ does not depend on the choice of the CONS $\{h_i\}$.

Let $\{\psi_i\}$ be another CONS in $\mathbf{H}$. Define $B_y\circ M^{\psi_i}(t)$ similarly to $B_y\circ M^{h_i}(t)$. We show that

$$\text{(2.12)} \qquad \lim_{n\to\infty} \mathbf{E} \sup_{t\le T} \left| \sum_{i=1}^{n} (B_y\circ M^{h_i}(t) - B_y\circ M^{\psi_i}(t))\right|^2 = 0.$$

From Doob's inequality (1.3.5) and § 1.3.15 it follows that

$$\mathbf{E} \sup_{t\le T}\left|\sum_{i=1}^{n}(B_y\circ M^{h_i}(t) - B_y\circ M^{\psi_i}(t))\right|^2 \le$$

$$\le 4\left(\mathbf{E} \sum_{i,j=1}^{n} \int_{[0,T]} (B'(s)y,\, h_i)(B'(s)y,\, h_j)d <M_i,\, M_j>_s -\right.$$

$$- 2 \sum_{i,j=1}^{n} \int_{[0,T]} (B'(s)y,\, h_i)(B'(s)y,\, \psi_i) \times$$

$$\times\, d<(M(\cdot),\, h_i),\, (M(\cdot),\, \psi_i)>_s +$$

$$+ \sum_{i,j=1}^{n} \int_{[0,T]} (B'(s)y,\, \psi_i)(B'(s)y,\, \psi_j) \times$$

$$\left.\times\, d<(M(\cdot),\, \psi_i),\, (M(\cdot),\, \psi_j)>_s\right) = 4(I_1(n) - I_2(n) + I_3(n)).$$

Theorem 1.13 implies that

$$I_2(n) = \mathbf{E} \int_{[0,t]} 2\left(Q_M(s) \sum_{i=1}^{n}(B'(s)y,\, h_i)h_i,\, \sum_{j=1}^{n}(B'(s)y,\, \psi_j)\psi_j\right)d<M>_s$$

and consequently

$$\lim_{n\to\infty} I_2(n) = \mathbf{E} \int_{[0,t]} 2(Q_M(s)\, B'(s)y,\, B'(s)y)d<M>_s.$$

It follows in the same way that

$$\lim_{n\to\infty} I_1(n) = \lim_{n\to\infty} I_3(n) = \mathbf{E} \int_{[0,t]} (Q_M(s)\, B'(s)y,\, B'(s)y)d<M>_s,$$

Thus we get the equality (2.12). For (2.11) and (2.12), we conclude that $B_y\circ M(t)$ does not depend on the choice of CONS in $\mathbb{H}$.

We next show that for $y \in \mathbf{X}$ fixed,

$$(2.13) \qquad (y,\ B \circ M(t))_{\mathbf{X}} = B_y \circ M(t), \quad \forall t, \qquad (\mathbf{P}\text{-}a.s.).$$

This equality obviously implies that the definition of the integral does not depend on the choice of CONS.

Note that

$$(2.14) \qquad \mathbb{E}|(y,\ B \circ M(t))_{\mathbf{X}} - B_y \circ M(t)|^2 \leq 2\mathbb{E}|(y,\ B \circ M(t) - B \circ M_{m,n}(t))_{\mathbf{X}}|^2 +$$

$$+ 2\mathbb{E}|(y,\ B \circ M_{m,n}(t))_{\mathbf{X}} - B_y \circ M(t)|^2 = V_1 + V_2.$$

From (2.9) it follows that

$$(2.15) \qquad \lim_{\substack{m \to \infty \\ n \to \infty}} V_1 = 0.$$

Evidentally,

$$(y,\ B \circ M_{m,n}(t)) = \sum_{i=1}^{n} \int_{[0,t]} \sum_{i=1}^{m} (B(s)h_i,\ e_j)_{\mathbf{X}} dM_i(s) \qquad (\mathbf{P}\text{-}a.s.),$$

and so

$$V_2 = \mathbb{E}\left|(y,\ B \circ M_{m,n}(t))_{\mathbf{X}} - \sum_{i=1}^{n} B_y \circ M^{h_i}(t)\right|^2 =$$

$$= \mathbb{E} \sum_{i,k=1}^{n} \int_{[0,t]} \left[\sum_{j=1}^{m} (B(s)\ h_i,\ e_j)_{\mathbf{X}}\ (y,\ e_j)_{\mathbf{X}} - (B'(s)y,\ h_i)\right] \times$$

$$\times \left[\sum_{j=1}^{m} (B(s)\ h_k,\ e_j)_{\mathbf{X}} (y,\ e_j)_{\mathbf{X}} - (B'(s)y,\ h_k)\right] d\langle M_i,\ M_k\rangle_s =$$

$$= \mathbb{E} \int_{[0,t]} \left(Q_M(s) \sum_{i=1}^{n} \left[\sum_{j=1}^{m} (b(s)h_i,\ e_j)_{\mathbf{X}}\ (y,\ e_j)_{\mathbf{X}} - (B'(s)y,\ h_i)\right] h_i,\right.$$

$$\sum_{k=1}^{n}\left[\sum_{j=1}^{m}(B(s)h_k, e_j)_X (y, e_j)_X - (B'(s)y, h_k)\right] h_k\Big) d<M>_s.$$

It is clear that

$$\lim_{\substack{m\to\infty \\ n\to\infty}} \left\|\sum_{i=1}^{n}\left[\sum_{j=1}^{m}(B(s)h_i, e_j)_X (y, e_j)_X - (B'(s)y, h_i)\right] h_i\right\| = 0,$$

and consequently

$$\lim_{\substack{m\to\infty \\ n\to\infty}} V_2 = 0. \tag{2.16}$$

Formula (2.13) follows from (2.11) and (2.14) - (2.16).

2.3. We state the main properties of the integral defined above.

(i) For every B_1, $B_2 \in \mathfrak{L}_2(<M>\circ \mathbf{P}, \mathfrak{L}_2(\mathbb{H},\mathbf{X}))$, $\alpha_1\beta \in \mathbf{R}^1$ *and* $t \in \mathbf{R}_+$

$$\int_{[0,t]} (\alpha\beta_1(s) + \beta B_2(s))dM(s) =$$

$$= \alpha \int_{[0,t]} B_1(s)dM(s) + \beta \int_{[0,t]} B_2(s)dM(s) \qquad (\mathbf{P}\text{-}a.s.).$$

(ii) For every s, u, $t \in \mathbb{R}_+$, such that $0 \le s \le u \le t$, and $B \in \mathfrak{L}_2(<M>\circ \mathbf{P}, \mathfrak{L}_2(\mathbb{H},\mathbf{X}))$

$$\int_{[0,t]} 1_{\{[s,u]\}}(r)B(r)dM(r) = \int_{[0,u]} B(r)dM(r) - \int_{[0,s]} B(r)dM(r) :=$$

$$= \int_{[s,u]} B(r)dM(r).$$

(iii) If B and $B_n \in \mathfrak{L}_2(<M>\circ \mathbf{P}, \mathfrak{L}_2(\mathbb{H},\mathbf{X}))$, $\forall_n \in \mathbf{N}$, and for some $T \in \mathbf{R}_+$

$$\lim_{n\to\infty} \mathbf{E} \int_{[0,T]} |||(B_n(s) - B(s))\, Q_M^{1/2}(s)|||^2 d<M>_s = 0,$$

then

$$\lim_{n\to\infty} \mathbb{E} \sup_{t\le T} \|\|B_n \circ M(t) - B\circ M(t)\|\|_{\mathbf{X}}^2 = 0.$$

(iv) For $B\in\mathfrak{L}_2(<M>\circ\mathbf{P}, \mathfrak{L}_2(\mathbb{H},\mathbf{X}))$ and every $A\in\mathfrak{L}(\mathbf{X},\mathbf{Y})$, where $\mathbf{Y}$ is a Hilbert space, the following equality holds

$$A \int_{[0,t]} B(s)dM(s) = \int_{[0,t]} AB(s)dM(s) \qquad (\mathbf{P}\text{-}a.s.).$$

(v) If $B\in\mathfrak{L}_2(<M>\circ\mathbf{P}, \mathfrak{L}_2(\mathbb{H},\mathbf{X}))$, $L\in\mathfrak{M}_2^c(\mathbb{R}_+,\mathbf{X})$ and $T\in\mathbb{R}_+$, then

$$\lim_{\substack{m\to\infty\\ n\to\infty}} \mathbb{E} \sup_{t\le T} \Bigg| <B\circ M, L>_t - \sum_{i=1}^{n}\sum_{j=1}^{m} \int_{[0,t]} (B(s)h_i, e_j)_{\mathbf{X}} d<M(\cdot), h_i), (L(\cdot), e_j)_{\mathbf{X}}>_s \Bigg| = 0,$$

In particular, if $C\in\mathfrak{L}_2(<M>\circ\mathbf{P}, \mathfrak{L}_2(\mathbb{H},\mathbf{X}))$ then from every z, $y\in\mathbf{X}$ and for all $t\in\mathbb{R}_+$

$$<B\circ M(\cdot), y)_{\mathbf{X}}, (C\circ M(\cdot), z)_{\mathbf{X}}>_t =$$

$$= \int_{[0,t]} (C(s)Q_M(s)\ B'(s)y, z)_{\mathbf{X}} d<M>_s$$

Properties (i), (ii) are obvious. Property (iii) follows from (2.10) and inequality (1.3.4).

Property (iv) follows easily form (2.13). To prove this note Theorem 1.11 ($\underline{\mathrm{v}}$) says that $AB\in\mathfrak{L}_2(<M>\circ\mathbf{P}, \mathfrak{L}_2(\mathbb{H},\mathbf{Y}))$. It is also clear for every $z\in\mathbf{Y}$, $B_{A'z} = (AB)_z$ (see notation of §2.2). From this, in view of (2.13), we have

$$\Big(z, A \int_{[0,t]} B(s)dM(s)\Big)_{\mathbf{Y}} = \Big(A'z, \int_{[0,t]} B(s)dM(s)\Big)_{\mathbf{X}} =$$

$$= \int_{[0,t]} B_{A'z}(s)dM(s) = \int_{[0,t]} (AB(s))_z dM(s) =$$

$$= \Big(z, \int_{[0,t]} AB(s)dM(s)\Big)_{\mathbf{Y}} \qquad (\mathbf{P}\text{-}a.s.)$$

The first part of the statement of (v) follows from (2.9) and Proposition 1.9. It is clear that Lemma 1.8 implies the equality

$$<B\circ M_{m,n}, L>_t = \sum_{j=1}^{m} <\sum_{i=1}^{n} B\circ M^{ij}, L_j>_t,$$

where $L_j(t) = (L(t), e_j)_{\mathbf{X}}$.

Hence, by property (v) of a martingale in $\mathbf{R}^1$ (see § 1.3.15),

$$<B\circ M_{m,n}(\cdot), L>_t =$$

$$= \sum_{i=1}^{n} \sum_{j=1}^{m} \int_{[0,t]} (B(s)h_i, e_j)_{\mathbf{X}} d<(M(\cdot), h_i), <(L(\cdot), e_j)_{\mathbf{X}}>_s \quad (\mathbf{P}\text{-}a.s.).$$

For the second part of the assertion, it suffices to verify it for $y, z\in\{e_j\}$. From the previous equality and Theorem 1.13 it follows that for every $t\in\mathbf{R}_+$ and e_i, $e_j \in \{e_j\}$

$$<(B\circ M_{m,n}(\cdot), e_i)_{\mathbf{X}}, (C\circ M_{m,n}(\cdot), e_j)_{\mathbf{X}}>_t =$$

$$= \sum_{l,k=1}^{n} \int_{[0,t]} (B'(s)\, e_i, h_k)\, (C'(s)\, e_j, h_l) d<M_k, M_l>_s =$$
$$= \int_{[0,t]} (Q_M(s) \textstyle\prod_n B'(s)\, e_i, \prod_n C'(s)\, e_j) d<M>_s \qquad (\mathbf{P}\text{-}a.s.).$$

Making use of Proposition 1.9 we pass to the limit as $m, n\to\infty$ and get

$$<(B_y\circ M(\cdot), e_i)_{\mathbf{X}}, (C\circ M(\cdot), e_j)_{\mathbf{X}}>_t = \int_{[0,t]} (Q_M(s)B'(s)\, e_i, C'(s)\, e_j) d<M>_s \qquad (\mathbf{P}\text{-}a.s). \ \square$$

We have proved the following statement.

Theorem 3. If $B\in\mathfrak{L}_2(<M>\circ\mathbf{P}, \mathfrak{L}_2(\mathbf{H},X))$ *and*

$$B\circ M^{ij}(t) := \int_{[0,t]} (B(s)\, h_i, e_j)_{\mathbf{X}} dM_i(s),$$

where h_i, e_j *are elements of CONS's* $\{h_i\}$ *and* $\{e_j\}$ *in* $\mathbb{H}$ *and* $\mathbb{X}$ *respectively, then there exists a random process* $B\circ M(t)\in\mathfrak{M}_2^c(\mathbb{R}_+,\mathbb{H})$ *such that for every* $T\in\mathbb{R}_+$,

$$\lim_{\substack{m\uparrow dim\mathbb{X}\\ n\uparrow dim\mathbb{H}}} \mathbb{E}\sup_{t\leq T}\|B\circ M(t) - \sum_{i=1}^{n}\sum_{j=1}^{m} e_j B\circ M^{ij}(t)\|_{\mathbb{X}}^2 = 0,$$

and

$$<B\circ M>_t = \int_{[0,t]} |||B(s)\, Q_M^{1/2}(s)|||^2 d<M>_s. \tag{2.17}$$

The process $B\circ M(t)$ *and the stochastic integral defined by the equality*

$$\int_{[0,t]} B(s)dM(s) = B\circ M(t), \qquad \forall t\in\mathbb{R}_+, \qquad (\mathbb{P}\text{-}a.s.),$$

have properties (i) - (v). □

2.4. Let $M(t)\in\mathfrak{M}_2^c(\mathbb{R}_+,\mathbb{H})$ and $T\in\mathbb{R}_+$. Define on $\mathcal{P}_T$ (see § 1.3.6) Dolean's measure $<M>\circ P_T$ in the same way as it was done in the one-dimensional case (see § 1.3.14), and denote by $\overline{\mathcal{P}}_T$ the completion of $\mathcal{P}_T$ with respect to this measure.

Consistent with the notation of § 1.2.13, $\mathbb{L}_2([0,T]\times\Omega;\ \overline{\mathcal{P}}_T;\ <M>\circ\mathbb{P}_T;\ \mathfrak{L}_2(\mathbb{H},\mathbb{X}))$ is the space (of classes) of $\overline{\mathcal{P}}_T$*-measurable functions* ϕ: $[0,T]\times\Omega\to\mathfrak{L}_2(\mathbb{H},\mathbb{X})$ such that

$$\mathbb{E}\int_{[0,T]} |||\Phi(t)|||^2 d<M>_t < \infty.$$

We denote this space by $\mathbb{L}_2^T$.

Every element ϕ of $\mathbb{L}_2^T$ has a $\mathcal{P}_T$-measurable representative $\tilde{\phi}$ (cf. § 1.2.7). The stochatic integral for $\tilde{\phi}$ with respect to $M(t)$ was defined in the previous section. Thus, we can define

$$\phi\circ M(t) := \tilde{\phi}\circ M(t)$$

and consider the stochastic integral with respect $M(t)$ as a linear strongly continuous operator from $\mathbb{L}_2^T$ to $\mathbb{L}_2(\Omega;\ \mathcal{F}_T;\ \mathbb{P}_T;\ \mathbb{X})$. It is also easy to prove that $\tilde{\phi}\circ M(t)$ has a version which is a linear strongly continuous operator

from $\mathbb{L}_2^T$ to $\mathbb{L}_2(\Omega;\ \mathcal{F}_T;\ \mathbb{P}_T;\ \mathbb{C}([0,T];\ \mathbb{H}))$.

From this we can extract the following result of great importance in this book.

Theorem 4. If $M(t)\in\mathfrak{M}_2^c(\mathbb{R}_+,\mathbb{H})$, $B^n\in\mathbb{L}_2^T$ *and* B^n *converges (weakly in* $\mathbb{L}_2^T$*) to* B *as* $n\to\infty$, *then*

$$\int_{[0,T]} B^n(s)dM(s) \to \int_{[0,T]} B(s)dM(s)$$

weakly in $\mathbb{L}_2(\Omega;\ \mathfrak{F}_T;\ \mathbf{P}_T;\ \mathbb{X})$ *and* $B^n\circ M(t) \to B\circ M(t)$ *weakly in* $\mathbb{L}_2(\Omega;\ \mathfrak{F}_T;\ \mathbf{P}_T;\ \mathbb{C}([0,T];\ \mathbb{X})$. □

The assertion of the theorem follows from the well-known fact that a strongly continuous linear operator in Banach spaces is also weakly continuous (see [25], Ch. V, §4, i. 15).

2.3. A Stochastic Integral with Respect to a Local Martingale.

3.0. Let Q be a self-adjoint, non-negative, nuclear operator $\mathfrak{L}(\mathbb{H})$. We denote by $\mathfrak{L}_Q(\mathbb{H},\mathbb{X})$ the set of all operators B (possibly unbounded) defined on $Q^{1/2}\mathbb{H}$ such that $\mathbb{B}$: $Q^{1/2}\mathbb{H}\to\mathbb{X}$ and $BQ^{1/2}\in\mathfrak{L}_2(\mathbb{H},\mathbb{X})$.

Let $\mathfrak{L}_2(<M>\circ\mathbf{P},\ \mathfrak{L}_{Q_M}(\mathbb{H},\mathbb{X}))$ denote the set of all mappings ϕ: $\mathbb{R}_+\times\Omega\to\mathfrak{L}_{Q_M}(\mathbb{H},\mathbb{X}$ such that $\phi(t,\omega)Q_M^{1/2}(t,\omega)$ is a predictable process taking value in $\mathfrak{L}_2(\mathbb{H},\mathbb{X})$ and

$$\mathbb{E}\int_{[0,t]} |||\Phi(s)Q_M^{1/2}(s)|||^2 d<M>_s < \infty,\quad \forall t\in\mathbb{R}_+.$$

In what follows, we use the notation

$$||\phi||_{Q_M}: = |||\phi Q_M^{1/2}|||$$

We will define a stochastis integral with respect to processes from $\mathfrak{M}_2^c(\mathbb{R}_+,\mathbb{H})$ and $\mathfrak{M}_{loc}^c(\mathbb{R}_+,\mathbb{H})$ for integrands from $\mathfrak{L}_2(<M>\circ\mathbf{P},\ \mathfrak{L}_{Q_M}(\mathbb{H},\mathbb{X}))$.

Definitions and assumptions from the previous section are still in force here.

3.1. To define the stochastic integral we have to recall some well-known results from functional analysis (see e.g. [110], § 106).

Lemma 1. *Let* Q *be a linear self-adjoint operator in Hilbert space* $\mathbf{H}$ *having lower and upper bounds (denoted by* m_0 *and* m_1 *respectively). Let also* $f(\lambda)$ *be a continuous non-negative real function on the interval* $[m_0, m_1]$, *and let* $\{P_n(\lambda) = a_0^n + a_1^n\lambda + ... + a_k^n\lambda^k\}$ *be a decreasing sequence of polynomials of finite degree converging to* $f(\lambda)$, *for every* $\lambda\subset[m_0, m_1]$, *as* $n\to\infty$. *Then the corresponding sequence of operators* $\{P_n(Q) = a_0^n + a_1^n Q + ... + a_k^n Q^k\}$ *converges (strongly) to the non-negative, bounded operator* $f(Q)$ *which does not depend on the choice of the approximating sequence* $\{P_n(\lambda)\}$. □

The following corollary is an easy consequence of the lemma.

Corollary 1. *If* h *is an eigen-function of the operator* Q *and* λ *is the corresponding eigen-value, then* h *is also an eigen-function of* $f(Q)$ *with corresponding eigen-value equal to* $f(\lambda)$. □

Let B be an element of $\mathfrak{L}_2(\langle M\rangle\circ\mathbf{P}, \mathfrak{L}_{Q_M}(\mathbf{H},\mathbf{X}))$. Denote

$$B_n(s) := B(s)\ Q_M^{1/2}(s)\ (n^{-1} + Q_M^{1/2}(s))^{-1},\quad n\in\mathbf{N}.$$

Obviously, the operator $(n^{-1} + Q_M^{1/2}(s))^{-1}$ is a bounded one and its norm does not exceed n for every $(s,w)\in\mathbf{R}_+\times\Omega$. Consequently

$$|||B_n(s)||| \leq n\ ||B(s)||_{Q_M}$$

From Theorem 1.13 and the lemma it follows that for every $x\in\mathbf{H}$, $Q_M^{1/2}(s)\ (n^{-1} + Q_M^{1/2}(s))^{-1}$ is a predictable $\mathbf{H}$-process, which implies that $\mathbf{B}_n(s)$ is a predictable $\mathfrak{L}_2(\mathbf{H},\mathbf{X})$-process. Thus $B_n\in\mathfrak{L}_2(\langle M\rangle\circ\mathbf{P}, \mathfrak{L}_2(\mathbf{H},\mathbf{X}))$, and consequently, the stochastic integral

$$\int_{[0,t]} B_n(s)\,dM(s).$$

is defined.

Let $\{\varphi_i(s,\omega)\}$ be a CONS consisting of the eigne-functions the operator $Q_M(s,w)$ and $\{\lambda_i(s,\omega)\}$ be the set of the corresponding eigen-values. We have from the corollary that

$$||B_n - B_m||^2_{Q_M} = \sum_{i=1}^{\infty} ||BQ_M^{1/2}\varphi_i||^2\ [\lambda_i^{1/2}(n^{-1}+\lambda_i^{1/2})^{-1} - \lambda_i^{1/2}(m^{-1}+\lambda_i^{1/2})^{-1}]^2.$$

Therefore, for every $(s,\omega)\in\mathbf{R}_+\times\Omega$

$$\lim_{\substack{m\to\infty\\ n\to\infty}} \|B_n(s) - B_m(s)\|\|^2_{Q_M(s)} = 0$$

and

$$\|B_n(s) - B_m(s)\|^2_{Q_M(s)} \leq 4\|B(s)\|^2_{Q_M(s)}.$$

From this, by inequality (1.3.4) and (2.17), and the Lebesgue theorem, we have that

$$\lim_{\substack{m\to\infty\\ n\to\infty}} \mathsf{E} \sup_{t\leq T} \|B_n\circ M(t) - B_m\circ M(t)\|^2_{\mathbf{X}} \leq$$

$$\leq \lim_{\substack{m\to\infty\\ n\to\infty}} \mathsf{E} \int_{[0,T]} \|B_n(t) - B_m(t)\|_{Q_M(t)} d{<}M{>}_t = 0.$$

Thus for every $B\in\mathfrak{L}_2({<}M{>}\circ\mathbf{P},\ \mathfrak{L}_{Q_M}(\mathbf{H},\mathbf{X}))$, we can define a stochastic process $B\circ M(t)\in\mathfrak{M}_2^c(\mathbf{R}_+,\mathbf{X})$ such that

$$\lim_{n\to\infty} \mathsf{E} \sup_{t\leq T} \|B\circ M(t) - B_n\circ M(t)\|^2_{\mathbf{X}} = 0 \tag{3.1}$$

for every $t\in\mathbf{R}_+$.

Similarly, we can prove that

$${<}B_n\circ M(t){>}_t = \int_{[0,t]} \|B_n(s)\|^2_{Q_M(s)} d{<}M{>}_s =$$

$$= \int_{[0,t]} \sum_{i=1}^{\infty} \|B(s)\, Q_M^{1/2}(s)\varphi_i(s)\|^2_{\mathbf{X}}\lambda_i(s)\, (n^{-1} + \lambda_i^{1/2}(s))^2 d{<}M{>}_s.$$

From this and Proposition 1.9 we find that

$${<}B\circ M{>}_t = \int_{[0,T]} \|B(s)\|^2_{Q_M(s)} d{<}M{>}_s. \tag{3.2}$$

Thus for $B\in\mathfrak{L}_2({<}M{>}\circ\mathbf{P},\ \mathfrak{L}_{Q_M}(\mathbf{H},\mathbf{X}))$, define a stochastic integral by the equality

$$\int_{[0,t]} B(s)dM(s) := B\circ M(t), \qquad \forall t\in\mathbb{R}_+, \qquad (\mathbb{P}\text{-}a.s.)$$

It is easy to show that for every $t\in\mathbb{R}_+$

$$\lim_{n\to\infty} \mathbb{E} \int_{[0,T]} \|B(s) - B_n(s)\|^2_{Q_M(s)} d{<}M{>}_s = 0.$$

Hence in view of (3.1) and property (iii) of the previous section both of the stochastic integrals defined in Theorem 1.3 and here coincide.

In view of inequality (1.3.4) and (3.2), the integral discussed above, does not depend on the method of approximation of B by elements of $\mathfrak{L}_2({<}M{>}\circ\mathbb{P}, \mathfrak{L}_2(\mathbb{H},\mathbb{X}))$. We shall denote by (i') - (iv') the properties obtained by replacing $\mathfrak{L}_2({<}M{>}\circ\mathbb{P}, \mathfrak{L}_2(\mathbb{H},\mathbb{X}))$ by $\mathfrak{L}_2({<}M{>}\circ\mathbb{P}, \mathfrak{L}_{Q_M}(\mathbb{H},\mathbb{X}))$.

It is easy to see that the integral defined in this section possesses properties (i') - (iv'). So we have proved the following result.

Theorem 1. *Let* $B\in\mathfrak{L}_2({<}M{>}\circ\mathbb{P}, \mathfrak{L}_{Q_M}(\mathbb{H},\mathbb{X}))$. *Then there exists a sequence* $B_n\in\mathfrak{L}_2({<}M{>}\circ\mathbb{P}, \mathfrak{L}_2(\mathbb{H},\mathbb{X}))$ *and a stochastic process* $B\circ M(t) \in \mathfrak{M}^c_2(\mathbb{R}_+,\mathbb{X})$ *such that for every* $T\in\mathbb{R}_+$,

$$(a) \quad \lim_{n\to\infty} \mathbb{E} \sup_{t\le T} \|B\circ M(t) - B_n\circ M(t)\|^2_{\mathbb{X}} = 0,$$

$$(b) \quad {<}B\circ M{>}_t = \int_{[0,T]} \|B(s)\|^2_{Q_M} d{<}M{>}_s \quad (\mathbb{P}\text{-}a.s.)$$

$$(c) \quad \lim_{n\to\infty} \mathbb{E} \int_{[0,T]} \|B(s) - B_n(s)\|^2_{Q_M(s)} d{<}M{>}_s = 0,$$

and $B\circ M(t)$ *does not depend on the choice of the sequence* $B_n\in\mathfrak{L}_2({<}M{>}\circ\mathbb{P}, \mathfrak{L}_2(\mathbb{H},\mathbb{X}))$ *with property (a). The process* $B\circ M(t)$ *and the stochastic integral defined by the equality*

$$\int_{[0,t]} B(s)dM(s) := B\circ M(t), \qquad \forall t\in\mathbb{R}_+$$

have properties (i') - (iv'). □

3.2. We consider a particular case of the integral defined above. Let $f(s)$ be a predictable $\mathbb{H}$-process such that

$$\mathbf{E} \int_{[0,t]} \|b(s)\|^2 d\langle M\rangle_s < \infty, \quad \forall t \in \mathbf{R}_+.$$

Denote by $\hat{b}(s)$ the linear operator from $\mathbb{H}$ to $\mathbb{R}^1$, defined by the equality $\hat{b}(s)h := (b(s), h)$, $\forall h \in \mathbb{H}$.

It is clear that $\hat{b} \in \mathfrak{L}_2(\langle M\rangle \circ \mathbf{P}, \mathfrak{L}_2(\mathbb{H}, \mathbb{R}^1))$ and by Theorem 1.3 for every $t \in \mathbb{R}_+$, ($\mathbb{P}$-*a.s.*)

$$\int_{[0,t]} \hat{b}(s) dM(s) = \lim_{n\to\infty} \sum_{i=1}^{n} \int_{[0,t]} (b(s), h_i) dM_i(s). \tag{3.3}$$

Warning 1. *In the future we shall denote* $\int_{[0,t]} \hat{b}(s)dM(s)$ *by* $\int_{[0,t]} (b(s), dM(s))$ *and call it the stochastic integral of* b *with respect to* M. □

The equality (3.3) shows that this notation is natural.

3.3. Define by $\mathfrak{L}_2(\langle M\rangle, \mathfrak{L}_{Q_M}(\mathbb{H},\mathbb{X}))$ the set of mappings $\Phi: \mathbf{R}_+ \times \Omega \to \mathfrak{L}_{Q_M}(\mathbb{H},\mathbb{X})$ such that $\Phi(t,\omega)Q_M^{1/2}(t,\omega)$ is a predictable stochastic process taking values in $\mathfrak{L}_2(\mathbb{H},\mathbb{X})$ and

$$\int_{[0,t]} \|\Phi(s)\|^2_{Q_M(s)} d\langle M\rangle_s < \infty, \qquad \forall t \in \mathbf{R}_+, \qquad (\mathbb{P}\text{-a.s.}).$$

In this section we shall define a stochastic integral with respect to a continuous local martingale $M(t)$ for integrands from $\mathfrak{L}_2(\langle M\rangle, \mathfrak{L}_{Q_M}(\mathbb{H}, \mathbb{X}))$.

Let $B \in \mathfrak{L}_2(\langle M\rangle, \mathfrak{L}_{Q_M}(\mathbb{H},\mathbb{X}))$. Let $\Gamma_n(t,\omega) = \left\{t,\omega: \int_{[0,t]} \|B(s)\|^2_{Q_M} d\langle M\rangle_s > n\right\}$,

$$\tau_n := \begin{cases} \inf_{t\in\mathbb{R}_+} \Gamma_n(t) & \text{if } \Gamma_n(t) \neq \emptyset \\ \infty & \text{if } \Gamma_n(t) = \emptyset \end{cases}$$

and $B^n(t) = B(t)\, 1_{\{[0,\tau_n]\}}(t)$ for every $n \in \mathbf{N}$.

Note that τ_n is a stopping time for every n (see § 1.3.7). It is also clear that $\tau_n \uparrow \infty$ $\mathbf{P}$-a.s. Next, the process $1_{\{[0,\tau_n]\}}(t)$ is a predictable one because it is left-continuous and $\mathfrak{F}_t$-adapted. Thus the process $B^n(t) \in \mathfrak{L}_2(<M> \circ \mathbf{P}, \mathfrak{L}_{Q_M}(\mathbb{H}, \mathbf{X}))$ for every n, and the stochastic integral $\int_{[0,t]} B^n(s)dM(s)$ is defined.

Let $\tau_0 = 0$ and

$$B \circ M(t) = \sum_{i=1}^{\infty} B^i \circ M(t)\, 1_{\{]\tau_{i-1},\, \tau_i]\}}(t) + 0\, 1_{\{t=0\}}(t).$$

It is easy to show that $B \circ M(t)$ is a continuous, $\mathfrak{F}_t$-adapted stochastic process and $B \circ M(t \wedge \tau_n) = B^n \circ M(t)$, $\forall n \in \mathbf{N}$, ($\mathbf{P}$-*a.s.*), which implies that $B \circ M(t) \in \mathfrak{M}^2_{loc}(\mathbf{R}_+, \mathbb{H})$ and

$$<B \circ M>_t = \int_{[0,t]} \|B(s)\|^2_{Q_M(s)}\, d<M>_s \qquad (\mathbf{P}\text{-}a.s.)$$

Define the stochastic integral by the equality

$$\int_{[0,t]} B(s)dM(s) := B \circ M(t) \qquad (\mathbf{P}\text{-}a.s.)$$

Let us denote by (i′′), (ii′′), (iv′′) the properties of the stochastic integral which are obtained from properties (i′), (ii′) and (iv′) by replacing $\mathfrak{L}_2(<M> \circ \mathbf{P}, \mathfrak{L}_2(\mathbb{H}, \mathbf{X}))$ and $\mathfrak{M}^c_2(\mathbf{R}_+, \mathbb{H})$ by $\mathfrak{L}_2(<M>, \mathfrak{L}_2(H, X))$ and $\mathfrak{M}^c_{loc}(\mathbf{R}_+, \mathbb{H})$ respectively. The stochastic integral defined in this section possesses these properties. Instead of the property (iii′) it has the following one:

(iii′′) Let $B \in \mathfrak{L}_2(<M>, \mathfrak{L}_{Q_M}(\mathbb{H}, \mathbf{X}))$ and $B_n \in \mathfrak{L}_2(<M>, \mathfrak{L}_{Q_M}(\mathbb{H}, \mathbf{X}))$, for every $n \in \mathbf{N}$, and for some $T \in \mathbf{R}_+$

$$\int_{[0,T]} \|B(s) - B_n(s)\|^2_{Q_M(s)} d<M>_s \to 0$$

in probability as $n \to \infty$.
Then

$$\sup_{t \le T} \|B^n \circ M(t) - B \circ M(t)\|_{\mathbf{X}} \to 0$$

in probability.

The validity of this statement follows easily from Corollary 1.7. Thus we have proved the following result.

Theorem 3. *Let* $B \in \mathfrak{L}_2(<M>, \mathfrak{L}_{Q_M}(\mathbf{H}, \mathbf{X}))$. *Then there exists a stochastic process* $B \circ M(t) \in \mathfrak{M}^2_{loc}(\mathbb{R}_+, \mathbf{X})$ *such that*

$$<B \circ M>_t = \int_{[0,t]} \|B(s)\|^2_{Q_M(s)} d<M>_s.$$

The stochastic integral defined by the equality

$$\int_{[0,t]} B(s)\,dM(s) := B \circ M(t), \quad (\mathbf{P}\text{-}a.s.)$$

possesses properties (i″) - (iv″). □

Example 3. Let G be a domain in $\mathbf{R}^d$ and $b(t) := b(t,x,w)$ be a predictable process, taking values in $\mathbf{L}_2(G)$, such that

$$\int_{[0,t]} \|b(t)\|^2_{\mathbf{L}_2(G)} dt < \infty, \quad \forall t \in \mathbf{R}_+ \qquad (\mathbf{P}\text{-a.s.})$$

Let $M(t) := w(t)$ be a one-dimensional standard Wiener process.

Suppose that $B(s)$ is an operator such that $B(s,w)h = b(s,\cdot,w)h$, $\forall h \in \mathbf{R}^1$, $s \in \mathbf{R}_+$ and $\omega \in \Omega$. Then, by the theorem the stochastic integral

$$\int_{[0,t]} B(s)\,dw(s)$$

is defined. □

Remark 3. *Making use of Theorem* 11.17 *from Ch. II of* [25], *it is easy to show that there exists a* $\mathcal{P} \otimes \mathcal{B}(G)$-*measurable function* $\hat{b}(t,x,w)$ *that coincides with the process* $b(t,\cdot,w)$ *from the previous section as an element of* $\mathbf{L}_2(G)$ *for* $l \times \mathbf{P}$-*a.a.* (t,w). *Moreover, this function satisfies the inequality*

$$\int_{[0,t]} \tilde{b}^2(s,x)\,ds < \infty, \quad \forall t \in \mathbf{R}_+, \qquad (l \times \mathbf{P}\text{-}a.s.).$$

Thus the stochastic integral $\int_{[0,t]} \tilde{b}(s,x)\,dw(s)$ *is defined for almost all* x (*cf.* § 1.4.1). *It is easy to see that after defining it on a* $\mathbf{P}$-*negligible set, it coincides (as a local martingale in* $\mathbb{L}_2(G)$*) with the stochastic integral defined in the example.* □

3.4. *Remark* 4. *Suppose* $b(t)$ *is a predictable* $\mathbb{H}$*-process such that*

$$\int_{[0,t]} \|b(s)\|^2 d\langle M\rangle_s < \infty, \quad \forall t \in \mathbb{R}_+, \qquad (\mathbf{P}\text{-}a.s.)$$

Then, arguing as in the §3.2, we define the stochastic integral of b *with respect to* M *by the equality*

$$\int_{[0,t]} (b(s),\, dM(s)) := \int_{[0,t]} \hat{b}(s)\, dM(s). \ \square$$

Warning 3. *In this section we have made a distinction between a stochastic integral and its continuous version (in particular, we have used for them different notation). From now on we shall consider only the continuous version of a stochastic martingale using for it the notation* $B \circ M(t)$ *as well as* $\int_{[0,t]} B(s)dM(s)$ *or* $\int_{[0,t]} (B(s),\, dM(s))$. □

3.5. *Remark* 5. *All the results of this chapter obtained for a process on* $\mathbb{R}_+$ *are still valid (sometimes with minor changes) for a process on the interval* $[T_0, T]$, $T_0,\ T \in \mathbb{R}_+$.

2.4. An Energy Equality in a Rigged Hilbert Space.

4.0. Imbedding theorems are of great importance in the theory of differential equations. We give now one of the simplest examples of such a theorem.

Let u: $[0,T] \to \overline{\mathbb{R}}_1$ and let $\frac{\partial}{\partial t} u$ be the generalized derivative of u.

Write $\mathbb{W}_2^1([0,t]) := \{u\colon u \in \mathbb{L}_2([0,T]);\ \frac{\partial}{\partial t} u \in \mathbb{L}_2([o,T])\}$.

Proposition 0. *If* $f \in \mathbb{W}_2^1([0,T])$ *then it can be modified so it is a continuous mapping from* $[0,t]$ *to* $\mathbb{R}^1$ *by a change on a set of zero measure.*

In other words the proposition means that in a sense the space $\mathbb{W}_2^1([0,T])$ *is imbedded into the space* $\mathbb{C}([0,T]);\ \mathbb{R}^1))$ *of continuous functions from* $[0,T]$ *to* $\mathbb{R}^1$. □

Definition 0. *We say that a Hilbert space* $\mathbb{Z}$ *is normally imbedded in another Hilbert space* $\mathbb{Z}'$ *if the imbedding is dense (in the topology of the space* $\mathbb{Z}'$*, generated by the norm* $\|\cdot\|_{\mathbb{Z}'}$*) and continuous, i.e. there a constant* N *such that*

$$\|z\|_{\mathbb{Z}'} \leq N\,\|z\|_{\mathbb{Z}} \qquad \forall z \in \mathbb{Z}.^{*} \ \square$$

In this section as well as above we denote the scalar product and the norm in $\mathbb{H}$ by $(\cdot,\cdot)$ and $\|\cdot\|$ respectively.

Let $\mathbf{X}$ and $\mathbb{H}$ be Hilbert spaces, and $\mathbf{X}$ be normally imbedded into $\mathbb{H}$. Identify $\mathbb{H}$ with its conjugate and denote by $\mathbf{X}^*$ the conjugate space to $\mathbf{X}$. Then $\mathbf{X} \subset \mathbb{H} \subset \mathbf{X}^*$. Suppose also that $hx = (x,h)$, $\forall x \in \mathbf{X}$, $h \in \mathbb{H}$, where hx is the value of the functional h on the element $x \in \mathbf{X}$.

The Hilbert space $\mathbb{H}$ equipped with the pair $\mathbf{X}$, $\mathbf{X}^*$ we shall call the rigged Hilbert space $\mathbb{H}$.

The following theorem, being a direct generalization of the proposition given above, is commonly used in the theory of partial differential equations.

Theorem 0 (*see e.g.* [89], *Ch.* 1, *i*.2.2). *Let* $u \in \mathbb{W} := (u\colon u \in \mathbb{L}_2([0,T];$ $\overline{\mathfrak{B}([0,T])};\, l;\, \mathbf{X}),\ \frac{\partial}{\partial t} u \in \mathbb{L}_2([0,T];\, \overline{\mathfrak{B}([0,T])};\, l;\, \mathbf{X}^*)$. *Then* u *has a version* (*with respect to* t), *which is a continuous mapping of* $[0,T]$ *in* $\mathbb{H}$. *Moreover, for all* $t \in [0,T]$

$$\|\tilde{u}(t)\|^2 = \|\tilde{u}(0)\|^2 + 2 \int_{[0,t]} \left(\frac{\partial}{\partial t} u(s)\right) u(s)\,ds. \ \square$$

The last formula is usually called the energy equality.

The aim of this section is an extension of this result to random functions $u(t)$. The assumption $\frac{\partial}{\partial t} u(t) \in \mathbb{L}_2([0,T]);\, \overline{\mathfrak{B}([0,T])};\, l,\, \mathbf{X}^*)$ will be replaced by the assumption that $u(t)$ is a semimartingale. The formula that generalizes the energy equality for $\|\tilde{u}(t)\|^2$ can be considered as the Ito formula for the square of the norm (in $\mathbb{H}$) of the continuous (in $\mathbb{H}$) version of the semimartingale $u(t)$. It is of great importance in the theory of evolution stochastic systems.

4.1. *Definition* 1. *The triple* $(\mathbf{X},\mathbb{H},\mathbf{X}')$ *will be called normal if* $\mathbf{X} \subset \mathbb{H} \subset \mathbf{X}'$, *and all the imbeddings are normal, and there exists a number* N *such that for all* $x \in \mathbf{X}$, $y \in \mathbb{H}$, $(x,y) \leq N\|x\|_{\mathbf{X}}\,\|y\|_{\mathbf{X}'}$. $\square$

Let $(\mathbf{X},\mathbb{H},\mathbf{X}')$ be a normal triple. Given $y \in \mathbf{X}'$ let us choose a sequence $\{y_n\}$ from $\mathbf{X}$ such that $\|y - y_n\|_{\mathbf{X}'} \to 0$ as $n \to \infty$. Write

---

*By an appropriate change of the norm of the one of the spaces we can always make N to be equal to 1. So from now on we shall suppose that N=1.

$$[x,y] := \lim_{n\to\infty}(x,y_n), \qquad \forall x\in \mathbf{X}.$$

Clearly, the limit in the right-hand side of the equality exists and does not depend on the choice of the sequence $\{y_n\}$ approximating y. Thus $[x,y]$ is well defined.

Obviously, $[x,y]$ is a bilinear form. We shall call it the canonical bilinear functional (CBF) of the triple

Let us collect some obvious properties of the CBF:

(i) $|[x,y]| \le N \, \|x\|_{\mathbf{X}} \cdot \|y\|_{\mathbf{X}'}$ for every $x\in\mathbf{X}$, $y\in\mathbf{X}'$, which implies that the CBF is continuous with respect to every variable;

(ii) $[x,y] = (x,y)$ for every $y\in\mathbb{H}$;

(iii) if $y(s)$ is an l-integrable function on the interval $[a,b]$ taking values in $\mathbf{X}'$, then

$$\left[x, \int_{[a,b]} y(s)\,ds\right] = \int_{[a,b]} [x,y(s)]\,ds, \qquad \forall x\in\mathbf{X}.$$

The proof of these properties is a simple exercise.

Remark 1. *It can be shown (see* [62], *Ch.* 4, § 5, *i.* 10) *that if a Hilbert space* $\mathbf{X}$ *is normally imbedded in a Hilbert space* $\mathbf{X}'$, *than there exists a space* $\mathbb{H}$ *so that* $(\mathbf{X},\mathbb{H},\mathbf{X}')$ *forms a normal triple. Moreover, the CBF of this triple produces the isometric isomorphism between* $\mathbf{X}^*$ *and* $\mathbf{X}'$ *(where* $\mathbf{X}^*$ *is the conjugate space for* $\mathbf{X}$*) by the formula*

$$x' \leftrightarrow [\cdot,x']\in\mathbf{X}^*, \quad \forall\, x' \in X'.$$

So, considering the CBF as a duality between $\mathbf{X}'$ *and* $\mathbf{X}^*$, *we can identify* $\mathbf{X}'$ *with* $\mathbf{X}^*$. □

4.2. Let us fix a normal triple $(\mathbf{X},\mathbb{H},\mathbf{X}')$, a standard probability space $\mathbb{F} = (\Omega, \mathfrak{F}, \{\mathfrak{F}_t\}_{t\in[0,T]}, \mathbb{P})$, and a continuous local martingale $M(t)$ (relative to $\{\mathfrak{F}_t\}$), taking values in $\mathbb{H}$, on this probability space. We suppose that the $\dim\mathbf{X} = \infty$. In the finite-dimensional case the results of this section are trivial.

Suppose we are given functions $x(t,w)$ and $x'(t,w)$ on $[0,T]\times\Omega$ taking values in $\mathbf{X}$ and $\mathbf{X}'$ respectively. Assume these functions are predictable and satisfy ($\mathbb{P}$-a.s.) the following inequality

$$\int_{[0,T]} (\|x(t,w)\|^2_{\mathbf{X}} + \|x'(t,w)\|^2_{\mathbf{X}'})\,dt < \infty.$$

Let $x(0)$ be an $\mathcal{F}_0$-measurable function on Ω taking values in **H**. These assumptions will be in force throughout this section. The main result is the following theorem.

Theorem 2. Let τ *be a stopping time and for every* $\eta \in \mathbf{X}$ $l \times \mathbf{P}$*-a.s. on the set* $\{(t,\omega)\colon t<\tau(w)\}$

$$(4.1) \qquad (\eta, x(t)) = (\eta, x(0)) + \int_{[0,t]} [\eta, x'(s)]ds + (\eta, M(t)).$$

Then there exist a set $\tilde{\Omega} \subset \Omega$ *and a function* $\tilde{x}(t,w)$ *that takes its values in* **H** *such that:*

(*i*) $\mathbf{P}(\tilde{\Omega}) = 1$, $\tilde{x}(t)$ *is* F_t*-measurable random variable on the set* $\{\omega\colon t<\tau(s)\}$ *for every* t, $\tilde{x}(t)$ *is continuous in* t *on* $[0,\tau(\omega)$ [*for every* $\omega \in \tilde{\Omega}$, *and* $x(t) = \tilde{x}(t)$ $(l \times \mathbf{P}$*-a.s.*) *on* $\{(t,\omega)\colon t<\tau(\omega)\}$ *and* $\eta \in \mathbf{X}$

(*ii*) *For* $w \in \tilde{\Omega}$, $t<\tau(w)$ *and* $\eta \in \mathbf{X}$

$$(4.2) \qquad (\eta, \tilde{x}(t)) = (\eta, x(0)) + \int_{[0,t]} [\eta, x'(s)]ds + (\eta, M(t));$$

(*iii*) *For* $\omega \in \tilde{\Omega}$, $t < \tau(\omega)$

$$||\tilde{x}(t)||^2 = ||x(0)||^2 + 2 \int_{[0,t]} [x(s), x'(s)]ds +$$

$$(4.3) \qquad + 2 \int_{[0,t]} (\tilde{x}(s), dM(s)) + \langle M\rangle_t;^*$$

(*iv*) *If* (4.1) *holds true for some* $t \geq 0$ *and every* $\eta \in \mathbf{X}$ $\mathbf{P}$*-a.s. on* $\{\omega\colon t<\tau(\omega)\}$, *then* $x(t) = \tilde{x}(t)$ $\mathbf{P}$*-a.s. on* $\{\omega\colon t<\tau(\omega)\}$. □

---

*As regards the stochastic integral $\int_{[0,t]} (\tilde{x}(s), dM(s))$ see Remarks 3.4 and 3.5.

The stochastic process $\tilde{x}(t)$ is predictable because it is continuous in **H** and $\mathcal{F}_t$-adapted.

Remark 2. To make the relation between this theorem and Theorem 0 clearer we note that if x(t) coincides ($l\times\mathbb{P}$-a.s.) with the $\mathbb{X}'$-semimartingale

$$y(t) = x(0) + \int_{[0,t]} x'(s)ds + M(t),$$

then (4.1) holds. □

In the case of $\mathbb{X}=\mathbb{H}=\mathbb{X}'$ formula (4.3) is the Ito formula for the square of the norm of $y(t)$.

Proof of Theorem 2. Write

$$r(t) = \|x(0)\|^2 + <M>_t + \int_{[0,t]} \|x(s)\|^2_{\mathbb{X}}\, ds + \int_{[0,t]} \|x'(s)\|^2_{\mathbb{X}}\, ds.$$

Since the process $r(t)$ is predictable, $\tau(n):= inf\{t\geq 0\colon r(t) \geq n\} \wedge \tau$ is a stopping time for every n and $\tau(n) \uparrow \tau$ as $n\rightarrow\infty$.

If the statement of the theorem, where τ is replaced by $\tau(n)$, is valid for all n, then it is also valid in its original form (we can define $\tilde{\Omega}:= \bigcap_n \tilde{\Omega}_n$, where $\tilde{\Omega}_n$ is $\tilde{\Omega}$ from the assertion of the theorem where τ is replaced by $\tau(n)$). Thus, without loss of generality we shall assume that there exists a constant N such that on the set $\{(t,\omega)\colon t<\tau(\omega)\}$, $\tau(t,\omega)<N$.

From this and Proposition 1.7 it follows that we can consider $M(t)$ as belonging to $\mathfrak{M}^c_2([0,T], \mathbb{H})$.

4.3. *Lemma* 3. *There exists a sequence of imbedded subdivisions* $\{0 = t^n_0 < t^n_1 < \dots < t^n_{k(n)+1} = T\}$ *of the interval* $[0,T]$, *with the diameter of subdivision tending to zero, and a set* $\Omega'\subset\Omega$ *having the following properties:*

1) $\mathbb{P}(\Omega') = 1$, *and for* $\omega\in\Omega'$, $t<\tau(\omega)$, $t\in I:= \{t^n_i,\ i=1,2,\dots k(n),\ n\in\mathbb{N}\}$ *and* $\eta\in\mathbb{X}$ *the equality* (4.1) *holds true.*

2) *Write* $x^1_n(t):= x(t^n_i)$ *for* $t\in[t^n_i, t^n_{i+1}[$ $i=1,2,..k(n)$, $x^1_n(t):= 0$ *for* $t\in[0, t^n_1[$, *and* $x^2_n(t):= x(t^n_{i+1})$ *for* $t\in[t^n_i, t^n_{i+1}[$, $i=0,1,..,(k(n)\text{-}1)$, $x^2_n(t):= 0$ *for* $t\in[t^n_{k(n)}, T[$.

Then

$$\text{(4.4)}\qquad \mathbb{E} \sup_{t\leq T} \|x^j_n(t)\|^2_{\mathbb{X}} < \infty,\ j=1,2,$$

and

$$\text{(4.5)}\qquad \lim_{n\rightarrow\infty} \mathbb{E} \int_{[0,T]} \|x(t) - x^i_n(t)\|^2_{\mathbb{X}}\, dt = 0,\ j=1,2.\ \square$$

Proof. We first prove the second statement. The proof proceeds by the well-known method of Doob (cf. [24] Ch. IX, §5).

Let f be an $\mathfrak{B}(\mathbb{R}^1)$-measurable function $\mathbf{R}^1$ with values in $\mathbf{X}$, vanishing outside a finite interval of $\mathbf{R}^1$ and such that $\int_{\mathbf{R}^1} \|f(s)\|^2_{\mathbf{X}}\, ds < \infty$. Then

$$\lim_{\delta \to 0} \int_{\mathbb{R}^1} \|f(s+\delta) - f(s)\|^2_{\mathbf{X}}\, ds = 0. \tag{4.6}$$

For each $\varepsilon < 0$, let $f_\varepsilon(s)$ be a continuous function taking values in $\mathbf{X}$, equal to zero outside some finite interval and such that

$$\int_{\mathbb{R}^1} \|f(s) - f_\varepsilon(s)\|^2_{\mathbf{X}}\, ds \leq \varepsilon^2.$$

From this inequality, , making use of the Minkowski inequality, we get

$$\overline{\lim_{\delta \to 0}} \left[\int_{\mathbb{R}^1} \|f(s+\delta) - f(s)\|^2_{\mathbf{X}}\, ds \right]^{1/2} \leq \overline{\lim_{\delta \to 0}} \left(\int_{\mathbf{R}^1} \|f_\varepsilon(s+\delta) - f_\varepsilon(s)\|^2_{\mathbf{X}}\, ds \right)^{1/2} + 2\varepsilon = 2\varepsilon.$$

Hence (4.6) is proved.
Define

$$\hat{x}(t,w) = \begin{cases} x(t,w), & t \in [0,T]; \\ 0, & t \in \mathbb{R}^1 \setminus [0,T]. \end{cases}$$

From (4.6) it follows that $\mathbb{P}$-a.s.

$$\lim_{\delta \to 0} \int_{\mathbb{R}^1} \|\hat{x}(t+\delta) - \hat{x}(t)\|^2_{\mathbf{X}}\, dt = 0.$$

Let $[a]$ be the integral part of the number a, $\chi^1(n,t) := 2^{-n}[2^n t]$ and $\chi^2(n,t) := \chi^1(n,t) + 2^{-n}$, *then*

$$\lim_{n \to \infty} \int_{\mathbb{R}^1} \|\hat{x}(\chi^j(n,t) + s) - \hat{x}(t+s)\|^2_{\mathbf{X}}\, ds = 0,\ j = 1,2 \quad (\mathbf{P}\text{-}a.s.)$$

Therefore, in view of boundness of the process $r(t)$, by dominated convergence theorem, we obtain

$$\lim_{n\to\infty} \mathbf{E} \int_{\mathbb{R}^1\times\mathbb{R}^1} \|\hat{x}(\chi^j(n,t) + s) - \hat{x}(t + s)\|_{\mathbf{X}}^2 \, dsdt = 0, \; j = 1,2.$$

From this if follows that there exists a sequence of integers r_n such that for l-a.a. s.

$$0 = \lim_{n\to\infty} \mathbf{E} \int_{\mathbb{R}^1} \|\hat{x}(\chi^j(r_n,t) + s) - \hat{x}(t + s)\|_{\mathbf{X}}^2 \, dt =$$

$$(4.7) \qquad = \lim_{n\to\infty} \mathbf{E} \int_{\mathbb{R}^1} \|\hat{x}(\chi^j(r_n,t - s) + s) - \hat{x}(t)\|_{\mathbf{X}}^2 \, dt, \; j = 1,2.$$

In view of separability of $\mathbf{X}$, by the Fubini theorem, there exists a set of $S\subset[0,T]$, where $[0,T]\backslash S$ is at most a countable set, such that for $t\in S$ and all $\eta\in\mathbf{X}$, the equality (4.1) holds true $\mathbf{P}$-a.s. on the set $\{\omega\colon t<\tau(\omega)\}$.

It is easy to show, that for l—a.s. $s\in[0,T]$, all the values of the function $\chi^j(r_n,t\text{-}s) + s$ hitting $[0,T]$, which it takes running over the interval $[0,T]$, belong to S. Thus we can choose s, which satisfies (4.7) and such that all the values of the function $\chi^j(r_n, t\text{-}s) + s$, for $j=1,2$, $n\leq 1$ and $t\in[0,T]$, belonging to $[0,T]$, belongs also to S. Define $\{t_i^n\}$ to be the set of the values $\chi^1(r_n, t\text{-}s) + s$ hitting $[0,T]$ for this s and $t\in[0,T]$ and amplified by points 0 and T. The subset of Ω, where equality (4.1) holds for all $\eta\in\mathbf{X}$ and $t:= t_i^n < \tau(w)$ for $i=1,2,..k(n)$, $n\in\mathbf{N}$ will be taken as Ω'

Evidently, for Ω' and I, chosen in such a way, the first part of the assertion of the lemma and equality (4.5) from the second part are valid. To prove equality (4.4) note that it is equivalent to the inequality

$$\mathbf{E} \int_{[0,T]} \|x_n^i(t)\|_{\mathbf{X}}^2 \, dt < \infty$$

This inequality for large n follows from (4.5). For small n, equality (4.4) is fulfilled so much the more because the subdivisions defined above are imbedded.

4.4. *Lemma* 4. *Let* $\widetilde{M}(t) := x(0) + M(t)$. *Then for* $\omega\in\Omega'$, t *and* $s\in I$ *and* $s\leq t<\tau(w)$,

$$(4.8) \qquad \|x(t)\|^2 = 2 \int_{[0,t]} [x(t), x'(u)]du + \|\widetilde{M}(t)\|^2 - \|x(t) - \widetilde{M}(t)\|^2,$$

$$\|x(t)\|^2 - \|x(s)\|^2 = 2\int_{[s,t]} [x(t), x'(u)]\,du + 2(x(s), M(t) - M(s)) +$$

$$+ \|M(t) - M(s)\|^2 - \|x(t) - x(s) - (M(t) - M(s))\|^2. \quad \square \tag{4.9}$$

Proof. Let us consider the equality (4.8). The equality (4.9) may be derived from (4.8) by simple algebraic transformation

From Lemma 3 it follows that for $\omega \in \Omega'$, $t \in I$ and $t < \tau(\omega)$, $x(t) \in X$. Hence in (4.1) we can take $x(t)$ as η. Thus we obtain that for $\omega \in \Omega'$, $t \in I$ and $t < \tau(\omega)$,

$$\|x(t)\|^2 = \int_{[0,t]} [x(t), x'(u)]\,du + (x(t), \widetilde{M}(t)).$$

From this it follows that for the same ω and t

$$\|\tilde{M}(t)\|^2 - \|x(t) - \tilde{M}(t)\|^2 = 2(x(t), \tilde{M}(t)) - \|x(t)\|^2 =$$

$$= \|x(t)\|^2 - 2\int_{[0,t]} [x(t), x'(u)]\,du,$$

which proves equality (4.8). $\square$

4.5. Write $\Omega'' := \Omega' \cap \{\omega: \sup_{t \in I,\, t < \tau} \|x(t)\|^2 < \infty\}$. From Lemma 3 we have that $\mathsf{P}(\Omega'') = 1$.

We also need the following lemma.

Lemma 5. For $\omega \in \Omega''$ and $t < \tau(\omega)$ the function $\tilde{x}(t)$, taking values in $\mathbf{H}$, is defined, is weakly continuous in $\mathbf{H}$ with respect to t, and satisfies (4.2) for all $\eta \in X$, $\omega \in \Omega''$ and $t < \tau(\omega)$. Moreover, $\tilde{x}(t)$ is F_t-measurable on the set $\{\omega: t < \tau(s)\}$, $\tilde{x}(t) = x(t)$ for $\omega \in \Omega''$, $t \in I$ and $t < \tau(w)$, and $\tilde{x}(t) = x(t)$ $l \times \mathsf{P}$-a.s. on the set $\{(t,\omega): t < \tau(\omega)\}$. $\square$

Proof. Since for $\omega \in \Omega''$, $t \in I$, $t < \tau$, and $\eta \in X$, the function $(\eta, x(t))$ coincides with the right-hand side of (4.1), it is continuous in t. Thus, for every $s \le \tau$, we can choose a sequence of points $t_i \in I$ not exceeding τ and such that $\lim_{i \to \infty} (\eta, x(t_i))$ exists. Since the function $\|x(t)\|$ is bounded on the set $I \cap \{t < \tau\}$, it follows that the sequence $\{x(t_i)\}$ has a weak limit in $\mathbf{H}$, which we shall denote by $\tilde{x}(s)$. It is clear that $\tilde{x}(t)$ satisfies (4.2) for all $\omega \in \Omega''$, $t < \tau(\omega)$ and $x \in X$. Now the assertion of the lemma follows easily. $\square$

4.6 Write $\tilde{x}(t) := \tilde{x}(t)$ for $t \geq \tau$ and $\tilde{x}(t) := 0$ for $\omega \notin \Omega''$. Since, for $t < \tau$ but not belonging to I, $\tilde{x}(t)$ is defined as a weak limit of $x(t_i)$ over some sequence of $t_i \in I$, and the norm of a weak limit does not exceed the lower limit of the norms of the approximating sequence, the following relation holds true:

$$(4.10) \qquad \sup_{t \leq T} \|\tilde{x}(t)\| = \sup_{t \in I,\ t < \tau} \|x(t)\| < \infty.$$

Next, it is obvious that the function $\tilde{x}(s)$ is predictable. This follows from the fact that it is weakly continuous and $\mathcal{F}_t$-adapted.

Hence by (4.10) it follows that the stochastic integral

$$\int_{[0,t]} (\tilde{x}(s),\ dM(s)),\ t \leq T,$$

is defined (see § 3.4).

Let $\tilde{x}_n(t) := \tilde{x}(t_i^n)$ for $t \in [t_i^n,\ t_{i+1}^n[$, $i = 0, 1, .. k(n)$. The next step of the proof is the following simple lemma.

Lemma 6. *The sequence*

$$(4.11) \qquad \sup_{t \leq T} \left| \int_{[0,t]} (x_n(s) - x(s),\ dM(s)) \right| \longrightarrow 0$$

in probability as $n \rightarrow \infty$. □

Proof. Let $\{h_i\}$, $i \in \mathbb{N}$, be a CONS in $\mathbb{H}$ and $\prod_r$ be a projection operator from $\mathbb{H}$ into the space generated by $h_1, \ldots, h_r$. By lemma 5, $x(s)$ is the weakly continuous function in $\mathbb{H}$ (for $\omega \in \Omega''$). Therefore, $\prod_r \tilde{x}(s)$ is the strongly continuous function in $\mathbb{H}$ for the same set of ω. Thus $\mathbb{P}$-a.s.

$$\lim_{n \rightarrow \infty} \int_{[0,t]} \|\textstyle\prod_r \tilde{x}_n(s) - \prod_r \tilde{x}(s)\|^2\ d\langle M \rangle_s = 0.$$

Therefore, to prove the lemma, it suffices to show that for every $\varepsilon > 0$,

$$\lim_{r \rightarrow \infty} \sup_n \mathbb{P} \left(\sup_{t \leq T} \left| \int_{[0,t]} ((\mathbb{I} - \textstyle\prod_r)\ \tilde{x}_n(s),\ dM(s)) \right| \geq \varepsilon \right) = 0$$

and

$$\lim_{r\to\infty} \mathbf{P}\left(\sup_{t\le T} \left| \int_{[0,t]} ((\mathbb{I} - \textstyle\prod_r)\, \tilde{x}(s),\ dM(s)) \right| \ge \varepsilon \right) = 0.$$

The second equality follows from (4.10) and Theorem 3.3 (property (iii$'''$)). For the first, note that

$$M_r^n(t) := \int_{[0,t]} ((\mathbb{I} - \textstyle\prod_r)\, \tilde{x}_n(s),\ dM(s)) =$$

$$\int_{[0,t]} (\tilde{x}_n(s),\ d(\mathbb{I} - \textstyle\prod_r)\, M(s)).$$

By Corollary 1.7 we have that for every $N,\ \delta \in \mathbf{R}_+$

$$\mathbf{P}\left(\sup_{t\le T} |\ M_r^n(t)\ | \ge \varepsilon\right) \le \varepsilon^{-1}\, 3\delta +$$

$$+ \mathbf{P}\left(\left(\int_{[0,T]} \|\tilde{x}_n(s)\|^2\ d\!<\left(\mathbb{I} - \textstyle\prod_r\right) M\!>_s\right)^{1/2} \ge \delta\right) \le$$

$$\le \varepsilon^{-1}\, 3\delta + \mathbf{P}\, (\sup_{t\le T} \|\tilde{x}_n(s)\| \ge N) +$$

$$+ \delta^{-1}\, N\, \mathbf{E}\|(\mathbb{I} - \textstyle\prod_r)\ M(T)\|^2)^{1/2}.$$

From (4.10) it follows that

$$\max_n \mathbf{P}\, (\sup_{t\le T} \|\tilde{x}_n(s)\| \ge N) = \mathbf{P}\, (\sup_{t\le T} \|\tilde{x}(s)\| \ge N),$$

the latter probability can be made small by choosing appropriate N. It is also clear that $\lim_{r\to\infty} \mathbf{E}\|(\mathbb{I} - \prod_r)\ M(t)\|^2 = 0$. Thus the lemma is proved. □

4.7. Now define the set $\tilde{\Omega}$ to meet the requirement of Theorem 2.

Let us choose a subsequence such that relation (4.11) holds $\mathbf{P}$-a.s. (From now on we shall identify it with the original sequence of x_n). Denote by Ω_1 the set of full probability, where this relation is valid.

Let H_k, $k \in \mathbf{N} \cup \{0\}$, be an expanding family of finite-dimensional subspaces of $\mathbb{H}$, $\dim \mathbb{H}_k = k$, and $\prod_k \mathbb{H} = \mathbb{H}_k$. By the results of §1.8 we have that for every k,t

(4.12) $$\lim_{n\to\infty} \sum_{t_{j+1}^n \le t} \|(\mathbb{I} - \Pi_k)(M(t_{j+1}^n) - M(t_j^n)\|^2 = \langle(\mathbb{I} - \Pi_k)\, M\rangle_t.$$

in probability.

Thus there exists a subsequence such that the convergence in (4.12) takes place $\mathbb{P}$-a.s. In the future we shall identify this subsequence with the original sequence. Denote

$$\Omega_2 := \bigcup_{k=0}^{\infty} \bigcap_{t\in I} \left\{\omega:\ \lim_{n\to\infty} \sum_{t_{j+1}^n \le t} \|(\mathbb{I} - \Pi_k)\,(M(t_{j+1}^n) - M(t_j^n))\|^2 = \right.$$

$$\left. = \langle(\mathbb{I} - \Pi_k)\, M\rangle_t\right\}.$$

From the above we see that $\mathbb{P}(\Omega_2) = 1$. Denote

$$\Omega_3 := \left\{\omega:\ \lim_{n\to\infty} \int_{[0,T]} \|\overset{j}{x}_n(t) - x(t)\|_{\mathbb{X}}^2\, dt = 0,\ j=1,2\right\},$$

$$\Omega_4 := \left\{\omega:\ \lim_{k\to\infty} \langle(\mathbb{I} - \Pi_k)\, M\rangle_T = 0\right\}$$

and $\tilde{\Omega} = \Omega'' \cap \bigcap_{i=1}^{4} \Omega_i$.

By Lemma 3 we have that $\mathbb{P}(\Omega_3) = 1$. Moreover,

$$\mathbb{E}\langle(\mathbb{I} - \Pi_k)\, M\rangle_T = \mathbb{E}\|(\mathbb{I} - \Pi_k)\, M(T)\|^2 \to 0.$$

Thus $\mathbb{P}(\Omega_4) = 1$ and consequently $\mathbb{P}(\tilde{\Omega}) = 1$.

Lemma 7. If $\omega\in\tilde{\Omega}$, t, $s\in I$ *and* $s<t<\tau(\omega)$, *then*

(4.13) $$\|\tilde{x}(t)\|^2 = \|x(0)\|^2 + 2\int_{[0,t]} [x(u),\, x'(u)]\,du +$$

$$+ 2\int_{[0,t]} (\tilde{x}(u),\, dM(u)) + \langle M\rangle_t,$$

$$\|\tilde{x}(t) - \tilde{x}(s)\|^2 = 2 \int_{[s,t]} [x(u) - x(s), x'(u)]du +$$

(4.14)
$$+ 2 \int_{[s,t]} (\tilde{x}(u) - \tilde{x}(s), dM(u)) + <M>_t - <M>_s. \quad \square$$

Proof. For $\omega \in \tilde{\Omega}$ and $t = t_i^n < \tau(\omega)$ we show that

$$\|\tilde{x}(t)\|^2 = \|x(0)\|^2 + 2 \int_{[0,t]} [x_n^2(u), x'(u)]du +$$

(4.15)
$$+ 2 \int_{[0,t]} (\tilde{x}_n(u), dM(u)) + \sum_{t_{j+1}^n \le t} \|M(t_{j+1}^n) - M(t_j^n)\|^2 -$$

$$- \sum_{t_{j+1}^n \le t} \|\tilde{x}(t_{j+1}^n) - \tilde{x}(t_j^n)) - (M(t_{j+1}^n) - M(t_j^n))\|^2.$$

To prove this, we note that for $j=2,\dots,k(n)$

$$\|\tilde{x}(t_j^n)\|^2 - \|\tilde{x}(t_{j-1}^n)\|^2 = 2 \int_{[t_{j-1}^n,\, t_j^n]} [x(t_j^n), x'(u)]du +$$

(4.16)
$$+ 2 (\tilde{x}(t_{j-1}^n), M(t_j^n) - M(t_{j-1}^n)) + \| M(t_j^n) - M(t_{j-1}^n)\|^2 -$$

$$- \|(\tilde{x}(t_j^n) - \tilde{x}(t_{j-1}^n)) - (M(t_j^n) - M(t_{j-1}^n))\|^2.$$

Next, by (4.8) taking into account that $\tilde{M}(0,w) = x(0,w) = \tilde{x}(0,w)$ for $\omega \in \Omega$, we have

(4.17)
$$\|\tilde{x}(t_1^n)\|^2 = 2 \int_{[0,t_1^n]} [x(t_1^n), x'(u)]du + \|\tilde{M}(0)\|^2 + 2(\tilde{x}(0), \tilde{M}(t_1^n)) -$$

$$- \tilde{M}(0)) + \|\tilde{M}(t_1^n) - \tilde{M}(0)\|^2 - \|(\tilde{x}(t_1^n) - \tilde{x}(0)) -$$

$$- (\tilde{M}(t_1^n) - \tilde{M}(0))\|^2.$$

Summing up equality (4.16) over all j and adding the resulting equality to (4.17) we get (4.15).
Next we show that the last term in the right-hand part of (4.15) tends to zero, as $n \to \infty$.

Denote

$$I_n := \sum_{t_{j+1}^n \le t} \|(\tilde{x}(t_{j+1}^n) - \tilde{x}(t_j^n) - (M(t_{j+1}^n) - M(t_j^n))\|^2.$$

Now represents I_n as follows:

$$I_n = \sum_{t_{j+1}^n \le t} (\tilde{x}(t_{j+1}^n) - \tilde{x}(t_j^n) - M(t_{j+1}^n) + M(t_j^n), \tilde{x}(t_{j+1}^n) -$$

$$- \tilde{x}(t_j^n) - \prod_k (M(t_{j+1}^n) - M(t_j^n))) - \sum_{t_{j+1}^n \le t} (\tilde{x}(t_{j+1}^n) - \tilde{x}(t_j^n) -$$

$$- M(t_{j+1}^n) + M(t_j^n), (\mathbb{I} - \prod_k) (M(t_{j+1}^n) - M(t_j^n))) = I_n^1 + I_n^2.$$

Since $\mathbf{X}$ is dense in $\mathbb{H}$ we can choose the CONS $\{x_i\}$, $i \in \mathbf{N}$, in $\mathbb{H}$ consisting of the elements of $\mathbf{X}$. Thus without loss of generality we shall assume that for all k, j, and n $\prod_k(M(t_{j+1}^n) - M(t_j^n)) \in \mathbf{X}$.

Write $\prod_k M(t) := v_k(t)$. Starting from $v_k(t)$, we construct the functions $v_{k,n}^j(t)$, $j=1,2$, exactly the same way as it was done in Lemma 3 for the function $x(t)$ and $x_n^j(t)$. From (4.1), where η is replaced by $v(t_{j+1}^n) - v(t_j^n) - \prod_k(M(t_{j+1}^n) - M(t_j^n))$, we get

$$I_n^1 = \int_{[0,t]} [x_n^2(u) - x_n^1(u), x'(u)]\,du -$$

(4.18)

$$- \int_{[0,t]} [v_{k,n}^2(u) - v_{k,n}^1(u), x'(u)]\,du.$$

The first term in this equality tends to zero as $n \to \infty$, since, in view of the properties of a normal triple and the Schwartz inequality, we have

$$\int_{[0,t]} \|[x_n^2(u) - x_n^1(u),\, x'(u)]\|\, du \le$$

$$\le N \int_{[0,t]} \|x_n^2(u) - x_n^1(u)\|_{\mathbf{X}} \cdot \|x'(u)\|_{\mathbf{X}'}\, du \le$$

$$\le N\Big(\int_{[0,T]} \|x_n^2(u) - x_n^1(u)\|_{\mathbf{X}}^2\, du\Big)^{1/2} \Big(\int_{[0,T]} \|x'(u)\|_{\mathbf{X}'}^2\, du\Big)^{1/2},$$

where the last term tends to zero as $n \to \infty$ since $w \in \Omega_3$.

The function $\prod_k M(t)$ is strongly continuous (in t) in $\mathbf{X}$. Hence $\|v_{k,n}^2(u) - v_{k,n}^1(u)\|_{\mathbf{X}} \to 0$ uniformly (relative to u) on $[0,T]$ as $n \to \infty$. Therefore, making use of estimates similar to the ones given above, it is easy to show that the second term in the right-hand side of (4.18) also tends to zero.

Hence

$$\lim_{n \to \infty} I_n^1 = 0.$$

On the other hand,

$$I_n^2 \le \Big(\sum_{t_{n+1}^n \le t} \|(\tilde{x}(t_{j+1}^n) - \tilde{x}(t_j^n)) - (M(t_{j+1}^n) - M(t_j^n))\|^2 \Big)^{1/2} \times$$

$$\times \Big(\sum_{t_{j+1}^n \le t} \|(\mathbb{I} - \prod_k)\,(M(t_{j+1}^n) - M(t_j^n)\|^2 \Big)^{1/2} =$$

$$= I_n^{1/2} < (\mathbb{I} - \prod_k)\, M >_t^{1/2}.$$

Taking into account that $<(\mathbb{I} - \prod_k)M>_t \to 0$ as $k \to \infty$, we get from the last inequality and (4.19) that $\lim_{n \to \infty} I_n = 0$, $\forall \omega \in \Omega$. Since $\omega \in \tilde{\Omega} \subset \Omega_3 \cap \Omega_1$, we have

$$\lim_{n \to \infty} \int_{[0,t]} [x_n^2(u),\, x'(u)]du = \int_{[0,t]} [x(u),\, x'(u)]du$$

and

$$\lim_{n\to\infty} \int_{[0,t]} (\tilde{x}_n(u), dM(u)) = \int_{[0,t]} (\tilde{x}(u), dM(u)).$$

Thus, passing to the limit in inequality (4.15) we get (4.13). Equality (4.14) can be derived from (4.13) by the application of the relations $(a\text{-}b)^2 = a^2\text{-}b^2\text{-}2b(a\text{-}b)$ and the identity

$$-\,2(\tilde{x}(s), (\tilde{x}(t) - \tilde{x}(s))) = -2 \int_{[s,t]} [x(s), x'(u)]du - 2 \int_{[s,t]} (\tilde{x}(s), dM(u)). \ \square$$

4.8. To complete the proof of the first three items of the statement of the theorem, it remains to prove that for $\omega \in \tilde{\Omega}$ the function $\tilde{x}(t,\omega)$ is strongly continuous in $\mathbb{H}$ with respect to t. Since a weakly continuous function with a continuous norm is strongly continuous, it is sufficient to prove the validity of (4.13) for all $t < \tau(\omega)$, where $\omega \in \tilde{\Omega}$. When $t=0$ equality (4.13) is clear. Let us fix $t>0$ such that $t < \tau(\omega)$ for $\omega \in \tilde{\Omega}$.

For every sufficiently large n we can find $j := j(n)$ such that $0 < t_j^n \le t < t_{j+1}^n$. Write $t(n) = t_{j(n)}^n$ and note that $t(n) \uparrow t$. Next,

$$\lim_{n\to\infty} \int_{[t(n),t]} \big|[x(u) - x(t(n)),\ x'(u)]\big|\ du \le$$

$$\le \lim_{n\to\infty} \int_{[0,T]} \|x(u) - x_n^1(u)\|_{\mathbf{X}} \cdot \|x'(u)\|_{\mathbf{X}'}\ du \le$$

$$\le \lim_{n\to\infty} \Big(\int_{[0,T]} \|x(u) - x_n^1(u)\|_{\mathbf{X}}^2\ du \Big)^{1/2} \Big(\int_{[0,T]} \|x'(u)\|_{\mathbf{X}'}^2\ du \Big)^{1/2} = 0,$$

$$\lim_{n\to\infty} (\langle M\rangle_t - \langle M\rangle_{t(n)}) = 0,$$

and

$$\lim_{n\to\infty} \Big| \int_{[t(n),s]} (\tilde{x}(u) - \tilde{x}(t(n)),\ dM(u)) \Big| =$$

$$= \lim_{n\to\infty} \sup_{s\le t} \Big| \int_{[0,s]} (\tilde{x}(u) - \tilde{x}_n(u), dM(u)) -$$

$$\int_{[0,t(n)]} (\tilde{x}(u) - \tilde{x}_n(u), dM(u)) \Big| \le$$

$$\le 2 \lim_{n\to\infty} \sup_{s\le t} \Big| \int_{[0,s]} (\tilde{x}(u) - \tilde{x}_n(u), dM(u)) \Big| = 0.$$

Thus there exists a subsequence $n(k)$ such that $s(k) = t(n(k))$, we have

$$\sum_{k=1}^{\infty} \Big\{ \Big(\int_{[s(k),s(k+1)]} |[x(u) - x(s(k)), x'(u)]| du \Big)^{1/2} +$$

$$+ (\langle M\rangle_{s(k+1)} - \langle M\rangle_{s(k)})^{1/2} +$$

$$+ \Big(\int_{[s(k),s(k+1)]} (\tilde{x}(u) - \tilde{x}(s(k)), dM(u)) \Big)^{1/2} < \infty.$$

Making use of (4.14), we obtain

$$\sum_{k=1}^{\infty} \| \tilde{x}(s(k+1)) - \tilde{x}(s(k)) \| < \infty.$$

Therefore $\tilde{x}(s(k))$ has a strong limit as $k\to\infty$. Since $s(k) \to t$, the function $\tilde{x}(s(k))$ converges in $\mathbb{H}$ weakly, and consequently strongly to, $\tilde{x}(t)$. Substituting in (4.13) the number $s(k)$ for t and passing to the limit as $k\to\infty$, we obtain (4.13) for a fixed t.

Item (iv′) of the statement of the theorem follows easily from the previous ones.

From (4.1) and (4.2) we obtain that $(\eta, \tilde{x}(t)) = (\eta, \tilde{x}(t))$, ($\mathbf{P}$-a.s.), on the set $\{\omega: t < \tau(\omega)\}$ for every $\eta \in \mathbf{X}$. From this, in view of separability of $\mathbf{X}$, it follows that $(\eta, \tilde{x}(t)) = (\eta, x(t))$ for all η simultaneously, ($\mathbf{P}$-a.s.), on the set $\{\omega: t < \tau(\omega)\}$. Since $\mathbf{X}$ is dense in $\mathbb{H}$, the statement of item (iv) of the theorem follows.

Chapter 3

LINEAR STOCHASTIC EVOLUTION SYSTEMS IN HILBERT SPACES.

3.0. Introduction.

0.1. Given a standard probability space $\mathbb{F}:=(\Omega, \mathcal{F}, \{\mathcal{F}_t\}_{t\in[0,T]}, \mathbf{P})$, where $T\in\mathbb{R}_+$, consider a Wiener process $W(t)$ taking values in $\mathbb{Y}$ with the covariance operator Q and a martingale $M(t)\in\mathfrak{M}_2^c([0,T];\,\mathbb{H})$.*

Denote, as above, by $\mathfrak{L}_Q(\mathbb{X},\mathbb{H})$, the set of all possibly unbounded, linear operators ϕ defined on $Q^{1/2}\mathbb{Y}$, mapping $Q^{1/2}\mathbb{Y}$ into $\mathbf{H}$ and such that $\phi Q^{1/2}$ belongs to $\mathfrak{L}_2(\mathbb{Y},\mathbb{H})$. Recall that $\mathfrak{L}_2(\mathbb{Y},\mathbb{H})$ is the space of Hilbert-Schmidt operators from $\mathbb{Y}$ to $\mathbb{H}$. The space $\mathfrak{L}_Q(\mathbb{Y},\mathbb{H})$ is a separable Hilbert space with respect to the scalar product $((\cdot,\cdot))$ corresponding to the norm $|||\phi|||:= \|\phi Q^{1/2}\|^2_{\mathfrak{L}_2(\mathbb{Y},\mathbb{H})}$. Let $(\mathbb{X},\mathbb{H},\mathbb{X}')$ be a normal triple. Throughout this chapter, unless otherwise stated, the notation introduced in Chapter 2 will be in force. In particular, we shall denote by $[\cdot,\cdot]$ the CBF of the normal triple $(\mathbb{X},\mathbb{H},\mathbb{X}')$, by $(\cdot,\cdot)$ the scalar product in $\mathbb{H}$, and by $\|\ \|$ the norm in $\mathbb{H}$.

We assume that linear operators

$$A(t,\omega):\ \mathbb{X}\rightarrow\mathbb{X}',\ B(t,\omega):\ \mathbb{X}\rightarrow\mathfrak{L}_Q(\mathbb{Y},\mathbb{H}),$$

and functions $\varphi(\omega)$, $f(t,\omega)$ taking values in $\mathbb{H}$ and $\mathbb{X}'$, respectively, are given for $t\in[0,T]$, $w\in\Omega$.

In this chapter we will consider linear evolution stochastic systems of the form

$$u(t)=\phi+\int_{[0,t]}(Au(s)+f(s))ds+\int_{[0,t]}Bu(s)\,d\mathbb{W}(s)+M(t).$$

System (0.1) is said to be coercive if it satisfies the condition (A) below.

---

*The space $\mathfrak{M}_2^c([0,T];\,\mathbb{H})$ is defined for martingales on [0,T] in the same way that the space $\mathfrak{M}_2^c(\mathbb{R}_+;\mathbb{H})$ for martingales on $\mathbb{R}_+$.

(A) There exist constants $K \in \mathbf{R}_+$ and $\delta > 0$ such that for all $(t,\omega) \in [0,T] \times \Omega$,

$$2[x, A(t,\omega)x] + |||B(t,\omega)x|||^2 + \delta\|x\|_{\mathbf{X}}^2 \le K\|x\|_{\mathbf{X}}^2, \qquad \forall x \in \mathbf{X}.$$

It is said to be dissipative if it satisfies the following condition:

(A′) There exists constant $K \in \mathbf{R}_+$ such that for all $(t,\omega) \in [0,T] \times \Omega$,

$$2[x, A(t,\omega)x] + |||B(t,\omega)x|||^2 \le K\|x\|^2 \qquad \forall x \in \mathbf{X}.$$

These two types of systems (coercive and dissipative*) are the subject of this chapter.

0.2. The structure of the chapter is as follows:

In Section 1 we consider coercive LSES. For such systems we prove existence and uniqueness theorems. Results about "improving the quality" of a solution are also given.

In the second section we define and discuss the important notation of a Hilbert scale. Next, we consider a dissipative LSES on a Hilbert scale.

In the third section we consider weaker conditions for the unicity of a solution of LSES (0.1) and investigate methods of approximating the solution of this system.

In the fourth section we illustrate the results of the previous section applying them to the Ito stochastic partial differential equation (of the arbitrary finite order). For this equation we consider the first boundary problem. In the same section we discuss briefly a variety of scales of Sobolev spaces which are imporant examples of Hilbert scales

3.1. Coercive Systems

1.0. In this section we deal with a LSES (0.1), where the operators A and B satisfy the coercivity condition. For a LSES of this type we prove the existence and uniqueness theorem.

In addition, we discuss conditions which ensure that a solution of the LSES

---

*An operator U: $\mathbf{X} \to \mathbf{X}^*$ is said to be coercive if there exists $\delta > 0$ such that $xUx \ge \delta\|x\|_{\mathbf{X}}^2$ for every $x \in \mathbf{X}$. If $xUx \ge 0$ for every $x \in \mathbf{X}$ it is callled dissipative.

Thus if $\mathbf{X}^* := \mathbf{X}$ and $[\cdot,\cdot]$ is the duality relation between $\mathbf{X}$ and $\mathbf{X}^*$, $\mathbf{Y} := \mathbf{R}^1$, and $B \equiv Q \equiv \mathbb{I}$, then condition (A) is equivalent to coercivity of the operator $S := 2^{-1}(K-1)\mathbb{I} - A$. Under the same assumptions condition (A′) is equivalent to dissipativity of S.

(0.1) belongs to a space imbedded into the one mentioned in the existence of a solution theorem. In applications, that means that a solution has some additional smoothness or other analytical properties and so its "quality" is higher.

1.1. Here, as in the other parts of the chapter we suppose that:
(i) the functions $A(t,\omega)x$, $B(t,\omega)x$ for every $x \in \mathbf{X}$ are predictable;
(ii) the random variable $\varphi(\omega)$ is $\mathfrak{F}_0$-measurable
(iii) the stochastic process $f(t,\omega)$ is predictable and (**P**-a.s.) l-integrable on the interval [0,T].

The family of the operators $A(t,\omega)$, mapping $\mathbf{X}$ and $\mathbf{X}'$, is said to be uniformly (relative to t,ω) continuous if the following condition holds:

(B). There exists constant $K \in \mathbf{R}_+$ such that for all $(t,\omega) \in [0,T] \times \Omega$, $\|A(t,\omega)x\|_{\mathbf{X}'} \leq K\|x\|_{\mathbf{X}}$ for every $x \in \mathbf{X}$.

Note that if the dissipativity condition holds and the family of the operators $A(t,\omega)$ is a family of the uniformly (relative to t,ω) continuous mappings from $\mathbf{X}$ to $\mathbf{X}'$, then the family $\mathbb{B}(t,\omega)$ is a family of uniformly (relative to t,ω) continuous operators from $\mathbf{X}$ into $\mathcal{L}_Q(\mathbf{Y},\mathbf{H})$, i.e. there exists a constant $K \in \mathbf{R}_+$ such that for every $(t,\omega) \in [0,T] \times \Omega$

$$(1.1) \qquad |||B(t,\omega)x||| \leq \mathbf{K}\|x\|_x .$$

This inequality is implied by (A), (B) and property (i) of CBF (see §2.4.1).

1.2. Let $\mathbb{Z}$ be a Hilbert space. We denote by $\mathbb{L}_2([0,T];\mathfrak{P};\mathbb{Z})$ the set of predictable representatives of the classes of functions from $\mathbb{L}_2([0,T] \times \Omega;\ \overline{\mathfrak{B}([0,T]) \otimes \mathfrak{F}};\ l \times \mathbf{P};\mathbb{Z})$ (see §1.2.13). Next, $\mathbb{L}_2^{\omega}([0,T];\mathfrak{P};\mathbb{Z})$ will denote the set of predictable processes $x(t,\omega)$ taking values in $\mathbb{Z}$ and such that

$$x(\cdot,\omega) \in \mathbb{L}_2([0,T];\ \overline{\mathfrak{B}([0,T])};\ l;\ \mathbb{Z}) \qquad (\mathbf{P}\text{-}a.s.).$$

By $\mathbb{C}([0,T];\mathfrak{P};\mathbb{H})$ we denote the set of strongly continuous in $\mathbf{H}$ predictable $\mathbf{H}$-processes.

Definition 2. A stochastic process $u(t)$ is said to be a solution of LSES (0.1) *if it belongs to* $\mathbb{L}_2^{\omega}([0,T]);\mathfrak{P};\mathbf{X})$ *and for all y from some dense (in the strong topology) subset of* $\mathbf{X}$, *$l \times \mathbf{P}$-a.s. on $[0,T] \times \Omega$ it satisfies the equality*

$$(y,\ u(t)) = (y,\phi) + \int_{[0,t]} [y,\ Au(s) + f(s)]ds +$$

$$+\Big(y, \int_{[0,t]} Bu(s)\, d\mathbf{W}(s) + M(t)\Big). \qquad \square \tag{1.2}$$

It is assumed, of course, that the integrals in the equality (0.1) are defined (in the sense of §1.2.11, §2.3.1. See also §2.1.13 and §1.4.1 (the third property of CBF)).

If conditions (A) and (B) are fulfilled, which in turn implies inequality (1.1), then as it is readily seen, the processes are predictable and the integrals in (1.2) are well-defined.

Remark 2. *If* $\mathbf{X}'=\mathbf{X}^*$ *and* $\mathbf{H}$ *is a Hilbert space rigged by the pair* $\mathbf{X}$, $\mathbf{X}^*$ *(see* §2.4.0), *then in accordance with the definition, the solution of the LSES* (0.1) *is a function* $u\epsilon\mathbf{L}_2^{\omega}([0,T]);\mathcal{P},\mathbf{X})$ *satisfying equality* (0.1) *as an equality in* $\mathbf{X}^*$, $(l\times\mathbf{P}$-*a.s.)* $\square$

1.3. *Definition* 3. *We say that a solution* u *of LSES* (0.1) *has a continuous version in* $\mathbf{H}$ *if there exists a function* $v\in\mathbf{C}([0,T]);\mathcal{P};\mathbf{H})$ *such that:*

(*i*) $v(t,\omega) = u(t,\omega)$, $(l\times\mathbf{P}$-*a.s.*);
(*ii*) *there exists a set* $\Omega'\subset\Omega$ *such that* $\mathbf{P}(\Omega') = 1$ *and for all* $(t,\omega)\in[0,T]\times\Omega'$, $y\in\mathbf{X}$,

$$(y,\ v(t,\omega)) = (y,\ \phi(\omega)) + \int_{[0,t]} [y,\ Av(s,\omega)]\, ds +$$

$$+\Big(y, \int_{[0,t]} Bv(s,\omega)\, d\mathbf{W}(s) + M(t,\omega)\Big). \qquad \square \tag{1.3}$$

Remark 3. *From Proposition* 1.2.9 *it follows that* $1_{\{v\in\mathbf{X}\}}$ *is a predictable (real) stochastic process. Thus, the integrals of* Av *and* Bv *in* (1.3) *considered as integrals of the functions* $1_{\{v\in\mathbf{X}\}}$ Av *and* $1_{\{v\in\mathbf{X}\}}$ Bv *are well-defined.* $\square$

It is evident that Theorem 2.4.2 yields the following result.

Proposition 3. *A solution of LSES* (0.1) *always has a continuous version in* $\mathbf{H}$. $\square$

Warning 3. *In what follows, a solution of a LSES will be identified with its continuous version.* $\square$

1.4. The main result of this section is the following theorem.

Theorem 4. Suppose that conditions (A) and (B) are fulfilled as well as condition

$$(C)\quad \mathbf{E}\|\phi\|^2 < \infty,\ \mathbf{E}\int_{[0,t]} \|f(t)\|^2_{\mathbf{X}'}\, dt < \infty,$$

*Then LSES (0.1) has a unique solution** $u \in \mathbf{L}_2([0,T];\mathcal{P};\mathbf{X})$, *and*

$$\mathbf{E}\sup_{t\le T}\|u(t)\|^2 + \mathbf{E}\int_{[0,T]}\|u(t)\|^2_{\mathbf{X}}\,dt \le \tag{1.4}$$

$$\le \mathbf{N}\,\mathbf{E}\left(\|\phi\|^2 + \int_{[0,T]}\|f(t)\|^2_{\mathbf{X}'}\,dt + \langle M\rangle_T\right),$$

where N depends only on K, T and δ. □

Proof. Let U_1 and U_2 be two solutions of the LSES (0.1). Denote $\mathcal{U} := u_1 - u_2$. Clearly, the process $\mathcal{U}$ is a solution in $(\mathbf{X},\mathbf{H},\mathbf{X}')$ of the LSES of the type (0.1) with $\varphi \equiv f \equiv M \equiv 0$.

Write $\Gamma_n(\omega) := \{t\colon t \le T,\ \|\mathcal{U}(t,\omega)\| > n\}$, $n \in N$,
and

$$\tau_n(\omega) = \begin{cases} \inf_t \Gamma_n(\omega), & \text{if } \Gamma_n(\omega) \ne \oslash; \\ T, & \text{if } \Gamma_n(\omega) = \oslash. \end{cases}$$

By Theorem 2.4.2 for all $t \in [0,T]$, ($\mathbf{P}$-*a.s.*), we have

$$\|\mathcal{U}(t\wedge\tau_n)\|^2 = \int_{[0,t\wedge\tau_n]} (2[\mathcal{U}(s),\ \mathbf{A}\mathcal{U}(s)] +$$

$$+ \||B\mathcal{U}(s)\||^2)ds + 2\int_{[0,t\wedge\tau_n]} (\mathcal{U}(s),\ d([B\mathcal{U}]\circ\mathbf{W}(s))).$$

*Here and in the sequal, if it does not cause misunderstanding, we will not refer to the normal triple connected with the solution.

Evidently, the stochastic integral in the right-hand part of equality (1.5) belongs to $\mathfrak{M}^c_{loc}([0,T];\mathbf{R}^1)$, and its quadratic variation is given by the integral

$$\int_{[0,t\wedge\tau_n]} \|\mathcal{U}(s)\|^2 \,|||B\mathcal{U}(s)|||^2 ds.$$

In view of (1.1), the last integral is dominated by

$$K^2 \int_{[0,t\wedge\tau_n]} \|\mathcal{U}(s)\|^4 ds \le K^2 n^4 T.$$

Thus Proposition 2.1.7 (iii) yields that the stochastic integral considered above is a martingale and its expectation is equal to zero. Hence taking expectations of both parts of equality (1.5) and applying condition (A), we find

$$\mathbf{E}\|\mathcal{U}(t\wedge\tau_n)\|^2 \le K \int_{[0,t]} \mathbf{E}\|\mathcal{U}(s\wedge\tau_n)\|^2 ds.$$

Using the Gronwall-Bellman lemma, we obtain that $\mathbf{E}\|\mathcal{U}(t\wedge\tau_n\|^2 = 0$ for all t and n. Since $\mathcal{U}\in\mathbf{C}([0,T];\mathcal{P};\mathbf{H})$ we have $\tau_n\uparrow T$ ($\mathbf{P}$-*a.s.*). Making use of this fact and Fatou's lemma it is easy to see that

$$\mathbf{P}(\sup_{t\le T} \|u_1(t)\text{-}u_2(t)\| > 0) = 0$$

and the uniqueness of the solution is proved.

We prove the existence of the solution by a Galerkin's method.

Let us fix CON's $\{h_i\}$ and $\{y_i\}$ in the spaces $\mathbf{H}$ and $\mathbf{Y}$, respectively. Since $\mathbf{X}$ is dense in $\mathbf{H}$ we may suppose that $h_i\in\mathbf{X}$.

Consider the following system of the Ito equations in $\mathbf{R}^n$:

$$u^i_n(t) = (h_i,\varphi) + \int_{[0,t]} \left[h_i,\ A(s)\sum_{j=1}^n h_j u^j_n(s) + f(s)\right]ds +$$

(1.6)

$$+ \sum_{k=1}^n \int_{[0,t]} \left(h_i,\left(B(s)\sum_{j=1}^n h_j u^j_n(s)\right)y_k\right)dw^k(s) + M_i(t),$$

where $w^k(s) := (w(s),y_k)_{\mathbf{Y}}$, $M_i(t) := (M(t),h_i)$.

By condition (B), for all $i\le n$,

$$\mathbb{E}(h_i,\varphi)^2 < \infty \quad and \quad \mathbb{E} \int_{[0,t]} [h_i,f(s)]^2 ds < \infty.$$

Thus system (1.6) has the unique continuous solution, and $\mathbb{E} \sup_{t \le T} \sum_{i=1}^{n} |u_n^i(t)|^2 < \infty$ (*see* §1.4.3).

1.5. For future use, we will require more precise estimates. Denote $u_n(t) := \sum_{i=1}^{n} u_n^i(t) h^i$.

Lemma 5. *There exists a constant* N, *independent of* n, *such that*

$$\mathbb{E} \sup_{t \le T} \|u_n(t)\|^2 = \mathbb{E} \int_{[0,T]} \|u_n(t)\|_{\mathbf{X}}^2 \, dt \le \tag{1.7}$$

$$\le N \, \mathbb{E}\Big(\|\varphi\|^2 + \int_{[0,T]} \|f(t)\|_{\mathbf{X}'}^2 \, dt + \langle M \rangle_t\Big). \qquad \square$$

Proof. Denote by $\prod_n$ the operator of orthogonal projection of $\mathbf{X}'$ on $\{h_1,\dots,h_n\}$ and by π_n the operator of orthogonal projection of $\mathbf{Y}$ on $\{y_1,\dots,y_n\}$.

Next, denote

$$\mathbb{L}_n(t) := \int_{[0,t]} \prod_n B u_n(s) \, \pi_n d\mathbf{W}(s)$$

and

$$\widehat{M}_n(t) := \mathbb{L}_n(t) + \prod_n M(t).$$

By formula (2.4.3), for all $t \in [0,T]$ and ω from the same set of full probability, we have

$$\sum_{i=1}^{n} |u_n^i(t)|^2 = \|u_n(t)\|^2 = \|\prod_n \varphi\|^2 +$$

$$+ 2 \int_{[0,t]} [u_n(s), Au_n(s) + f(s)] ds + \tag{1.8}$$

$$+ 2 \int_{[0,t]} (u_n(s), d\widehat{M}_n(s)) + \langle \widehat{M}_n \rangle_t.$$

Also, it is clear that

$$<L_n>_t = \int_{[0,t]} |||\Pi_n B u_n(s)\pi_n|||^2 ds$$

by Theorem 1.3.16 and inequality (1.1)

$$\mathbb{E}\left(\int_{[0,T]} \|u_n(s)\|^2 d <\widehat{M}_n>_s \right)^{1/2} \le$$

$$\le N\, \mathbb{E} \sup_{t\le T} \|u_n(s)\| \left(\int_{[0,T]} |||\Pi_n B u_n(s)\pi_n|||^2 ds + <\Pi_n M>_T \right)^{1/2} \le$$

$$\le 2N\, \mathbb{E}\left(\sup_{t\le T} \|u_n(t)\|^2 + \int_{[0,T]} |||Bu_n(s)|||^2 ds + <M>_T \right) < \infty.$$

Thus the stochastic integral in (1.8) is a martingale (see §2.2.3). Making use of this fact, we obtain from(1.8) that

$$\begin{aligned} \mathbb{E}\|u_n(t)\|^2 &= \mathbb{E}\|\Pi_n\varphi\|^2 + \mathbb{E} \int_{[0,t]} (2[u_n(s), Au_n(s), + f(s)] + \\ &+ |||\Pi_n B u_n(s)\ \pi_n|||^2) ds + 2\mathbb{E}<\mathbf{L}_n, \Pi_n M>_t + \mathbb{E}<\Pi_n M>_t. \end{aligned} \tag{1.9}$$

Applying property (i) of a CBF (see §2.4.1) and the elementary inequality

$$2ab \le \varepsilon a^2 + \frac{1}{\varepsilon}\, b^2\ (\varepsilon > 0), \tag{1.10}$$

it is easy to see that

$$2\mathbb{E}[u_n(s), f(s)] \le \varepsilon \mathbb{E}\|u_n(s)\|^2_{\mathbf{X}} + \varepsilon^{-1}\mathbb{E}\|f(s)\|^2_{\mathbf{X}'}, \tag{1.11}$$

and

$$\begin{aligned} 2\mathbb{E}<\mathbf{L}_n, \Pi_n M>_t &= 2\mathbb{E}(\mathbf{L}_n(t), \Pi_n M(t) \le \\ &\le \varepsilon \int_{[0,t]} |||\Pi_n B u_n(s)\pi_n|||^2 ds + \varepsilon^{-1}\mathbb{E}\|\Pi_n M(t)\|^2. \end{aligned} \tag{1.12}$$

Since a norm of a projection operator does not exceed 1, and in view of the coercivity property (A), we find from (1.9), (1.11), and (1.12) that

$$\mathbf{E}\|u_n(t)\|^2 + \delta\mathbf{E}\int_{[0,t]}\|u_n(s)\|^2_{\mathbf{X}}ds \le \mathbf{E}\|\varphi\|^2 + \varepsilon\mathbf{E}\int_{[0,t]}\|u_n(s)\|^2_{\mathbf{X}}ds +$$

(1.13)

$$+\ \varepsilon\mathbf{E}\int_{[0,t]}|||Bu_n(s)|||^2ds + \varepsilon^{-1}\mathbf{E}\int_{[0,t]}\|f(s)\|^2_{\mathbf{X}'}ds + (1+\varepsilon^{-1})\mathbf{E}\|M(t)\|^2.$$

From (1.1) we obtain

$$\varepsilon\mathbf{E}\int_{[0,t]}|||B(s)u_n(s)|||^2ds \le \varepsilon K\,\mathbf{E}\int_{[0,t]}\|u_n(s)\|^2ds. \tag{1.14}$$

Combining (1.13), (1.14) and taking ε sufficiently small, we find that for some $\delta_1>0$ and all $t\in[0,T]$

$$\mathbf{E}\|u_n(t)\|^2 + \delta_1\mathbf{E}\int_{[0,t]}\|u_n(s)\|^2_{\mathbf{X}}\,ds \le$$

(1.15)

$$\le N\,\mathbf{E}\Big(\|\varphi\|^2 + \int_{[0,t]}\|f(s)\|^2_{\mathbf{X}'}\,ds + \|M(t)\|^2 + \int_{[0,t]}\|u_n(s)\|^2ds\Big).$$

From this, by the Gronwall-Bellman lemma, we see that

$$\sup_{t\le T}\mathbf{E}\|u_n(t)\|^2 \le N\Big(\|\varphi\|^2 + \int_{[0,T]}\|f(t)\|^2_{\mathbf{X}'}\,dt + \langle M\rangle_t\Big). \tag{1.16}$$

which along with (1.15) implies the lemma. □

1.6. By definition, $u_n(t)$ is a predictable stochastic process in $\mathbf{X}$. Moreover, from Lemma 5, it follows that some subsequence of u_n converges weakly in the space $\mathbb{L}_2([0,T]\times\Omega,\ \overline{\mathcal{P}}_T;l\times\mathbf{P};\mathbf{X})$ (see the notation of §1.3.6). We shall identify this subsequece with the original sequence. From Proposition 1.2.7 it follows that the limit u has some $\mathcal{P}$-measurable version, which will be denoted in the same way. Hence $u\in\mathbb{L}_2([0,T];\mathcal{P};\mathbf{X})$.

Let η be an arbitrary bounded random variable on $(\Omega,\mathcal{F})$ and ψ be an arbitrary bounded *l-measurable function on* $[0,T]$.

From 1.6, for $n\in\mathbb{N}$ and $h_i\in\{h_i\}$, where $i\le n$, we have

$$\mathbf{E}\int_{[0,T]}\eta\psi(t)\,(h_i,u_n(t))dt = \mathbf{E}\int_{[0,T]}\eta\psi(t)\,\Big\{(h_i,\varphi) +$$

$$+ \int_{[0,t]} [h_i, Au_n(s) + f(s)]ds +$$

$$+ \int_{[0,t]} (h_i, Bu_n(s)\pi_n d\mathbf{W}(s)) + M_i(t)\Big\}dt.$$

Evidently,

$$\mathbb{E} \int_{[0,T]} \eta\psi(t)\, (h_i, u_n(t))dt \to \mathbb{E} \int_{[0,T]} \eta\psi(t)\, (h_i, u(t))dt. \tag{1.17}$$

Next, in view of condition (B) and Lemma 5,

$$\mathbb{E}\left|\eta \int_{[0,t]} \eta[h_i, Au_n(s)]ds\right| < N < \infty, \tag{1.18}$$

where N does not depend on n.

It is also clear that for every $t\in[0,T]$

$$\mathbb{E} \int_{[0,t]} \eta[h_i, Au_n(s)]ds \to \mathbb{E} \int_{[0,t]} \eta[h_i, Au(s)]ds. \tag{1.19}$$

From (1.18), (1.19) and Fubini's theorem we have by the dominated convergens theorem

$$\mathbb{E} \int_{[0,T]} \eta\psi(t) \int_{[0,t]} [h_i, Au_n(s)]dsdt = \int_{[0,T]} \psi(t)\, \mathbb{E} \int_{[0,t]} \eta[h_i,$$

$$Au_n(s)]dsdt \to \int_{[0,T]} \psi(t)\, \mathbb{E} \int_{[0,t]} \eta[h_i, Au(s)]dsdt. \tag{1.20}$$

From (1.1) and Lemma 5 it follows that

$$(1.21) \qquad \mathbb{E}\left|\eta \int_{[0,t]} (h_i, \prod_n Bu_n(s)\pi_n d\mathbb{W}(s))\right| < N < \infty,$$

where N does not depend on n.

By Theorem 2.2.4 we find that for every $t\in[0,T]$

$$(1.22) \qquad \mathbb{E}\eta \int_{[0,t]} (h_i, \prod_n Bu_n(s)\pi_n d\mathbb{W}(s)) \to \mathbb{E}_n \int_{[0,t]} (h_i, Bu(s)d\mathbb{W}(s)).$$

Therefore,

$$\mathbb{E} \int_{[0,T]} \eta\psi(t) \int_{[0,t]} (h_i, Bu_n(s)\pi_n d\mathbb{W}(s))dt \to$$

$$(1.23) \qquad \mathbb{E} \int_{[0,T]} \eta\psi(t) \int_{[0,t]} (h_i, Bu(s)d\mathbb{W}(s))dt.$$

It is also clear that

$$(1.24) \qquad \mathbb{E} \int_{[0,T]} \eta\psi(t)\, (h_i, \prod_n M(t))dt \to \mathbb{E} \int_{[0,T]} \eta\psi(t)\, (h_i, M(t))dt.$$

Combining (1.17), (1.20), and (1.23), (1.24) we obtain that $l\times\mathbf{P}$-*a.s.*

$$(h_i, u(t)) = (h_i,\varphi) + \int_{[0,t]} [h_i, Au(s) + f(s)]ds +$$

$$+ \int_{[0,t]} (h_i, Bu(s)d\mathbb{W}(s)) + (h_i, M(t)).$$

Thus the existence of the solution for system (0.1) is proved. In view of Proposition 3 we can consider this solution as a continuous function of t in $\mathbb{H}$.

To complete the proof of the theorem it only remains to justify inequality (1.4).

Let

$$L(t) := 2 \int_{[0,t]} Bu(s)d\mathbb{W}(s) \text{ and } \widehat{M}(t) := L(t) + M(t).$$

By Theorem 2.4.2, for all $(t,\omega) \in [0,T]\times\Omega'$, where $\mathbf{P}(\Omega') = 1$, we have

$$\|u(t)\|^2 = \|\varphi\|^2 + 2 \int_{[0,t]} [u(s), Au(s) + f(s)]ds + \tag{1.25}$$
$$+ 2 \int_{[0,t]} (u(s), d\widehat{M}(s)) + <\widehat{M}>_t.$$

Define

$$\Gamma_n(\omega) := \{t := t \le T, \|u(t,\omega)\|^2 \ge n\}$$

and

$$\tau_n(\omega) := \begin{cases} \inf_t \Gamma_n(\omega), & \text{if } \Gamma_n(\omega) \neq \oslash, \\ T, & \text{if } \Gamma_n(\omega) = \oslash. \end{cases}$$

Evidently,

$$\mathbb{E} \int_{[0,t\wedge\tau_n]} \|u(s)\|^2 d<\widehat{M}>_s \le N\,\mathbb{E} \int_{[0,t\wedge\tau_n]} \|u(s)\|^2\, |||Bu(s)|||^2 ds +$$
$$+ N\,\mathbb{E} \int_{[0,t\wedge\tau_n]} \|u(s)\|^2 d<M>_s < \infty.$$

Thus, by Proposition 2.1.7,

$$\mathbb{E} \int_{[0,t\wedge\tau_n]} (u(s), d\widehat{M}(s)) = 0.$$

Applying this, we obtain form (1.25) that

$$\mathbf{E}\|u(t\wedge\tau_n)\|^2 = \mathbf{E}\|\varphi\|^2 + 2\mathbb{E} \int_{[0,t\wedge\tau_n]} [u(s), Au(s) + f(s)]ds +$$
$$+ \mathbb{E}<\widehat{M}>_{t\wedge\tau_n}.$$

From the last equality, exactly in the same way as in Lemma 5, we easily derive that

$$\sup_{t\le T}\mathbb{E}\,\|u(t\wedge\tau_n)\|^2+\mathbb{E}\int_{[0,t\wedge\tau_n]}\|u(s)\|_{\mathbf{X}}^2ds\le$$

(1.26)

$$\le N\,\mathbb{E}\Big(\|\varphi\|^2+\int_{[0,T]}\|f(s)\|_{\mathbf{X}'}^2\,ds+\langle M\rangle_T\Big).$$

Since $\|u(t)\|$ is a continuous function of t, $\tau_n\uparrow T$ as $n\to\infty$. Thus passing to the limit in (1.26), we obtain, by Fatou's lemma and the monotone convergence theorem, that

$$\sup_{t\le T}\mathbb{E}\|u(t)\|^2+\mathbb{E}\int_{[0,T]}\|u(s)\|_{\mathbf{X}}^2\,ds\le$$

(1.27)

$$\le N\,\mathbb{E}\Big(\|\varphi\|^2+\int_{[0,T]}\|f(s)\|_{\mathbf{X}'}^2\,ds+\langle M\rangle_T\Big).$$

1.7. It only remains to prove that a similar estimate is also valid for $\mathbb{E}\sup_{t\le T}\|u(t)\|^2$.

It is clear that

(1.28) $$\langle\widehat{M}\rangle_t=\langle M\rangle_t+\int_{[0,t]}|||Bu(s)|||^2ds+2\Big\langle M,\int_{[0,\cdot]}Bu(s)d\mathbf{W}(s)\Big\rangle.$$

Thus from (1.25), in view of condition (A), it follows that

(1.29) $$\mathbb{E}\sup_{t\le T}\|u(t\wedge\tau_n)\|^2\le 2\,\mathbb{E}\int_{[0,T]}\Big|[u(s),f(s)]+$$

$$+K\|u(s)\|^2\,\Big|ds+2\,\mathbb{E}\sup_{t\le T}\Big|\int_{[0,t\wedge\tau_n]}(u(s),d\widehat{M}(s))\Big|+$$

$$+2\mathbb{E}\Big(M(T),\int_{[0,T]}Bu(s)d\mathbf{W}(s)\Big)+\mathbb{E}\langle M\rangle_T.$$

By the Burkholder-Davis inequality (§2.1.7) we find that

$$\mathsf{E}\sup_{t\le T}\left|\int_{[0,t\wedge\tau_n]}(u(s),\,d\widehat{M}(s))\right| \le 3\mathsf{E}\left(\int_{[0,t\wedge\tau_n]}\|u(s)\|^2 d<\widehat{M}>_s\right)^{1/2} \le$$

$$\le 3\mathsf{E}(\sup_{t\le T}\|u(t\wedge\tau_n)\| <\widehat{M}>_T^{1/2}) \le \tag{1.30}$$

$$\le \tfrac{3}{2}\,\varepsilon\mathsf{E}\sup_{t\le T}\|u(t\wedge\tau_n)\|^2 + \tfrac{\varepsilon}{2\varepsilon}\,\mathsf{E}<\widehat{M}>_T.$$

The last inequality, obtained via application of the inequality (1.10) is valid for every $\varepsilon>0$.

From (1.1) it follows that

$$\mathsf{E}\Big(M(t),\int_{[0,T]}Bu(s)d\mathsf{W}(s)\Big) \le \tfrac{1}{2}\Big(\mathsf{E}<M>_T + K\mathsf{E}\int_{[0,T]}\|u(s)\|^2_{\mathbf{X}}ds\Big). \tag{1.31}$$

Making use of the Kunita-Watanabe inequality and (1.1), we obtain the inequality

$$\mathsf{E}<\widehat{M}>_T \le \Big(\mathsf{E}<M>_T + \mathsf{E}\int_{[0,T]}\|u(s)\|^2_{\mathbf{X}}ds\Big). \tag{1.32}$$

Next combine (1.29) and (1.32) and then choose ε to be small enough. As a result, we obtain

$$\mathsf{E}\sup_{t\le T}\|u(t\wedge\tau_n)\|^2 \le N\,\mathsf{E}\Big(<M>_T + \int_{[0,T]}\|u(s)\|^2_{\mathbf{X}}ds\Big).$$

Estimate (1.4) follows from (1.27) and the inequality we'll obtain if we pass to the limit as $n\to\infty$ in the above inequality. □

Proposition 7. *Suppose the LSES (0.1) in* $(\mathbf{X},\mathbf{H},\mathbf{X}')$ *has a solution from* $\mathbf{L}_2([0,T]\times\Omega;\mathcal{P};X)$ *and possesses the dissipativity property* A'. *Suppose also that conditions B and C of the theorem are fulfilled. Then there exists a constant* N, *which depends only on* K *and* T, *such that*

$$\mathbf{E}\sup_{t\leq T}\|u(t)\|^2 \leq N\,\mathbf{E}\Big(\|\varphi\|^2 + \int_{[0,T]}\|f(t)\|^2_{\mathbf{X}'}\,dt + <M>_T\Big). \quad \square$$

1.8. In this paragraph conditions are studied which guarantee that a solution of LSES (0.1) belongs to some space normally imbedded in the space X. In applications to stochastic partial differential equations this may correspond to additional smoothness or some other additional analytic properties of the solution.

Let $(\mathbf{V},\mathbf{U},\mathbf{V}')$ be a normal triple. Denote by $[\cdot,\cdot]_{\mathbf{U}}$ its CBF.

Throughout this paragraph it is assumed that $\mathbf{V}\subset\mathbf{X}$, $\mathbf{U}\subset\mathbf{H}$ and $\mathbf{V}'\subset\mathbf{X}'$. All these imbeddings are assumed to be normal (see definition 2.4.0).

Later we will refer to the following classic result in functional analysis (see e.g. [62], §1.10).

Proposition 8. *Let* $\mathbb{H}_1$ *and* $\mathbb{H}_0$ *be Hilbert spaces, and suppose that* $\mathbb{H}_1$ *is normally imbedded in* $\mathbb{H}_0$. *Then there exists a unique positive, self-adjoint operator* Λ_0 *on* $\mathbb{H}_0$ *with the domain* $\mathbb{H}_1$ *such that* $\|x\|_{\mathbb{H}_1} = \|\Lambda_0\|_{\mathbb{H}_0}$ *for every* $x\in\mathbb{H}_1$. $\square$

Denote by Λ_1 the operator Λ_0 from the above proposition, when $\mathbb{H}_1 \equiv \mathbf{U}$ and $\mathbb{H}_0 \equiv \mathbb{H}$.

Throughout this item it is supposed that

$$\mathbb{Z} := \{x : x\in\mathbf{V}\cap\mathbf{D}(\Lambda_1^2);\ \Lambda_1^2 x\in\mathbf{X}\} \neq \emptyset$$

Lemma 8. *For every* $x\in\mathbb{Z}$, *and* $y\in\mathbf{V}'$, $[x,y]_{\mathbf{U}} = [\Lambda_1^2 x, y]$. $\square$

Proof. Let $\{y_n\}$, $n\in\mathbf{N}$, be a sequence of elements of $\mathbf{V}$ converging strongly in $\mathbf{V}'$ to y. Then by the proposition we have

$$[x,y]_{\mathbf{U}} = \lim_{n\to\infty}(x,y_n)_{\mathbf{U}} = \lim_{n\to\infty}(\Lambda_1^2 x,\ y_n) = [\Lambda_1^2 x, y]. \quad \square$$

In addition to the assumptions made in the introduction and in §1.1, before the formulation of the theorem, we suppose in this item that for all $(t,\omega) \in [0,T]\times\Omega$, $A(t,\omega)$: $\mathbf{V}\to\mathbf{V}'$, $\mathbf{B}(t,\omega)$: $\mathbf{V}\to\mathfrak{L}_Q(\mathbf{Y},\mathbf{U})$, and $\varphi(\omega)\in\mathbf{U}$, $M(t) \in \mathfrak{M}_2^c([0,T];\mathbf{U})$.

The norm in $\mathfrak{L}_Q(\mathbf{Y},\mathbf{U})$ similar to the norm $|||\cdot|||$ in $\mathfrak{L}_Q(\mathbf{Y},\mathbf{H})$ will be denoted by $|||\cdot|||_{\mathbf{U}}$.

Theorem 8. *Suppose that* $\Lambda_1^2\mathbb{Z}$ *is dense in* $\mathbf{X}$ (*with respect to the strong topology*) *and for all* $(t,\omega) \in [0,T]\times\Omega$, *some* $K'\in\mathbf{R}_+$ *and* $\delta'>0$ *the following conditions are fulfilled*:

$$(A_1) \qquad 2[x,\ A(t,\omega)x]_{\mathbf{U}} + |||B(t,\omega)|||^2_{\mathbf{U}} + \delta'\|x\|^2_{\mathbf{V}} \leq K'\ \|x\|^2_{\mathbf{U}},\ \forall x\in\mathbf{V};$$

(B_1) $\quad ||A(t,\omega)x||_{\mathbf{V}'} \leq K'||x||_{\mathbf{V}}, \ \forall x \in \mathbf{V};$

(C_1) $\quad \mathsf{E}||\varphi||^2_{\mathbf{U}} < \infty$ *and* $\mathsf{E}\int_{[0,T]} ||f(t)||^2_{\mathbf{V}'}\,dt < \infty.$

Then the system (0.1) *has a solution* u *in* $(\mathbf{X},\mathbf{H},\mathbf{X}')$ *belonging to* $\mathbf{L}_2([0,T];\mathcal{P};\mathbf{V}) \cap \mathbf{C}([0,T];\mathcal{P};\mathbf{U})$, *and there exists a constant* $N \in \mathbf{R}_+$, *depending only on* K', δ' *and* T, *such that*

$$\mathsf{E}\sup_{t\leq T}||u(t)||^2_{\mathbf{U}} + \mathsf{E}\int_{[0,T]} ||u(t)||^2_{\mathbf{V}}\,dt \leq \tag{1.33}$$

$$\leq N\,\mathsf{E}\Big(||\varphi||^2_{\mathbf{U}} + \int_{[0,T]} ||f(t)||^2_{\mathbf{V}'}\,dt + ||M(t)||^2_{\mathbf{U}}\Big). \qquad \square$$

Proof. Consider the LSES (0.1) in $(\mathbf{V},\mathbf{U},\mathbf{V}')$. Observe that the processes $A(t)x$, $f(t)$, and $B(t)x$ for very $x \in \mathbf{V}$, are predictable as processes in $\mathbf{V}'$ and $\mathfrak{L}_Q(\mathbf{Y},\mathbf{U})$, respectively and the real variable φ is $\mathcal{F}$-measurable in $\mathbf{U}$. Consider, for example, the process $B(t)x$. It suffices to show that for every y and h from dense subsets of $\mathbf{Y}$ and $\mathbf{H}$, respectively, the process $(B(t)xQ^{1/2}y,h)_{\mathbf{U}}$ is a predictable one (see §1.2.8 and §1.2.6). We take as these sets $\mathbf{Y}$ itself and $\mathbf{D}(\Lambda_1^2)$. It follows from the proposition that

$$(B(t)x)y',h)_{\mathbf{U}} = (B(t)x)y',\ \Lambda_1^2 h).$$

Since $B(t)x$ is a predictable $\mathfrak{L}_Q(\mathbf{Y},\mathbf{H})$-process, the right-hand part of the last equality is predictable. Thus Theorem 4 is applicable to the LSES (0.1) considered as system in $(\mathbf{V},\mathbf{U},\mathbf{V}')$. Therefore, it has a solution U in $(\mathbf{V},\mathbf{U},\mathbf{V}')$ satisfying inequality (1.33). It remains to prove that this solution is also a solution of the LSES (0.1) in $(\mathbf{X},\mathbf{H},\mathbf{X}')$.

From item (i) of Proposition 1.2.9 and since by assumption, the imbedding $\mathbf{V}$ in $\mathbf{H}$ is normal, we obtain that $u \in \mathbf{L}_2([0,T]\times\Omega;\mathcal{P};\mathbf{X})$. Furthermore, by Theorem 4, for all $y \in \mathbf{V}$ and $l\times\mathbf{P}$-*a.a.* t,ω the following equality holds:

$$(y,u(t))_{\mathbf{U}} = (y,\varphi)_{\mathbf{U}} + \int_{[0,t]} [y,\ Au(s) + f(s)]_{\mathbf{U}}ds +$$

$$+ (y,\ \tilde{M}(t))_{\mathbf{U}},\ \text{where}\ M(t) := M(t) + \int_{[0,T]} Bu(s)d\mathbf{W}(s).$$

From this, making use of the proposition and the lemma of this item, we find that for all $y \in \mathbf{V}$, ($l\times\mathbf{P}$-*a.s.*),

$$(\Lambda_1^2 y, u(t)) = (\Lambda_1^2 y, \varphi) + \int_{[0,t]} [\Lambda_1^2 y, Au(s) + f(s)]ds + (\Lambda_1^2 y, \tilde{M}(t)).$$

Since $\Lambda_1^2\mathbb{Z}$ is dense in $\mathbf{X}$, the last equality completes the proof of the theorem. □

3.2. Dissipative Systems.

2.0. In this section we consider the LSES (0.1) in a scale of Hilbert processes. A family of Hilbert spaces $\{\mathbb{H}^\alpha\}$, $\alpha\in\mathbf{R}^1$, is called a Hilbert scale if these spaces possess the following properties:

(i) For $\beta>\alpha$ the space $\mathbb{H}^\beta$ is normally imbedded in $\mathbb{H}^\alpha$.
(ii) For every $\alpha<\beta<\gamma$ the following inequality is valid*

$$||x||_\beta \le ||x||_\alpha^{(\gamma-\beta)/(\gamma-\alpha)} \cdot ||x||_\gamma^{(\beta-\alpha)/(\gamma-\alpha)}, \ \forall x\in\mathbb{H}^\gamma.$$

System (0.1) is not assumed to be coercive in the normal triple of the elements of the Hilbert scale, where it is considered, but it is assumed that this system is a dissipative one in another scale "imbedded" in the original one. Under such assumptions we shall prove the existence and uniqueness of a solution.

2.1. Let $\mathbf{X}$ and $\mathbb{H}$ be a pair of Hilbert spaces and suppose $\mathbf{X}$ is normally imbedded in $\mathbb{H}$.

There exists a Hilbert scale connecting these spaces, that is, there exists a Hilbert scale such that one of its elements coincides with $\mathbf{X}$ and another with $\mathbb{H}$. This fact is well known (see e.g. [62], Ch. IV, §1, i. 10 or [42], Ch. IV, §9, i. 1). Nevertheless, for the readers' comfort, we give in this paragraph an outline of the construction of such a Hilbert scale. We also will give some additional information about the properties of this Hilbert scale.

Denote by Λ the operator Λ_0 from Proposition 1.8, where $\mathbb{H}_1$ is replaced by $\mathbf{X}$ and $\mathbb{H}_2$ by $\mathbb{H}$. Making use of the spectral decomposition of the identity E_λ corresponding to the operator Λ, we define powers of this operator by the formula

$$\Lambda^\alpha x = \int_{[1,\infty]} \lambda^\alpha dE_\lambda x, \ \alpha\in\mathbf{R}^1. \tag{2.1}$$

*Here and below we use notation $||\cdot||_\alpha := ||\cdot||_{\mathbb{H}^\alpha}$.

We denote by $\mathbb{H}^\alpha$, for $\alpha>0$, the domain of Λ^α. It is well known that $\mathbb{H}^\alpha$ is a Hilbert space with respect to the scalar product

$$(2.2) \qquad (\cdot,\cdot)_\alpha = (\Lambda^\alpha\cdot;\ \Lambda^\alpha\cdot).$$

For $\alpha>0$, we define the space $\mathbb{H}^\alpha$ as the completion of $\mathbb{H}$ with respect to the norm $\|\cdot\|_\alpha = \|\Lambda^\alpha\cdot\|$.

Making use of the representation (2.1) it is easy to show that the family of Hilbert spaces $\{\mathbb{H}^\alpha\}$ discussed above is a Hilbert scale. This scale will henceforth be called the Hilbert scale connecting $\mathbf{X}$ and $\mathbb{H}$. Note two important properties of the Hilbert scale connecting $\mathbf{X}$ and $\mathbb{H}$:

(a) this scale is unique,
(b) the space $\mathbb{H}^\infty := \bigcap_{\alpha\in\mathbb{R}^1} \mathbb{H}^\alpha$ is dense in $\mathbb{H}^\alpha$ (in the strong topology of the latter space) for every $\alpha\in\mathbb{R}^1$.

Let us choose $\alpha,\beta\in\mathbb{R}^1$ such that $\alpha>\beta$, and set $\gamma := 2\beta-\alpha$. Consider the triple of spaces $(\mathbb{H}^\alpha,\mathbb{H}^\beta,\mathbb{H}^\gamma)$. Evidently for every $x\in\mathbb{H}^\alpha$ and $y\in\mathbb{H}^\beta$, we have

$$|(x,y)_\beta| = |(\Lambda^\beta x,\Lambda^\beta y)_0| = |(\Lambda^{\beta-\alpha}\Lambda^\alpha y),\Lambda^\beta y)_0| = \|(\Lambda^\alpha x,\Lambda^\gamma y)_0\| \leq \|x\|_\alpha\, \|y\|_\gamma.$$

Thus the triple $(\mathbb{H}^\alpha,\mathbb{H}^\beta,\mathbb{H}^\gamma)$ is a normal one. CBF of this triple will be denoted by $[\cdot,\cdot]_{\alpha,\beta}$. By property (i) of CBF (§2.4.1) the mapping $\mathfrak{J}$: $y\rightarrow[\cdot,y]_{\alpha,\beta}$ is a mapping of $\mathbb{H}^\gamma$ into some subspace of $(\mathbb{H}^\alpha)^*$.

In fact the following stronger statement is valid (see [40], Ch. IV, §1, i. 10).

Proposition 1. *The mapping $\mathfrak{J}$ is an isometric isomorphism of the spaces $\mathbb{H}^\gamma$ and $(\mathbb{H}^\alpha)^*$.* □

Remark 1. *We have established that CBF $[\cdot,\cdot]_{\alpha,\beta}$ brings the spaces $\mathbb{H}^\alpha$ and $\mathbb{H}^{2\beta-\alpha}$ into a duality. In particular, in the Hilbert scale $\{\mathbb{H}^\alpha\}$ the CBF $[\cdot,\cdot]_{\alpha,0}$ brings into duality the spaces $\mathbb{H}^\alpha$ and $\mathbb{H}^{-\alpha}$ for every α.*

It is also clear how, given a pair of Hilbert spaces $\mathbf{X}$, $\mathbb{H}$, where the first one is normally imbedded in the second, to form a normal triple $(\mathbf{X},\mathbb{H},\mathbf{X}')$ such that its CBF brings $\mathbf{X}$ and $\mathbf{X}'$ into duality. This possibility was mentioned above (see Remark 2.4.1). For this purpose we construct the scale connecting $\mathbf{X}$ and $\mathbb{H}$, letting $\mathbb{H}^1 := \mathbf{X}$ and $\mathbb{H}^0 := \mathbb{H}$, and setting $\mathbf{X}' := \mathbb{H}^{-1}$.

2.2. Let $\{\mathbb{H}^{\alpha}\}$, $\alpha \in \mathbb{R}^1$, be the Hilbert scale connecting $\mathbf{X}$ and $\mathbf{H}$, where $\mathbf{H}^1 := \mathbf{X}$, $\mathbb{H}^0 := \mathbb{H}$, and $\mathbb{H}^{-1} := \mathbf{X}'$. Let us fix a number $\lambda \geq 1$. Suppose that $M(t) \in \mathfrak{M}^c_{loc}([0,T];\ \mathbb{H}^{\lambda+1})$, $A(t,\omega)$: $\mathbb{H}^{\lambda+1} \to \mathbb{H}^{\lambda-1}$, and $\mathbb{B}(t,\omega)$: $\mathbf{H}^{\lambda+1} \to \mathfrak{L}_Q(\mathbf{Y},\mathbf{H}^{\lambda})$, $f(t,\omega)$ and $\varphi(\omega) \in \mathbb{H}^{\gamma}$ for all t,ω.

Recall that the assumptions made in the Introduction are still in force.

Denote by $[\cdot,\cdot]_{\alpha}$, the CBF of the triple $(\mathbf{H}^{\alpha+1}, \mathbf{H}^{\alpha}, \mathbf{H}^{\alpha-1})$ and by $||\cdot||$ the norm in $\mathfrak{L}_Q(\mathbf{Y},\mathbf{H}^{\alpha})$ similar to the norm $|||\cdot|||$ in $\mathfrak{L}_Q(\mathbf{Y},\mathbf{H})$.

Theorem 2. Suppose that there exists a constant $K \in \mathbf{R}_+$ such that for all $(t,\omega) \in [0,T] \times \Omega$ the following assumptions hold.

$$(A_2) \qquad 2[x,\ A(t,\omega)x]_{\lambda} + |||B(t,\omega)x|||^2_{\lambda} \leq K||x||^2_{\lambda}, \quad \forall x \in \mathbf{H}^{\lambda+1};$$

$$(B_2) \qquad ||A(t,w)x||_{\alpha-1} \leq K||x||_{\alpha+1}, \quad \forall x \in \mathbb{H}^{\lambda+1},\ \alpha = \lambda,\ 0;$$

$$|||B(t,\omega)x|||_{\alpha-1} \leq K||x||_{\alpha}, \quad \forall x \in \mathbb{H}^{\lambda+1},\ \alpha = \lambda,\ 1;$$

$$(C_2) \qquad \mathbb{E}||\varphi||^2_{\lambda} < \infty,\ \mathbb{E}\int_{[0,T]} ||f(t)||^2_{\lambda}\, dt < \infty.$$

Then the LSES (0.1) *has a solution u in* $(\mathbb{H}^1,\mathbb{H}^0,\mathbb{H}^{-1})$. *This solution belongs to* $\mathbb{L}_2([0,T];\mathcal{P};\mathbb{H}^{\lambda})$ *and for some* $N \in \mathbb{R}_+$, *depending only on K and T, satisfies the following inequality:*

$$(2.3) \qquad \mathbb{E}\int_{[0,T]} ||u(t)||^2_{\lambda}\, dt \leq N\, \mathbb{E}\Big(||\varphi||^2_{\lambda} + \int_{[0,T]} ||f(t)||^2_{\lambda} dt + ||M(t)||^2_{\lambda+1}\Big). \qquad \square$$

Proof. First of all note that

$$(2.4) \qquad ||x||^2_{\lambda+1} = [x,\ \Lambda^2 x]_{\lambda},\ \forall x \in \mathbb{H}^{\lambda+1}.$$

Indeed, let $x \in \mathbb{H}^{\lambda+2}$, *then* $\Lambda^2 x \in \mathbb{H}^{\lambda}$ and by property (ii) of CBF (§2.4.1)

$$[x,\ \Lambda^2 x]_{\lambda} = (x,\ \Lambda^2 x)_{\lambda} = ||x||^2_{\lambda+1}.$$

We choose a sequence $\{x_n\}$ from $\mathbb{H}^{\lambda+2}$ strongly converging to x in $\mathbf{H}^{\lambda+1}$.

Clearly,

$$\lim_{n\to\infty} ||\Lambda^2 x_n - \Lambda^2 x||_{\lambda-1} = \lim_{n\to\infty} ||\Lambda^{\lambda+1}(x_n - x)||_0 = \lim_{n\to\infty} ||x_n - x||_{\lambda+1} = 0.$$

Making use of this fact, it is easy to prove the validity of (2.4) for all $x \in \mathbb{H}^{\lambda+1}$.

Let $A_\varepsilon(t,\omega) := A(t,\omega) + \varepsilon\, \Lambda^2$, $\varepsilon > 0$. From the assumption (A_2) and (2.4) it follows that for all $(t,\omega) \in [0,T]\times\Omega$ and every $x \in \mathbb{H}^{\lambda+1}$,

(2.5) $$2[x, A_\varepsilon(t,\omega)x]_\lambda + |||B(t,\omega)x|||_\lambda^2 \le K\|x\|_\lambda^2 - \varepsilon\|x\|_{\lambda+1}^2$$

and

$$\|A_\varepsilon(t,\omega)x\|_{\lambda-1} \le (K+\varepsilon)\,\|x\|_{\lambda+1}.$$

Exactly in the same way as it was done in §1.8 we can show that $A_\varepsilon(t,\omega)x$, $f(t)$ and $B(t)x$, $(x \in \mathbb{H}^{\lambda+1})$, are predictable stochastic processes in $\mathbb{H}^{\lambda-1}$ and $\mathfrak{L}_Q(\mathbf{Y},\mathbb{H}^\lambda)$, respectively, and that φ is an $\mathcal{F}$-measurable random variable taking values in $\mathbb{H}^\alpha$.

Hence for every $\varepsilon > 0$, the LSES

(2.6) $$u(t) = \varphi + \int_{[0,t]} [A_\varepsilon u(s) + f(s)]\,ds + \int_{[0,t]} Bu(s)\,d\mathbf{W}(s) + M(t)$$

satisfied the assumptions of Theorem 1.4 and consequently has the unique solution $u^\varepsilon(t) \in \mathbb{L}_2([0,T];\mathcal{P};\mathbb{H}^{\lambda+1}) \cap \mathbb{C}([0,T];\mathcal{P};\mathbb{H}^\lambda)$. Moreover, there exists a constant $N \in \mathbb{R}_+$ such that

(2.7) $$\mathbb{E}\sup_{t\le T}\|u^\varepsilon(t)\|_\lambda^2 + \mathbb{E}\int_{[0,t]}\|u^\varepsilon(t)\|_{\lambda+1}^2\,dt \le$$
$$\le N\,\mathbb{E}\Big(\|\varphi\|_\lambda^2 + \int_{[0,T]}\|f(s)\|_{\lambda-1}^2\,ds + \|M(t)\|_\lambda^2\Big) < \infty.$$

The constant N in the inequality depends, generally speaking, on ε. In the next item, it will be shown that a similar estimate is valid with the constant independent on ε.

2.3. *Lemma* 3. *The following estimate, where the constant N depends only on K and T, is valid for the solution* u^ε *of the LSES* (2.6). □

Proof. From Theorem 2.4.2 it follows that for all $t \in [0,T]$ and ω from some set of probability 1

(2.8) $$\sup_{t\le T}\mathbb{E}\|u^\varepsilon(t)\|_\lambda^2 \le N\,\mathbb{E}\Big(\|\varphi\|_\lambda^2 + \int_{[0,T]}\|f(s)\|_\lambda^2\,ds + \|M(t)\|_{\lambda+1}^2\Big).$$

By estimate (2.7) and Proposition 2.1.7 it is easy to prove that the stochastic integrals in (2.9) are continuous martingales. Thus from (2.9) we obtain

$$\|u^{\varepsilon}(t)\|_{\lambda}^{2} = \|\varphi\|_{\lambda}^{2} + 2\int_{[0,t]} [u^{\varepsilon}(s),\ A_{\varepsilon}u^{\varepsilon}(s) + f(s)]_{\lambda}ds +$$

(2.9)

$$+\ 2\int_{[0,t]} (u^{\varepsilon}(s),\ d((Bu^{\varepsilon})\circ \mathbb{W}(t) + M(t))_{\lambda} + <(Bu^{\varepsilon})\circ\mathbb{W} + M>_{t}.$$

Since $f(s,\omega) \in \mathbb{H}^{\lambda}$ for all (s,ω), we have $[u^{\varepsilon}(s),\ f(s)]_{\lambda} = (u^{\varepsilon}(s),\ f(s))_{\lambda}$. This equality implies that

$$(2.11)\quad \mathbb{E}\int_{[0,t]} [u^{\varepsilon}(s),f(s)]_{\lambda}ds \le \tfrac{1}{2}\Big(\mathbb{E}\int_{[0,t]} \|u^{\varepsilon}(s)\|_{\lambda}^{2}ds + \mathbb{E}\int_{[0,t]} \|f(s)\|_{\lambda}^{2}ds\Big).$$

Next, it is obvious that

$$\mathbb{E}\|(Bu^{\varepsilon})\circ\mathbb{W}(t) + M(t)\|_{\lambda}^{2} = \mathbb{E}\int_{[0,t]} |||Bu^{\varepsilon}(s)|||_{\lambda}^{2}ds +$$

(2.12)

$$+\ 2\mathbb{E}((Bu^{\varepsilon})\circ\mathbb{W}(t),\ M(t))_{\lambda} + \mathbb{E}\|M(t)\|_{\lambda}^{2}.$$

From (2.2) it follws that

$$\Big(\int_{[0,t]} Bu^{\varepsilon}(s)d\mathbb{W}(s),\ M(t)\Big)_{\lambda} = \Big(\int_{[o,t]} \Lambda^{\lambda-1}Bu^{\varepsilon}(s)d\mathbb{W}(s),\ \Lambda^{\lambda+1}M(t)\Big)_{0} \le$$

(2.13)

$$\le \tfrac{1}{2}\Big(\int_{[0,t]} \Lambda^{\lambda-1}Bu^{\varepsilon}(s)d\mathbb{W}(s)\|_{0}^{2} + \|M(t)\|_{\lambda-1}^{2}\Big).$$

By the assumption (B_2) we find

$$(2.14)\qquad \mathbb{E}\int_{[0,t]} |||\Lambda^{\lambda-1}Bu^{\varepsilon}(s)|||_{0}^{2}ds = \mathbb{E}\int_{[0,t]} |||Bu^{\varepsilon}(s)|||_{\lambda-1}^{2}ds \le$$

$$\le K\,\mathbb{E}\int_{[0,T]} \|u^{\varepsilon}(s)\|_{\lambda}^{2}ds.$$

Collecting (2.11) - (2.14) and making use of (2.5), we obtain from (2.10) that for every $t\in[0,T]$,

$$\mathbf{E}\|u^{\varepsilon}(t)\|_{\lambda}^{2} \le N\,\mathbf{E}\Big(\|\varphi\|_{\lambda}^{2} + \int_{[0,T]} \|f(t)\|_{\lambda}^{2} dt + \|M(t)\|_{\lambda+1}^{2} + \int_{[0,t]} \|u^{\varepsilon}(s)\|_{\lambda}^{2} ds\Big),$$

where the constant N does not depend on ε. From this, by the Gronwall-Bellman lemma we obtain (2.8).

2.4. Since u^{ε} is a solution of the LSES (2.6) in $(\mathbf{H}^{\lambda+1},\mathbf{H}^{\lambda},\mathbf{H}^{\lambda-1})$, the following equality holds true on $[0.T]\times\Omega$, ($\ell\times\mathbf{P}$-*a.s.*), *for every* $y\in\mathbf{H}^{\infty}$.

$$(y,u^{\varepsilon}(t))_{\lambda} = (y,\varphi)_{\lambda} + \int_{[0,t]} [y, A_{\varepsilon}u^{\varepsilon}(s) + f(s)]_{\lambda} ds + \tag{2.15}$$
$$+ \Big(y, \int_{[0,t]} Bu^{\varepsilon}(s)d\mathbf{W}(s) + M(t)\Big)_{\lambda}.$$

In Lemma 1.8, let $\mathbf{V}:=\mathbb{H}^{\lambda+1}$, $\mathbf{U}:=\mathbb{H}^{\lambda}$, $\mathbf{V}':=\mathbf{H}^{\lambda-1}$ and, $\Lambda_1:=\Lambda^{\lambda}$. Since $\mathbb{H}^{\infty}$ is dense in every $\mathbb{H}^{\lambda}$, it follows from this lemma that for all s,ω, $[A_{\varepsilon}u^{\varepsilon}(s) + f(s),y]_{\lambda} = [A_{\varepsilon}u^{\varepsilon}(s) + f(s), \Lambda^{2\lambda}y]_{0}$. From this and (2.2), making use of self-adjointness of Λ^{λ}, we obtain that equality (2.15) is equivalent to the following one:

$$(\Lambda^{2\lambda}y, u^{\varepsilon}(t))_{0} = (\Lambda^{2\lambda}y, \varphi)_{0} + \int_{[0,t]} [\Lambda^{2\lambda}y, A_{\varepsilon}u^{\varepsilon}(s) + f(s)]ds + \tag{2.16}$$
$$+ \Big(\Lambda^{2\lambda}y, \int_{[0,t]} Bu^{\varepsilon}(s)d\mathbf{W}(s) + M(t)\Big)_{0}.$$

From Proposition 1.2.9 (a) it follows that $u^{\varepsilon}\in\mathbf{L}_2([0,T];\mathcal{P};\mathbf{H}^1)$. Thus it is proved that u^{ε} is a solution of the LSES (0.1) in $(\mathbf{H}^1,\mathbf{H}^0,\mathbf{H}^{-1})$. Recall that $\{\Lambda^2 y\}$, $y\in\mathbb{H}^{\infty}$, is a dense subset of $\mathbb{H}^1$.

Lemma 3 implies that there exist $u\in\mathbf{L}^2([0,T];\mathcal{P};\mathbf{H}^1)$ and some subsequence $\{\varepsilon_n\}$ converging to zero as $n\to\infty$ *such that* $u^{\varepsilon_n}\to u$ weakly in $\mathbf{L}_2([0,T]\times\Omega; \mathcal{B}([0,T])\otimes\mathcal{F}; \ell\times\mathbf{P}; \mathbb{H}^1)$.

Arguing in the same way as in Theorem 1.4, we pass to the limit in (2.16) (over the sequence u^{ε_n}) and prove that u is the solution of the LSES (0.1) in $(\mathbb{H}^1,\mathbb{H}^0,\mathbb{H}^{-1})$.

Since the norm of a weak limit is bounded by the limit of the norms of an approximating system, (2.8) implies the inequality in the statement of the theorem. Thus the theorem is proved. □

3.3. Uniqueness and the Markov Property.

3.0. The assumptions of Theorem 1.4 ensure the existence and uniqueness of a solution of the LSES (0.1) in a normal triple $(\mathbf{X},\mathbb{H},\mathbf{X}')$. However, if we consider the problem of uniqueness as it is, these assumptions are too burdensome. In this section we prove the uniqueness theorem under less restrictive conditions, in particular, ones which dissipative systems satisfy.

In addition to the uniqueness theorem, we also discuss in this section, possibilities for the approximation of the solution of the LSES (0.1) by solutions of other systems of the same type. We also prove that if the coefficients, initial and external values of the LSES (0.1) do not depend on ω, then the solution of this system possesses the Markov property.

3.1. *Theorem* 1. *Let* $K(t,\omega)$ *and* $N(t.\omega)$ *be* $\mathcal{F}_t$*-adapted stochastic processes taking values in* $\mathbb{R}_+$ *such that*

$$\sup_{t\leq T} K(t,\omega) < \infty \quad and \int_{[0,T]} N(t,\omega)dt < \infty.$$

Suppose that for all $(t,\omega) \in [0,T]\times\Omega$ *the following assumptions are valid:*

A_ω. $2\,[x,\, A(t,\omega)x] + |||B(t,\omega)x|||^2 \leq N(t,\omega)\,||x||^2, \quad \forall x\in\mathbf{X},$

B_ω. $||A(t,\omega)x||_{\mathbf{X}'} \leq K(t,\omega)\,||x||_{\mathbf{X}}, \quad \forall x\in\mathbf{X}.$

Then a solution of the LSES (0.1) *(if it does exist) is uniquely determined.* □

Proof. We first observe that under the assumptions made above, the integrals in (0.1) are well-defined. From property (i) of a CBF (see §2.4.1) and conditions (A_ω), (B_ω) it follows that for all $(t,\omega) \in [0,T]\times\Omega$

$$(3.1) \qquad |||B(t,\omega)x|||^2 \leq K(t,\omega)\,||x||^2_{\mathbf{X}} + N(t,\omega)\,||x||^2,\ \forall x\in\mathbf{X}.$$

This implies, in particular, that for every $(t,\omega) \in [0,T]\times\Omega$, $A(t,\omega) \in \mathbb{L}(\mathbf{X},\mathbf{X}')$ and $B(t,\omega) \in \mathbb{L}(\mathbf{X},\mathfrak{L}_Q(\mathbb{Y},\mathbb{H}))$. Thus the operators A and B transform predictable processes into predictable ones. Moreover, from the assumptions of the theorem and (3.1), it follows that for every $x \in \mathbf{L}_2^{\omega}([0,T];\mathcal{P};\mathbf{X})$,

$$\int_{[0,T]} (||Ax(t,\omega)||_{\mathbf{X}'} + |||Bx(t,\omega)|||^2)\,dt < \infty \quad (\mathbf{P}\text{-}a.s.).$$

Therefore the integrals in (0.1) are actually defined (see §1.2.11 and §2.3.3).

Let u_1 *and* u_2, be solutions of the system (0.1) in $(\mathbf{X},\mathbf{H},\mathbf{X}')$ and $\bar{u} = u_1 - u_2$. It is clear that $\bar{u}$ is also a solution of a LSES of the same type with $\varphi \equiv f \equiv M \equiv 0$.

Denote

$$r(t,\omega) := ||\bar{u}(t,\omega)||^2 + \int_{[0,t]} (N(s,\omega) + K(s,\omega)\, ||\bar{u}(s,\omega)||_{\mathbf{X}}^2))\,ds,$$

$$\Gamma_n(\omega) := \{t\colon t \le T,\ r(t,\omega) \ge n\},\ n \in \mathbf{N},\ \textit{and}$$

$$\tau_n(\omega) := \begin{cases} \inf_t \Gamma_n(\omega) & \text{if } \Gamma_n(\omega) \ne \oslash; \\ T, & \text{if } \Gamma_n(\omega) = \oslash. \end{cases}$$

Evidently, τ_n is a stopping time, $\tau_n \uparrow T$ as $n \to \infty$, and

$$\int_{[0,t\wedge\tau_n]} B\bar{u}(s)\,dW(s) \in \mathfrak{M}_2^c([0,T],\ H).$$

The latter relation follows from the inequality

$$(3.2) \qquad \int_{[0,t\wedge\tau_n]} |||B\bar{u}(s)|||^2 ds \le N_n < \infty,\ \forall n \in \mathbf{N},\ (\mathbf{P}\text{-}a.s.),$$

where the constant N_n does not depend on ω, which is valid in view of (3.1).

Thus $(B\bar{u}) \circ W(t)$ is a continuous local martingale with respect to $\{\tau_n\}$.

From Theorem 2.4.2 it follows that for all $t \le T$ on one and the same ω-set of probability 1,

$$(3.3) \qquad \begin{aligned} &||\bar{u}(t\wedge\tau_n||^2\, exp\Big\{ - \int_{[0,t\wedge\tau_n]} N(s)||\bar{u}(s)||^2 ds\Big\} = \\ &= \int_{[0,t\wedge\tau_n]} exp\Big\{- \int_{[0,s]} N(\tau)||\bar{u}(\tau)||^2 d\tau\Big\} \times \end{aligned}$$

$$\times\ (2[\bar u(s),\ A\bar u(s)] + |||B\bar u(s)|||^2 - N(s)||\bar u(s)||^2)ds +$$

$$+\ 2\int_{[0,t\wedge\tau_n]}\Big(exp\Big\{-\int_{[0,s]}N(\tau)||\bar u(\tau)||^2 d\tau\Big\}\ \bar u(s),\ d[(B\bar u)\circ\mathbf{W}(s)]\Big).$$

Next, in view of (3.2),

$$\mathsf{E}\Big(\int_{[0,t\wedge\tau_n]} exp\Big\{-2\int_{[0,s]}N(\tau)||\bar u(\tau)||^2 d\tau\Big\}\ ||\bar u(s)||^2\ \times$$

$$\times\ |||B(s)\bar u(s)|||^2 ds\Big)^{1/2} \le$$

$$\le \mathsf{E}\sup_{[s\le t\wedge\tau_n]}||\bar u(s)||\Big(\int_{[0,t\wedge\tau_n]}|||B\bar u(s)|||^2 ds\Big)^{1/2} \le NN_n < \infty.$$

Hence the stochastic integral in (3.3) is a martingale (see Proposition 2.1.7) and so that its expectation is equal to zero. In view of this and condition (A_ω), we obtain by taking expectations of both parts of the inequality (3.3) that

$$\mathsf{E}||\bar u(t\wedge\tau_n)||^2\ exp\Big\{-\int_{[t\wedge\tau_n]}N(s)||u(s)||^2 ds\Big\} \le 0,$$

and consequently

$$\mathsf{P}(||\bar u(t\wedge\tau_n)|| = 0) = 1.$$

By passing to the limit in the last equality as $\tau_n \uparrow \mathrm{T}$ (this is possible in view of continuity of $||\bar u(t)||$ with respect to t), we find that

$$\mathsf{P}(\sup_{t\le T}||u_1(t) - u_2(t)\}\} = 0) = 1. \qquad \square$$

3.2. Suppose that the collection $\{A(t,\omega),\ B(t,\omega),\ \varphi(\omega),\ f(t,\omega),\ M(t,\omega)\}$ satisfies, as always in this section, the assumptions given in the Introduction and in §1.1 before the statement of the theorem. We shall also suppose that a sequence of collections $\{A_n(t,\omega),\ B_n(t,\omega),\ \varphi(\omega),\ f_n(t,\omega),\ M_n(t,\omega)\}$ satisfying the same assumptions (maybe with different constants) is given.

Consider the following LSES:

$$u^n(t) = \varphi_n + \int_{[0,t]} (A_n u^n(s) + f_n(s))ds +$$

(3.4)

$$+ \int_{[0,t]} B_n u^n(s) d\mathbb{W}(s) + M_n(t).$$

Suppose that the LSES's (0.1) and (3.4) (the latter for every ($n \in \mathbb{N}$) have solutions from $\mathbb{L}_2([0,T];\mathcal{P};\mathbf{X})$.

Theorem 2. Suppose that the LSES (3.4) *satisifes the assumptions of Proposition* 1.7 *for all* $n \in \mathbb{N}$ *with constants which do not depend on n. We assume also that the following conditions hold true.*

a) $\lim_{n\to\infty} \mathbb{E}(||\varphi_n - \varphi||^2 +$

$$+ \int_{[0,T]} ||f_n(t) - f(t)||^2_{\mathbf{X}'} dt + <M_n - M>_T = 0;$$

b) for every $x \in \mathbb{L}_2([0,T];\mathcal{P};\mathbf{X})$

$$\lim_{n\to\infty} \mathbb{E} \int_{[0,T]} (||(A_n - A)x(t)||^2_{\mathbf{X}'} + |||(B_n - B)x(t)|||^2 dt = 0.$$

Then,

$$\lim_{n\to\infty} \mathbb{E} \sup_{t\le T} ||u^n(t) - u(t)||^2 = 0.$$

If in addition to the assumptions made above the LSES's (0.1) *and* (3.4) *are coercive for all n with constants that do not depend on n, then,*

$$\lim_{n\to\infty} \mathbb{E} \int_{[0,T]} ||u^n(t) - u(t)||^2_{\mathbf{X}} dt = 0. \quad \square$$

Proof. We prove only the first statement, since the second one can be proved in the same way.

Write $v^n := u^n - u$. Clearly, v^n is a solution (in $(\mathbf{X},\mathbf{H},\mathbf{X}')$) of the following the LSES:

$$v^n(t) = \tilde{\varphi}_n + \int_{[0,t]} (A_n v^n(s) + \tilde{f}_n(s))ds +$$

(3.5)

$$+ \int_{[0,t]} B_n v^n(s)d\mathbb{W}(s) + \tilde{M}_n(t),$$

where

$$\tilde{\varphi}_n := \varphi_n - \varphi, \ \tilde{f}_n(s) := (A_n - A)\ u(s) +$$

$$+ \tilde{f}_n(s) - f(s),\ \tilde{M}_n(t) := M_n(t) - M(t) +$$

$$+ \int_{[0,t]} (B_n - B)\ u(s)d\mathbb{W}(s).$$

Evidently, the LSES (3.5) also satisfies the assumptions of Proposition 1.7, and consequently there exists a constant N independent on n and such that

$$\mathbb{E} \sup_{t \leq T} ||v^n||^2 \leq N\ \mathbb{E}(||\tilde{\varphi}_n||^2 + \int_{[0,T]} ||\tilde{f}_n(t)||^2_{\mathbf{X}'}\ dt + ||\tilde{M}_n(T)||^2).$$

This inequality implies the first part of the statement. □

3.3. In this paragraph we shall assume that:

(a) The operators A,B and the function f do not depend on ω.

(b) $M(t) := \int_{[0,t]} g(s)d\mathbb{W}(s),$

where $g(s) \in \mathbb{L}_2([0,T],\ \overline{\mathfrak{B}([0,T])},\ l,\ \mathfrak{L}_Q(\mathbf{Y},\mathbb{H})).$

We shall also suppose that the LSES (0.1) satisfies the conditions of Theorem 1.4 or Theorem 2.2 and Proposition 1.7, which ensure the existence, uniqueness and continuous dependence on the initial values of the system.

It will be shown that under these assumptions a solution of the LSES (0.1) possesses the Markov property, that is, for every Borel set Γ from $\mathbb{H}$ and arbitrary $s,\ t \in [0,T]$ such that $s \leq t$, ($\mathbb{P}$-a.s.),

(3.6) $$\mathbb{P}(u(t) \in \Gamma | \mathfrak{F}_s(u)) = \mathbb{P}(u(t) \in \Gamma | u(s)).$$

Here and below, $\mathcal{F}_s(u)$ is the σ-algebra generated by the solution $u(r,\omega)$ of the LSES (0.1) for $r\in[0,s]$. □

Let $y(s) \in \mathbb{L}_2(\Omega;\mathcal{F}_s;\mathbb{P};\mathbb{H})$. Consider the LSES

$$u(t,y(s),s) = y(s) + \int_{[s,t]} (A(r)u(t,y(s),s) +$$

(3.7)

$$+ f(r))dr + \int_{[0,t]} (B(r)u(r,y(s),r) + g(r))d\mathbb{W}(r).$$

for $t\in[s,T]$.

As in Definition 1.2, we call a predictable process $u(t,y(s),s)$ belonging to $\mathbb{L}_2([s,t];\ \overline{\mathfrak{B}([s,t])};\ l;\ \mathbf{X})$, ($\mathbb{P}$-*a.s.*), a solution of the LSES (3.7) in $(\mathbf{X},\mathbb{H},\mathbf{X}')$, if for all x from some dense (in the strong topology) subset of $\mathbf{X}$ and $l\times\mathbb{P}$-*a.a.* t, ω from $[s,t]\times\Omega$ the following equality holds:

$$(x,\ u(t,y(s),s)) = (s,\ u(s)) +$$

$$+ \int_{[s,t]} [x,\ A(r)\ u(r,y(s),s) + f(r)]dr +$$

$$+ \left(x,\ \int_{[s,t]} (B(r)\ U(r,y(s),s) + g(r))d\mathbb{W}(r)\right).$$

Remark 3. *It is easy to see that all the results obtained for the LSES* (0.1) *carry over to the case of a similar LSES considered on the interval* $[T_0,T]$ *with the* $\mathcal{F}_{T_0}$*-measurable initial condition* φ. □

In the future, all the results obtained for the LSES (0.1) are without any references, carried over to the LSES (3.7).

Theorem 3. Let f be a bounded Borel function on $\mathbb{H}$ taking values in $\mathbf{R}^1$ and $y(s,\omega)\equiv x\in\mathbb{H}$. Then for every s, $t\in[0,T]$, such that $s\leq t$, $\phi(x):= \mathbb{E}f(u(t,x,s))$ is a Borel function on $\mathbb{H}$ and
$\mathbb{P}$-a.s.,

$$\mathbb{E}[f(u(t))|\mathcal{F}_s(u)] = \Phi(u(s)). \quad □$$

Proof. It suffices to prove the statement for a continuous function *f*. Henceforce, we assume that f is continuous. From Theorem 2, it follows that $\mathbb{E}f(u(t,x,s))$ is a continuous function and consequently Borel.

Note that $u(t,x,s)$ does not depend on the σ-algebra $\mathcal{F}_s$ (recall that x does not depend on ω). To prove this, observe that $u(t,x,s)$ is a unique solution of the LSES (0.1), and from the analysis of the proofs of Theorems 1.4 and 2.2 it follows that the solution can be obtained as the limit of solutions of finite-dimensional systems. For them the above property is valid (see e.g. [68], Ch. II, §9), and consequently it is valid for $u(t,x,s)$.

Let $\{u_n(s,\omega)\}$, $n\in\mathbb{N}$, be a sequence of $\mathcal{F}_s$-measurable step-functions in $\mathbb{H}$ converging uniformly in the strong topology of $\mathbb{H}$ to $u(s,\omega)$. Let also Γ_n be the set of values of $u_n(s,\omega)$. It is easy to verify that

$$v(s) = \sum_{x\in\Gamma_n} 1\{u_n(s) = x)\ u(t,x,s)$$

is a solution of the LSES (3.7), where $y(s) = u_n(s)$. In view of uniqueness of the solution, $v(s,x)$ coincides with $u(t,u_n(s),s)$. Since $u(t,x,s)$ does not depend on $\mathcal{F}_s$, this yields

$$\mathbb{E}[f(u(t,u_n(s),s))|\mathcal{F}_s(u)] = \mathbb{E}\Big(\sum_{x\in\Gamma_n} 1\{u_n(s) = x\}\ f(u(t,x,s))\Big|\ \mathcal{F}_s(u)] =$$

$$= \sum_{x\in\Gamma_n} 1\{u_n(s) = x\}\ \mathbb{E}\ f(u,(t,x,s)) = \Phi(u_n(s)).$$

From the external equality, making use of continuity of f and Φ, we obtain (3.8) by passing to the limit. □

3.4. The First Boundary Problem for Ito's Partial Differential Equations.

4.0. Denote by x^i the i-th coordinate of the vector $x\in\mathbf{R}^d$. In this section we denote by α, β and γ, d-dimensional multi-indices, that is, d-dimensional vectors with coordinates from $\mathbb{N}\cup\{0\}$. Given a d-dimensional multiindex Z, we define its length $|Z|$ by the formula $|Z| = Z^1+...+Z^d$. A real number (function) with upper multiindex will denote an element of a set of real numbers (functions) indexed by multiindices. A function with lower multiindex will denote the derivative defined as follows:

$$f_\alpha := \frac{\partial^{\alpha^1+...+\alpha^d}}{\partial(x^1)^{\alpha^1}...\partial(x^d)^{\alpha^d}},\ \alpha := (\alpha^1,...,\alpha^d).$$

Let G be a domain in $\mathbf{R}^d$. In this section we consider the following boundary problem in the cylinder $[0,T]\times G$:

$$u(t,x,\omega) = \varphi(x,\omega) - \int_{[0,t]} \sum_{|\alpha|\le m} (-1)^{|\alpha|} \Big(\sum_{|\beta|\le m} a^{\alpha\beta}(s,x,\omega)u_\beta(s,x,\omega) +$$

(4.1)

$$+ f^\alpha(s,x,\omega)\Big)_\alpha \, ds + \int_{[0,t]} \sum_{l=1}^{d_1} \sum_{|\beta|\le m} b^\beta(s,x,\omega) \times$$

$$u_\beta(s,x,\omega)dw^l(s,\omega) + M(t,x,\omega),$$

(4.2) $$u_\gamma(t,x,\omega)|_S = 0$$

for all γ such that $|\gamma| \le d$-1, where S is the lateral side of the cylinder $[0,T]\times G$, $M(t) \in \mathfrak{M}_2^c([0,T]; \mathsf{L}_2(G))$, and $w(t)$ is a Wiener process in $\mathbf{R}^{d_1}$ with the unit correlation operator, where $w^i(t)$ is the i-th coordinate of $w(t)$ (i=1,2,...,d_1).

Making use of Theorem 1.4, we show that under general assumptions concerning the coefficients and the initial and external values this problem has a unique solution in a Sobolev space.

4.1. Sobolev spaces are of great importance in the thoery of partial differential equations. So we begin with a short review of some important results concerning spaces of this type. Full proofs of the results given below in items 1 to 7 can be found e.g. in [103], [131].

Suppose that numbers $m\in\mathbb{N}$, $p\in[1,\infty[$ are fixed. Denote $dx := dx^1....dx^d$.

Remark 1. *This notation will be used in what follows.* □

Definition 1. *The space (of classes) of real functions on G belonging together with their derivatives up to the order of m to* $\mathsf{L}_p(G)$ *is called the Sobolev space* $\mathbb{W}_p^m(G)$. □

Theorem 1. *The Sobolev space* $\mathbb{W}_p^m(T)$ *endowed by the norm*

$$\|u\|_{G,m,p} = \Big(\sum_{|\alpha|\le m} \int_G |u_\alpha(x)|^p \, dx \Big)^{1/p},$$

*is a separable, reflexive Banach space**. *For p=2 this space*

is a Hilbert space with respect to the scalar product $(\cdot,\cdot)_{G,m,2}$ *generated by the norm* $||\cdot||_{G,,m,p}$. □

Warning 1. *If* $G = \mathbf{R}^d$, *we shall omit the index G in the notation of the scalar product and the norm.* □

4.2. For $p=2$ and $G = \mathbf{R}^d$ there exists another very helpful definition of a Sobolev space, which we give below.

Let Δ be Laplace operator $\sum_{i=1}^{d} \frac{\partial^2}{\partial(x^i)^2}$. Denote $\Lambda^2 := I-\Delta$.

Clearly, Λ^2 is a positive, self-adjoint, and unbounded operator on $\mathbf{H}^0 := \mathbf{L}_2(\mathbf{R}^d)$. Let us define the Hilbert space $\{\mathbf{H}^s\}$, $s \in \mathbf{R}^1$, by the

operator $\Lambda = \sqrt{\Lambda^2}$ and the space $\mathbf{H}^0$, as it was done in §3.1. Namely, define $\mathbf{H}^s$, for $s>0$, as the domain of the operator Λ^s, and for $s \leq 0$, as the completion of $\mathbf{H}^0$ in the norm $||\cdot||_s := ||\Lambda^s \cdot||_{\mathbf{L}_2(\mathbf{R}^d)}$.

It was stated above that spaces constructed in such a way are Hilbert ones with respect to the scalar product $(\cdot,\cdot)_s = (\Lambda^s \cdot, \Lambda^s \cdot)_{\mathbf{L}_2(\mathbf{R}^d)}$.

Theorem 2. *For* $m \in \mathbf{N}$ *the space* $\mathbf{W}_2^m(\mathbf{R}^d)$ *and* $\mathbf{H}^m$ *are equivalent, that is,* $\mathbf{H}^m = \mathbf{W}_2^m(\mathbf{R}^d)$ *as sets, and the norms* $||\cdot||_m$ *and* $||\cdot||_{m,2}$ *are equivalent, i.e. there exist constants* N_1, N_2 *such that*

$$N_1||u||_m \leq ||u||_{m,2} \leq N_2||u||_m, \ \forall u \in \mathbf{H}^m. \quad □$$

Warning 2. *In view of the last theorem we can and shall identify the spaces* $\mathbf{W}_2^m(\mathbf{R}^d)$ *and* $\mathbf{H}^m$ *for* $m \in \mathbf{N}$. *As a rule we shall use the notation* $\mathbf{H}^m$ *for this space as well as the corresponding notation* $||\cdot||_m$ *for the norm and* $(\cdot,\cdot)_m$ *for the scalar product.* □

From the properties of a Hilbert scale we obtain the following result.

Corollary 2. *For every* $k,m \in \mathbf{N}$ *the spaces* $\mathbf{H}^{m+k}$, $\mathbf{H}^m$, *and* $\mathbf{H}^{m-k}$ *form a normal triple, and the mapping*

*Here and below in this section unless otherwise stated derivatives are assumed to be generalized ones, in the sense of S.L. Sobolev.

$$\mathbb{H}^{m-k} \ni x \leftrightarrow [\cdot,x]_{m-k,m} \in (\mathbb{H}^{m+k})^*$$

where $[\cdot,\cdot]_{m-k}$ is the CBF of this triple, is an isometric isomorphism of the spaces $(\mathbb{H}^{m+k})^*$ and $\mathbb{H}^{m-k}$. □

From the construction of a Hilbert scale it is easy to obtain the following statement.

Proposition 2. *(i) For every nonnegative m and k,* $\Lambda^k \mathbb{H}^m = \mathbb{H}^{m-k}$ *and* $(\Lambda^{2k}x,y)_m := (x,y)_{m+k}$ *for* $x \in \mathbb{H}^{m+2k}$ *and* $y \in \mathbb{H}^{m+k}$.

(ii) The space $C_0^\infty(\mathbb{R}^d)$ *is dense in* $\mathbb{H}^m$ *(in the strong topology of the latter space) for every integer m.* □

Remark 2. *Sobolev spaces* $\mathbb{H}^s$ *with fractional s are not used in the book. So we do not consider them in detail in this section.*

4.3. *Definition* 3. *The boundary Γ of the domain* $G \subset \mathbb{R}^d$ *is said to be regular if there exist a finite open covering* $\{\Gamma_i\}$ *(of the boundary), a finite set of open, finite cones* K_j, *and a number* $\varepsilon > 0$ *such that the following conditions are valid:*

(a) For every point from Γ, the sphere with the center at this point and radius ε lies entirely in some set Γ_i *from the covering.*

(b) For every point from $\Gamma_i \cap G$, *the cone with the apex at point, obtained by the parallel transfer of some* K_j, *lies entirely in G.* □

Lemma 3. *Let G be a bounded domain in* $\mathbb{R}^d$ *and* $\mathbb{R}(G)$ *be the set of functions which are restrictions of* $C_0^\infty(\mathbb{R}^d)$ *functions to G. Then R(G) is dense in* $W_p^m(G)$ *(in the strong topology of the last space).* □

4.4. The theorem presented below is among the most important results of the theory of Sobolev spaces.

Theorem 4. *Let G be a bounded domain with regular boundary and* $2(d-mp) \leq pd$. *Then the space* $W_p^m(G)$ *is dense and continuously imbedded in* $\mathbb{L}_2(G)$ *(see §1.2.9). If for* $n \in \mathbb{N} \cap [0\}$, $n \leq m$ *and* $p(m-n) > d$, *then* $W_p^m(G)$ *is dense and continuously imbedded in* $C^n(G \cup \Gamma)^*$. □

Remark 4. *The theorem given above is a special case of the well-known "Sobolev's imbedding theorem".* □

---

*Here and below, $C^n(S)$ is the space of n times continuously differentiable functions on $S \subset \mathbb{R}^d$.

4.5. Let G be a bounded domain with regular boundary. This assumption will be in force throughout the section.

Definition 5. *The closure of* $C_0^\infty(G)$ *with respect to the norm* $\|\cdot\|_{G,1,2}$ *will be called the (Sobolev) space* $\mathring{\mathbb{H}}^1$. □

Theorem 5. *The space* $\mathring{\mathbb{H}}^s$ *is a separable space is normally imbedded in* $\mathbb{L}_2(G)$. □

The Hilbert space connecting $\mathbb{X} = \mathring{\mathbb{H}}^1$ and $\mathbb{H} = \mathbb{L}_2(G)$ will be denoted by $\{\mathring{\mathbb{H}}^s\}$, $s \in \mathbb{R}^1$. Existence of this scale follows from Theorem 5 and Remark 2.1.

Proposition 5. *The space* $\mathring{\mathbb{H}}^s$, $s \in \mathbb{R}_+$, *is the closure of* $\mathbb{C}_0^\infty(G)$ *with respect to the norm* $\|\cdot\|_s$. *The norms* $\|\cdot\|_{\mathring{\mathbb{H}}^m}$ *and* $\|\cdot\|_{G,2,m}$ *are equivalent.* □

Warning 5. *The norm, the scalar product, and the CBF for spaces from the scale* $\{\mathring{\mathbb{H}}^s\}$, $s \in \mathbb{R}^1$, *will be denoted in the same way as corresponding objects from the scale* $\{\mathbb{H}^s\}$, $s \in \mathbb{R}^1$. □

From the properties of a Hilbert space we obtain the following result.

Corollary 5. *For all integers* $k, m \in \mathbb{N}$ *the spaces* $\mathring{\mathbb{H}}^{m+k}$, $\mathring{\mathbb{H}}^m$ *and* $\mathring{\mathbb{H}}^{m-k}$ *form a normal triple, and the mapping*

$$\mathring{\mathbb{H}}^{m-k} \ni x \leftrightarrow [\cdot, x]_{m-k,m} \in (\mathring{\mathbb{H}}^{m+k})^*$$

is an isometric isomorphism between $\mathring{\mathbb{H}}^{m-k}$ *and* $(\mathring{\mathbb{H}}^{m+k})^*$. □

The following result has proven to be very useful.

Lemma 5. *(Friedrichs). There exists a constant* $\mathbb{N}$ *such that*

$$\|u\|_m \leq N \sum_{|\alpha|=m} \|u_\alpha\|_0, \ \forall u \in \mathring{\mathbb{H}}^m, \ m \in \mathbb{N}. \quad □$$

4.6. *Proposition* 6. *Let* $m \in \mathbb{N} \cup \{0\}$, *then*

(*i*) *There exists a constant* N *such that*

$$\left\| \frac{\partial}{\partial x^i} u \right\|_{m-1} \leq \mathbb{N} \|u\|_m, \ \forall u \in \mathbb{H}^m \ (\mathring{\mathbb{H}}^m).$$

(ii) A function $u \in \mathbb{H}^{m-1}(\overset{\circ}{\mathbb{H}}{}^{m-1}$, *respectively) if and only if this function can be represented as follows:* $u = \sum_{|\alpha| \leq 1} (v^{\alpha})_{\alpha}$ *for* $v^{\alpha} \in \mathbb{H}^m$ ($\overset{\circ}{\mathbb{H}}{}^m$, *respectively).*

(iii) If v belongs to $\mathbb{C}_b^1(\mathbb{R}^d)$ *or* $\mathbb{H}^1$ *and* $u \in \mathbb{H}^1$, *then*

$$\int_{\mathbb{R}^d} v(x) \frac{\partial}{\partial x^i} u(x)\,dx = - \int_{\mathbb{R}^d} \left(\frac{\partial}{\partial x^i} v(x)\right) u(x)\,dx. \quad \square$$

4.7. When it necessary to consider bounded or "mildly" increasing functions, the spaces $\mathbb{W}_p^m(\mathbb{R}^d)$ are of no use. In such a situation their modifications, the so-called weighted Sobolev spaces have proved to be useful.

In this book we make use of one of such modifications.

Denote by $\mathbb{L}_p(r,\mathbb{R}^d)$, $r \in \mathbb{R}^1$, $p \geq 1$, the space consisting of the classes of $\overline{\mathcal{B}(\mathbb{R}^d)}$-measurable functions on $\mathbb{R}^d$ such that

$$\int_{\mathbb{R}^d} |(1 + |x|_d^2)^{r/2} f(x)|^2\,dx \leq \infty.$$

Next, let $\mathbb{W}_p^m(r,\mathbb{R}^d)$, $m \in \mathbb{N} \cup \{0\}$ be the space of real functions on $\mathbb{R}^d$ belonging together with their generalized derivatives up to the order m to $\mathbb{L}_p(r)$.

Theorem 7. The space $\mathbb{W}_p^m(r,\mathbb{R}^d)$ *endowed with the norm*

$$||u||_{m,p,r} = \left(\int_{\mathbb{R}^d} \sum_{|\alpha| \leq m} |(1 + |x|_d^2)^{r/2} u_{\alpha}(x)|^p\,dx \right)^{1/p},$$

is a separable Banach space, and $\mathbb{W}_2^m(r,m)$ *is a Hilbert space with respect to the scalar product generated by the norm* $||\cdot||_{m,2,r}$. *The set* $\mathbf{C}_0^{\infty}(\mathbf{R}^d)$ *is dense in* $\mathbb{W}_p^m(r,\mathbb{R}^d)$ *(in the strong topology of the latter space).* $\square$

Warning 7. Throughout what follows the space $\mathbf{W}_p^m(r,\mathbf{R}^d)$ *will be denoted by* $\mathbb{H}^m(r,\mathbb{R}^d)$. $\square$

Lemma 7. (i) Let f be a $\mathcal{B}(R^d)$*-measurable real function and for some s,* $n \in \mathbf{R}_+$

$$\sum_{|\alpha|\leq m} |f_\alpha(x)| \leq N(1+|x|_d^2)^{s/2}.$$

Then for $r < -p^{-1}(d+sp)$ *and arbitrary* $p\geq 1$, $f(x) \in \mathbf{W}_p^m(r,\mathbb{R}^d)$ *and* $\|f\|_{m,p,r} < N_1 < \infty$, *where* N_1 *does not depend on f.*

(*ii*) *If* s, $r\in\mathbb{R}^1$, *then the operator S defined by the equality* $Sf(x) =$

$(1+|x|_d^2)^{s/2} f(x)$ *belongs to*

$$\mathfrak{L}(\mathbf{W}_p^n(r,\mathbb{R}^d),\ \mathbf{W}_p^m(r\text{-}s,\ \mathbb{R}^d)).\quad \square$$

The reader will prove this lemma easily.

4.8. Now we proceed to a more explicit consideration of problem (4.1), (4.2). We shall suppose that $a^{\alpha\beta}(t,x,\omega)$, $b^{l\beta}(t,x,\omega)$, *and* $f^\alpha(t,x,\omega)$ are $\overline{\mathcal{B}[0,T]\times G)\otimes\mathcal{F}}$-measurable real functions on $\overline{[0,T]\times G\times\Omega}$, which are predictable for every $x\in G$. The function $\varphi(x,\omega)$ is assumed to be $\overline{\mathcal{B}(G)\otimes\mathcal{F}_0}$-measurable. Suppose also that $a^{\alpha\beta}$, $b^{l\beta}$ are the uniformly bounded functions and $\varphi(\cdot,\omega)$, $f^\alpha(\cdot,t,\omega)$ belongs to $\mathbf{L}_2(G)$ for all t, ω, and

$$\mathbf{E}\|\varphi\|_0^2 < \infty,\ \mathbf{E} \int_{[0,T]} \sum_\alpha \|f^\alpha(t)\|_0^2\, dt < \infty.$$

Definition 8. *A function* $u\in\mathbf{L}_2^\omega([0,T];\mathcal{P};\mathring{\mathbf{H}}^m)$ *is called a generalized solution of problem* (4.1), (4.2), *if for every* $y\in\mathbf{C}_0^\infty(G)$, *the following equality holds true* $l\times\mathbf{P}$-*a.s.*):

$$(y,u(t))_0 = (y,\varphi)_0 - \int_{[0,t]} \sum_{|\alpha|\leq m} \Big(y_\alpha, \sum_{|\alpha|\leq m} a^{\alpha\beta} u_\beta(s) + f^\alpha(s)\Big)_0 ds +$$

(4.3)

$$+\sum_{l=1}^{d_1} \int_{[0,t]} (y, \sum_{|\beta|\leq m} b^{l\beta} u_\beta(s))_0 dw^l(s) + (y,\ M(t))_0.\quad \square$$

Note that the integrals in (4.3) are well-defined, since in view of item (i) of the Proposition 6, $\sum_{|\beta|\leq m} a^{\alpha\beta} u_\beta(s)$ and $\sum_{|\beta|\leq m} b^{l\beta} u_\beta(s)$ belongs to $\mathbf{L}_2^\omega([0,T];$ $\mathcal{P};\mathbf{L}_2(G))$.

Remark 8. *Equality* (4.3) *can be obtained formally by multiplying* (4.1) *by* y *and integrating by parts the equality we have arrived at.*

If we manage to prove that the generalized solution of the problem (4.1), (4.2) *belongs to* $\mathbf{L}_p([0,T];\ \overline{\mathcal{B}([0,T])};\ l;\ \mathbf{W}_p^{2m+1}(\mathbf{R}^d))$ *for a sufficiently large* p, *then it can be shown that under some additional assumptions concerning the "smoothness" of the boundary of the domain, the solution has a version taking values in the space* $\mathbf{C}^{2m}(G\cup\Gamma)$, *being continuous in* (t,x) ($\mathbf{P}$-*a.s.*) *in this space and such that all its derivatives up to the order of* m-1 *are equal to zero on the boundary. Such a version satisfies equalities* (4.1), (4.2) *for all* t,x, ($\mathbf{P}$-*a.s.*), *and can therefore be called the classical solution of the Dirichlet's problem* (4.1), (4.2).

It is not the author's aid to consider in this section the problem when the generalized solution possesses the classical version, but in the chapter to follow a similar question concerning the Cauchy problem for the equation of the type (4.2) *will be treated in detail.* □

Let us continue our consideration of the notion of the generalized solution.

Note that for every y, $v\in\overset{\circ}{\mathbf{H}}{}^m$

$$\left|\sum_{|\alpha|\leq m}(y_\alpha,\sum_{|\beta|\leq m}a^{\alpha\beta}(t,\cdot,\omega)v_\beta)_0\right|\leq \tag{4.4}$$

$$\leq N\sum_{|\alpha|\leq m}\|y_\alpha\|_0\sum_{|\beta|\leq m}\|v_\beta\|\leq N_1\|y\|_m\,\|v\|_m,$$

where N_1 does not depend on (t,ω).

Thus for every fixed $v\in\overset{\circ}{\mathbf{H}}{}^m$, for every $(t,\omega)\in[0,T]\times\Omega$, there exists a linear operator $A(t,\omega)$: $\overset{\circ}{\mathbf{H}}{}^m\to\overset{\circ}{\mathbf{H}}{}^{-m}$ such that

$$-\sum_{|\alpha|\leq m}(y_\alpha,\sum_{|\beta|\leq m}a^{\alpha\beta}(t,\omega)v_\beta)_0=[y,\mathbf{A}(t,\omega)v].^* \tag{4.5}$$

From (4.4) and (4.5) in view of the isometry of $\overset{\circ}{\mathbf{H}}{}^{-m}$ and $(\overset{\circ}{\mathbf{H}}{}^m)^*$ it follows that for all t,ω,

---

*Here and below in this section we denote by $[\cdot,\cdot]$ the CBF of the normal triple $(\overset{\circ}{\mathbf{H}}{}^m,\ \mathbf{L}_2(G),\ \overset{\circ}{\mathbf{H}}{}^{-m})$ (see §5).

(4.6) $$\|\mathbb{A}(t,\omega)v\|_{-m} \le N\|v\|_m, \quad \forall v\in\overset{\circ}{\mathbb{H}}{}^m,$$

where the constant N does not depend on (t,ω).

By the Pettis theorem (§1.2.8) and (4.5) we obtain that for every $v\in\overset{\circ}{\mathbb{H}}{}^m$, $A(t,\omega)v$ is a predictable $\overset{\circ}{\mathbb{H}}{}^{-m}$-process.

Let us also denote

(4.7) $$B(t,\omega)v := \Big(\sum_{|\beta|\le m} b^{1\beta}(t,\omega)v_\beta, \dots, \sum_{|\beta|\le m} b^{d_1\beta}(t,\omega)v_\beta\Big).$$

It is clear that for all (t,ω), $B(t,\omega)$: $\overset{\circ}{\mathbb{H}}{}^{-m} \to \mathfrak{L}_2(\mathbb{R}^{d_1}, \mathbb{L}_2(G))$ and there exists a constant N that does not depend on t,ω such that

$$\|B(t,\omega)v\|_{\mathfrak{L}_2(\mathbb{R}^{d_1}, \mathbb{L}_2(G))} \le N\|v\|_m, \quad \forall v\in\overset{\circ}{\mathbb{H}}{}^m.$$

Evidently, for every $v\in\overset{\circ}{\mathbb{H}}{}^m$, $B(t,\omega)v$ is a predictable process in $\mathbb{L}_2(\mathbb{R}^{d_1}, \mathbb{L}_2(G))$.

At last define $f(t)$ by the equality

(4.8) $$f(t) := \sum_{|\alpha|\le m} f_\alpha^\alpha(t,x,\omega).$$

From this definition and Proposition 6 (ii) we find that f(t) is a predictable $\overset{\circ}{\mathbb{H}}{}^{-m}$-process.

Thus a generalized solution of problem (4.1), (4.2) (provided this solution does exist) is a solution of the LSES (0.1) (with the operators A, B and the function f defined by equalities (4.5), (4.7) and (4.8)) in the normal triple ($\overset{\circ}{\mathbb{H}}{}^m$, $\mathbb{L}_2(g)$, $\overset{\circ}{\mathbb{H}}{}^{-m}$) and vice versa. This yields in particular that a generalized solution of the problem (4.1) (4.2) has a version from $\mathbb{C}([0,T];\mathcal{P};\mathbb{L}_2(G))$, (see Proposition 1.3). So, from now on we shall identify a generalized solution of problem (4.1), (4.2) with its version from $\mathbb{C}([0,T],\mathcal{P};\mathbb{L}_2(G))$.

Theorem 8. Let the following condition be valid:

(A) There exist constants $K\in\mathbb{R}_+$ *and* $\delta>0$ *such that for every collection of real numbers* $\{\xi^\alpha, |\alpha|\le m\}$ *for all* $(t,x,\omega) \in [0,T]\times G\times\Omega$,

$$2\sum_{\substack{|\alpha|\le m\\|\beta|\le m}} a^{\alpha\beta}(t,x,\omega)\,\xi^\alpha\xi^\beta - \sum_{l=1}^{d_1}\left|\sum_{|\beta|\le m} b^{\ell\beta}(t,x,\omega)\xi^\beta\right|^2 \ge$$

$$\ge \delta \sum_{|\alpha|\le m} |\xi^\alpha|^2 - K|\xi^0|^2.$$

Then the problem (4.1) (4.2) *has a unique generalized solution, and there exists a constant N such that*

$$\mathbb{E}\sup_{t\le T}\|u(t)\|_0^2 + \mathbb{E}\int_{[0,T]}\|u(s)\|_m^2\,ds \le \tag{4.9}$$

$$\le N\,\mathbb{E}\Big(\|\varphi\|_0^2 + \int_{[0,T]}\|f(s)\|_{-m}^2\,ds + \|M(t)\|_0^2\Big). \quad \square$$

Proof. We now show that the LSES in $(\mathring{\mathbb{H}}^m, \mathbb{L}_2(G), \mathring{\mathbb{H}}^{-m})$ which is equivalent to the problem (4.1), (4.2), satisfies the conditions of Theorem 1.4.

Conditions (B) and (C) of this theorem are fulfilled in view of (4.6) and the assumptions made in the beginning of this paragraph. Thus it suffices to prove that this LSES is coercive.

Let $v\in\mathring{\mathbb{H}}^m$, then by (4.5) and (4.7), we find that for all t,ω,

$$2[v,Av] + \|Bv\|^2_{\mathfrak{L}_2(\mathbb{R}^{d_1},\mathbb{L}_2(G))} = -2\sum_{\substack{|\alpha|\le m\\|\beta|\le m}}(v_\alpha, a^{\alpha\beta}v_\beta)_0 +$$

$$+\sum_{l=1}^{d_1}\Big\|\sum_{|\beta|\le m} b^{l\beta}v_\beta\Big\|_0^2 = \int_G\Big(-2\sum_{\substack{|\alpha|\le m\\|\beta|\le m}} a^{\alpha\beta}v_\alpha v_\beta(x) +$$

$$+\sum_{l=1}^{d_1}\Big|\sum_{|\beta|\le m} b^{l\beta}v_\beta(x)\Big|^2\Big)dx.$$

In view of condition (A) of the theorem the latter expression does not exceed

$$-\delta \sum_{|\alpha|=m} \|v_\alpha\|_0^2 + K\|v\|_0^2,$$

which, in its turn, by the Fridrichs lemma (§5) is majorized by the expression $-\delta\, N\|v\|_m^2 + K\|v\|_C^2$. Thus the validity of condition (A) (coercivity) of Theorem 1.4. is proved, which completes the proof of the theorem. □

Remark 8. *It can be readily seen that the conditions*

$$\mathbb{E}\|\varphi\|_0^2 < \infty \quad and \quad \mathbb{E} \int_{[0,T]} \sum_{\alpha} \|f^\alpha(t)\|_0^2\, dt < \infty,$$

are superfluous for the uniqueness of the problem (4.1), (4.2), *since the difference of two generalized solutions is also a generalized solution of the same problem with* $\varphi \equiv f^\alpha \equiv 0$. □

The theorem of this item and Theorem 3.3.3 yield the following result.

Corollary 8. *A generalized solution of the problem* (4.1), (4.2) *possesses the Markov property, that is, for every Borel set* $A \in \mathbb{L}_2(G)$ *and every* s, $t \in [0,T]$ *such that* $s \leq t$ *the following equality holds* ($\mathbb{P}$-*a.s.*)

$$\mathbb{P}(u(t) \in \mathcal{A} | \mathfrak{F}_s(u)) = \mathbb{P}(u(t) \in \mathcal{A} | u(s)),$$

where $\mathfrak{F}_s(u)$ *is the* σ-*algebra generated by* $u(\tau, x, \omega)$ *for* $\tau \leq s$. □

4.9. *Remark* 9. *All the results of this chapter are easily carried over to the case of complex valued operators, initial conditions and external forces. In this case, we have to replace in the coercivity and dissipativity for the CBF, by condition for its real part.*

Chapter 4

ITO'S SECOND ORDER PARABOLIC EQUATIONS

4.0. Introduction.

0.1. Let us fix T_0, $T\in\mathbb{R}_+$ with $T_0 \leq T$, and d, $d_1\in\mathbb{N}$. Suppose that a standard probability space $\mathbb{F} = (\Omega, \mathcal{F}, \{\mathcal{F}_t\}_{t\in[0,T]}, \mathbb{P})$ and a standard Wiener process $w(t)$ in $\mathbb{R}^{d_1}$ on this probability space are given.

Warning 1. *Throughout what follows:*

(*a*) *upper dices* i,i',j,l,k *denote the numbers of coordinates; the same indices put below denote differentiation with respect to the coordinates with corresponding numbers (e.g.* $g_{ij} = \frac{\partial^2}{\partial x^i \partial x^j} g(x)$);

(*b*) *repeated indices* i,i',j,l,k *in monomials are summed over (e.g.* $\int_{[0,t]} h^l(s)dw^l(s):= \sum_{l=1}^{d_1} \int_{[0,t]} h^l(s)dw^l(s)$);

(c) *the notations of section* 3.4 *are still in force;*

(*d*) *in the designations of Sobolev spaces* $\mathbf{W}_p^m(\mathbf{R}^d)$, $\mathbf{H}^p(r,\mathbf{R}^d)$ *and* $\mathbf{W}_p^m(r,\mathbf{R}^d)$, $\mathbf{L}_p(r,\mathbf{R}^d)$ *the argument will be omitted.*

0.2. The subject of this chapter is the equation

$$du(t,x,\omega) = [a^{ij}(t,x,\omega)a^{ij}(t,x,\omega) + b^i(t,x,\omega)u_i(t,x,\omega) +$$

$$+ c(t,x,\omega)u(t,x,\omega) + f(t,x,\omega)]dt + [\alpha^{ij}(t,x,\omega)u_i(t,x,\omega) +$$

$$+ h^l(t,x,\omega)u(t,x,\omega) + g^l(t,x,\omega)]\, dw^l(t),$$

$$(t,x,\omega)\in]\, T_0,T] \times \mathbf{R}^d\times\Omega.$$

We investigate the problem of the solvability of forward and backward Cauchy problems in generalized and classical settings, that is, in Sobolev spaces and in the space of continuously differentiable functions, respectively.

From the formal point of view it is natural to consider a slightly more general equations then (0.1), namely, the equation containing operators of the second order in the dw term. However, the author is unaware of any "physical" model involving equations of this type.

Definition 2. *An equation of the type* (0.1) *will be called parabolic* (*superparabolic, respectively*) *if for all* $(t,x,\omega) \in [T_0,T]\times\Omega\times\mathbf{R}^d$ *the following inequality holds*

(0.2) $$2a^{ij}(t,x,\omega)\,\xi^i\xi^j - \sum_{i=1}^{d_1}|\alpha^{ij}(t,x,\omega)\xi^i|^2 \geq 0, \quad \forall\xi\in\mathbf{R}^d$$

(0.3) $$(2a^{ij}(t,x,\omega)\xi^i\xi^j - \sum_{i=1}^{d_1}|\alpha^{ij}(t,x,\omega)\xi^i|^2 \geq$$
$$\geq \delta\sum_{i=1}^{d}|\xi^i|^2, \quad \forall\xi\in\mathbf{R}^d,$$

where $\delta > 0$ *and does not depend on* t,ω,x,ξ). □

Evidently, the parabolic condition holds if and only if the matrix

$$(A^{ij}) := 2a^{ij} - \sigma^{il}\sigma^{il},$$

is non-negative definite, and the superparabolic condition is valid iff this matrix is uniformly positive definite.

When $\sigma^{il} \equiv 0$, conditions (0.2), (0.3) turn into the usual conditions of the theory of partial differential equations (see [104], [4.1]). For example, if $\sigma^{il} \equiv 0$ and some additional minor assumptions are satisfied, condition (0.3) is equivalent to the uniform ellipticity of the operator $\mathcal{L}(\cdot) := (a^{ij}(\cdot)_i)_j + b^i\times(\cdot)_i + c\times(\cdot)$.

Note that if the coefficients of equation (0.1) are uniformly bounded, then the superparabolic condition (0.3) is a special case of condition (A) of Theorem 3.4.8 for m=1.

Considering the Cauchy problem for equation (0.1) as a LSES in some normal triple of Sobolev spaces, we shall prove below that the superparabolic condition ensures that the LSES is coercive and the parabolic condition guarantees its dissipativity.

Getting ahead of the story, we may say that the "behavior" of a superparabolic Ito's equation is very much like the "behavior" of a deterministic parabolic equation, and there is a great resemblance between properties of a parabolic Ito's equation and a deterministic degenerate parabolic equation (elliptic-parabolic equation).

It is important to understand to what extent the parabolic condition is a restrictive one for an Ito's partial differential equation. The following example shows that this condition cannot be omitted if we are looking for a solution of equation (0.1) in the class of square integrable functions.

Example. Consider the following special case of equation (0.1)

$$u(t,x) = \varphi(x) + \int_{[0,t]} \alpha^2 \frac{\partial^2}{\partial x^2} u(s,x)\,ds + \int_{[0,t]} \sigma \frac{\partial}{\partial x} u(s,x)\,dw(s), \tag{0.4}$$

where $\varphi \in \mathbb{L}_2$.

The function $u(t,x,\omega) \in \mathbb{L}_2\ ([0,T], \overline{\mathfrak{B}([0,T])}, \ell;\ \mathbf{W}_2^2)$ will be said to be a solution of equation (0.4) if it is continuous in t, continuously differentiable (in a classical sense) up to the second order and satisfies equation (0.4) for all $t \in [0,T]$, ($\mathbf{P}$-*a.s.*).

The Fourier transforms of the functions $\varphi(x)$ and $u(t,x)$ will be denoted by $\hat{\varphi}(y)$ and $\hat{u}(t,x)$, respectively.

Passing to Fourier transforms, we obtain from (0.4),

$$\hat{u}(t,y) = \hat{\varphi}(y) - \int_{[0,t]} (y\alpha)^2 \hat{u}(s,y)\,ds + i \int_{[0,t]} (\sigma y)\hat{u}(s,y)\,dw(s).$$

This equation has the unique solution of the form

$$\hat{\varphi}\ exp\{-\tfrac{1}{2}(2a^2-\sigma^2)y^2 t + i\sigma y w(t)\}.$$

(*see* [36], *Ch.* 4, §1). Thus by Parseval's equality,

$$\|u(t)\|^2_{\mathbb{L}_2} = \int_{\mathbf{R}^1} |\hat{\varphi}(y)|^2\ exp\{-(2a^2-\sigma^2)y^2 t\}\,dy.$$

If the parabolic condition does not hold, i.e. $2a^2-\sigma^2<0$, then the integral on the right-hand side of the latter equality converges only under very special assumptions on φ.

If we drop the requirement of the classical solvability of equation (0.4) and consider it as a LSES in $(\mathbb{H}^1,\mathbb{L}_2,\mathbb{H}^1)$ in the spirit of definition 3.4.8., then the result remains the same: if the parabolic condition does not hold a solution (to be more specific, a generalized solution) exists only for very special initial condition

φ (*see*[70], *Ch. III*, §3).

0.3. Before proceeding to the main part of the chapter we introduce some notation which will be used extensively in what follows.

Let $\mathbf{X}$ be a separable Banach space and p be a number from $[1,\infty[$.

Denote

$$\mathbb{L}_p([T_0,T];\mathbf{X}) := \mathbb{L}_p([T_0,T];\ \overline{\mathfrak{B}([T_0,T])};\ l;\ \mathbf{X});$$

$$\mathbb{L}_p([T_0,T]\times\Omega;\ \mathbf{X}) := \mathbb{L}_p([T_0,T]\times\Omega;\ \overline{\mathfrak{B}([T_0,T])\otimes\mathfrak{F}};\ l\times\mathbf{P},\mathbf{X});$$

$$\mathbb{L}_p(\Omega;\tilde{\mathfrak{F}};\mathbf{X}) := \mathbb{L}_p(\Omega;\tilde{\mathfrak{F}};\tilde{\mathbf{P}};\mathbf{X}),$$

where $\tilde{\mathfrak{F}}$ is a sub-σ-algebra of $\mathfrak{F}$ and $\tilde{\mathbf{P}}$ is the restriction of $\mathbf{P}$ on $\mathfrak{F}$,

$$\mathbb{L}_p(\Omega;\mathbf{X}) := \mathbb{L}_p(\Omega;\mathfrak{F};\mathbf{P};\mathbf{X});$$

$$\mathbb{L}_p(\Omega;\mathbb{C}(\{T_0,T];\mathbf{X})) := \mathbb{L}_p(\Omega;\mathfrak{F};\mathbf{P};\mathbb{C}([T_0,T],\mathbf{X}))$$

$$\mathbb{L}_p([T_0,T]\times\Omega;\ \mathbb{C}_w\mathbf{X})$$

is the set of all representatives of $\mathbb{L}_p([T_0,T])\times\Omega;\ \mathbf{X})$ weakly continuous in t, $\mathbf{P}$-a.s.) as a function in $\mathbf{X}$;

$$\mathbb{L}_p^{\omega}([T_0,T]);\mathbf{X})$$

is the set of all $\overline{\mathfrak{B}([T_0,T])\otimes\mathfrak{F}}$ measurable mappings f: $[T_0,T]\times\Omega \to \mathbf{X}$ such that $f(\cdot,\omega)\in L_p([T_0,T];\ \mathbf{X})$, ($\mathbf{P}$-*a.s.*);

$$\mathbb{L}_p([T_0,T];\mathfrak{P};\mathbf{X})$$

is the set of all predictable representatives of $\mathbb{L}_p(T_0,T]\times\Omega;\mathbf{X})$;

$$\mathbb{L}_p^{\omega}([T_0,T];\mathfrak{P};\mathbf{X})$$

is the set of all predictable representatives of $\mathbb{L}_p^{\omega}([T_0,T];\mathbf{X})$;

$$\mathbb{C}([T_0,T];\mathfrak{P};\mathbf{X})$$

is the set of all $\mathbf{X}$-processes strongly continuous from $[T_0,T]$ into $\mathbf{X}$.

Warning 3. *Similar notation will be in force for backward predictable functions. For such functions the symbol* $\mathfrak{P}$ *in the notation will be replaced by* $\overleftarrow{\mathfrak{P}}$.

4.1. The Cauchy Problem for Superparabolic Ito's Second Order Equations.

1.0. In this section we consider the Cauchy problem

$$du(t,x,\omega) = [(a^{ij}(t,x,\omega)u_i(t,x,\omega))_j + b^i(t,x,\omega)u_i(t,x,\omega) +$$

$$+ c(t,x,\omega)u(t,x,\omega) + f(t,x,\omega)]\,dt +$$

$$+ [\sigma^{ij}(t,x,\omega)u(t,x,\omega) + h^i(t,x,\omega)u(t,x,\omega) + \tag{1.1}$$

$$+ g^l[t,x,\omega)]\,dw^l(t),$$

$$(t,x,\omega)\in]\ T_0,T]\times\mathbf{R}^d\times\Omega,$$

$$u(T_0,x,\omega) = \varphi(x,\omega)\ (x,\omega)\in\mathbf{R}^d\times\Omega. \tag{1.2}$$

Equation (1.1) differs from (0.1) the term containing higher derivatives. In the case that the coefficients α^{il} are differentiable in x, equations (0.1) and (1.1) are equivalent since

$$(\alpha^{ij}u_i = \alpha^{ij}u_{ij} + \alpha^{ij}_i u_j.$$

On the other hand, provided the superparabolic condition is fulfilled, problem (1.1), (1.2) as it will be shown below is solvable without any assumptions on differentiability or even continuity of the coefficients of the equation. A similar statement for the case of problem (0.1), (1.2) is not valid. The results concerning problem (0.1), (1.2) can be obtained from the corresponding results for problem (1.1), (1.2) by reducing equation (0.1) to the form (1.1).

Throughout this section it will be supposed that the coefficients of equation

(1.1) a^{ij}, b^i, c, σ^{il}, h^l where i, j=1,2,...d, l=1,2,...d_1, are $\overline{\mathcal{B}([T_0,T]\times\mathbf{R}^d)\otimes\mathcal{F}}$-measurable, bounded, predictable (for every $x\in\mathbf{R}^d$), real functions, and φ is an $\mathcal{F}_{T_0}$-measurable function taking values in $\mathbf{L}_2$.

Apart from that it is assumed that $f\in L_2^\omega([T_0,T];\mathcal{P};\mathbf{H}^{-1})$ and $g^l\in\mathbf{L}_2^\omega([T_0,T];\mathcal{P};\mathbf{L}_2)$ for every l=1,2,...,d_1.

In the next item it will be proved that problem (1.1), (1.2) has a unique solution in the class $\mathbf{L}_2([T_0,T];\mathcal{P};\mathbf{H}^1)\cap\mathbf{L}_2(\Omega;\mathbf{C}([T_0,T];\mathbf{H}^0))$.

Below we shall investigate the dependence of analytical properties of the solution on analytical properties of the coefficients of the equation, the initial condition φ and the external forces f, g^l. In particular it will be proved that under appropriate assumptions problem (1.1), (1.2) has a solution in $\mathbb{L}_p([T_0,T]\times\Omega; \mathbf{C}_w\mathbf{W}_p^m) \cap \mathbb{L}_2([T_0,T]; \mathcal{P}; \mathbb{H}^{m+1}) \cap \mathbb{L}_2(\Omega; \mathbf{C}([T_0,T]); \mathbf{H}^m))$ and a "classic" solution.

1.1. *Definition* 1. *A function* $u\in\mathbb{L}_2^\omega([T_0,T];\mathcal{P};\mathbb{H}^1)$ *is called a generalized solution of problem* (1.1), (1.2) *if for every* $y\in\mathbf{C}_0^\infty(\mathbb{R}^d)$, *the following equality holds true* $l\times\mathbb{P}$-*a.s.*:

$$(u(\ ),y)_0 = (\varphi,y)_0 + \int_{[T_0,t]} (-(a^{ij}u_i(s),\ y_i)_0 +$$

(1.3)

$$+ (b^i u_i(s) + cu(s),y)_0 + [f(s),y]_0)ds +$$

$$+ \int_{[T_0,T]} (\sigma^{ij}u_i(s) + h^l(s)\ u(s) + g^l(s)y)_0 dw^l(s),$$

where $[\cdot,\cdot]_0$ *is the CBF of the normal triple* $(\mathbb{H}^1,\mathbb{L}_2,\mathbb{H}^{-1})$. □

Making use of the Schwarz inequality and of the first property of a CBF (§ 2.4.1), we find easily that all the integrals in (1.3) are well-defined.

Applying the same Schwarz inequality we obtain that for every $y\in\mathbf{C}_0^\infty(\mathbb{R}^n)$

$$|-(a^{ij}v_i,\ y_j)_0 + (b^i v_i + cv,\ y)_0| \le N\|v\|_1\ \|y\|_1. \tag{1.4}$$

where the constant N does not depend on (t,ω).

Hence for every $v\in\mathbb{H}^1$, the expression in the left-hand side of inequality (1.4) is a linear continuous functional on $\mathbb{H}^1$. Thus since the CBF $[\cdot,\cdot]_0$ brings $(\mathbb{H}^1)^*$ and $\mathbb{H}^1$ into isometric isomorphism (see Proposition 3.2.1), there exists a linear operator $A(t,\omega)$: $\mathbb{H}^1 \to \mathbb{H}^{-1}$ such that

$$-(a^{ij}(t,\omega)v_i,y_j)_0 + (b^i(t,\omega)v_i + c(t,\omega)v,y)_0 = [y,A(t,\omega)v]_0. \tag{1.5}$$

From (1.4) and (1.5) in view of the isometry of $(\mathbb{H}^1)^*$ and $\mathbb{H}^{-1}$ mentioned above, it follows that

$$\|A(t,\omega)v\|_{-1} \le N\|v\|_1,\quad \forall v\in\mathbb{H}^1.$$

Define also $\mathcal{M}^l(t,\omega)v := \sigma^{il}(t,\omega)v_i + h^l(t,\omega)v$, and

$$B(t,\omega)v := (\mathcal{M}^1(t,\omega)v,\ldots,\mathcal{M}^{d_1}(t,\omega)v). \tag{1.6}$$

Clearly, $B(t,\omega)$: $\mathbb{H}^1 \to \mathfrak{L}_2(\mathbb{R}^{d_1},\mathbb{L}_2)$ for every (t,ω). Making use of the Pettis theorem it is easy to verify that $A(t,\omega)v$ is a predictable $\mathbb{H}^{-1}$-process and $\mathbb{B}(t,\omega)v$ is a predictable $\mathfrak{L}_2(\mathbb{R}^{d_1},\mathbb{L}_2)$-process.

Thus problem (1.1), (1.2) can be considered as a LSES of the type (3.0.1) with the operators A and B defined by the equality (1.5), (1.6) and

$$\mathbb{M}(t) := \sum_{l=1}^{d_1} \int_{[T_0,t]} g^l(s)\, dw^l(s).$$

As can be easily seen, the notion of a generalized solution of problem (1.1), (1.2) is equivalent to the notion of a solution of the LSES just mentioned above, given by the Definition 3.1.2 (see also Remark 3.3.3). Hence Proposition 3.1.3 yields that a generalized solution u of problem (1.1), (1.2) has a version from $\mathbb{C}([T_0,T];\mathcal{P};\mathbb{L}_2)$. From now on this version will be identified with u.

Remark 1. *From the arguments presented above it follows that the generalized solution u of problem* (1.1), (1.2) *satisfies equality* (1.3) *for all* $t\in[T_0,T]$ *and* $w\in\Omega'\subset\Omega$, *where* $\mathbb{P}(\Omega')=1$. □

Theorem 1. *Let the superparabolic condition* (0.3) *be fulfilled; then a generalized solution of problem* (1.1), (1.2) (*provided it exists*) *is unique.*

If, in addition, $\varphi\in\mathbb{L}_2(\Omega;\mathbb{L}_2)$, $f\in\mathbb{L}_2([T_0,T]\times\Omega;\ \mathbb{H}^{-1})$, *and* $g^l\in\mathbb{L}_2(T_0,T]\times\Omega;\mathbb{L}_2)$ *for every* $l=1,2,\ldots,d_1$, *then the problem has a generalized solution u from the class* $\mathbb{L}_2([T_0,T];\mathcal{P};\mathbb{H}^1) \cap \mathbb{C}([T_0,T];\mathcal{P};\mathbb{L}_2)$, *and there exists a constant* $N\in\mathbb{R}_+$ *depending only on* T_0,T,δ,d,d_1, *and* $\max_{\ell,i,j,t,x,\omega} (|a^{ij}|, |b^i|, |\sigma^{il}|, h^l|)$ *such that*

$$\mathbb{E} \sup_{t\in[T_0,T]} \|u(t)\|_0^2 + \mathbb{E} \int_{[T_0,T]} \|u(s)\|_1^2 ds \le$$

$$\le N\mathbb{E}\left(\|\varphi\|_0^2 + \int_{[T_0,T]} \left(\|f(t)\|_{-1}^2 + \sum_{l=1}^{d_1} \|g^l(t)\|_0^2 \right) dt \right). \ \square$$

Proof. To begin with, we show that the coercivity condition the normal triple $(\mathbb{H}^1,\mathbb{L}_2,\mathbb{H}^{-1})$ (condition A of §3.0.1) is fulfilled for the operators A,B defined above. This will be sufficient to prove the theorem because it justifies application of theorems 3.1.4 and 3.3.1 to problem (1.1), (1.2).

Clearly, for every $v\in\mathbb{H}^1$ and all t,ω

$$2[v, \mathbb{A}v]_n + \|Bv\|^2_{\mathfrak{L}_2(\mathbb{R}^d,\mathbb{L}_2)} = -2(a^{ij}v_{i},v_{j})_0 +$$

$$+ \sum_{l=1}^{d_1} \|\sigma^{il}v_{i}\|_0^2 + 2(b^iv_i + cv,v)_0 +$$

$$+ 2\sum_{l=1}^{d_1} (\sigma^{il}v_{i}, h^lv)_0 + \sum_{l=1}^{d_1} \|h^lv\|_0^2.$$

From the superparabolic condition (0.3) it follows that

$$- 2(a^{ij}v_{i}, v_{j})_0 + \sum_{l=1}^{d_1} \|\sigma^{il}v_{i}\|_0^2 =$$

$$(1.7) \qquad = \int_{\mathbb{R}^d}\left[- 2a^{ij}v_{i}v_{j}(x) + \sum_{l=1}^{d_1}|\sigma^{il}v_{i}(x)|^2\right]dx \le$$

$$\le - \delta \int_{\mathbb{R}^d} \sum_{i=1}^{d} |v_{i}(x)|^2 dx \le - \delta\|v\|^2_{1,2} + \delta\|v\|_0^2.$$

On the other hand, with the help of the Schwartz inequality we find that

$$U:= \Bigg|2(b^iv_i + cv,v)_0 + 2\sum_{l=1}^{d_1} (\sigma^{il}v_{i}, h^lv)_0 +$$

$$+ \sum_{l=1}^{d_1} \|h^lv\|_0^2\Bigg| \le N(\|v\|_0^2 + \sum_{i=1}^{d}\|v_i\|_0\, \|v\|_0.$$

From Proposition 3.4.6 it follows that

$$\|v_i\|_0 \leq N\|v\|_{1,2}.$$

Thus making use of the inequality 3.1.10 it is easy to verify that the following inequality holds:

$$U \leq 2^{-1} \varepsilon N\|v\|_{1,2}^2 + N(\varepsilon^{-1}) \|v\|_0^2, \tag{1.8}$$

where the constant $N(\varepsilon^{-1})$ depends on ε.

Choosing ε to be sufficiently small, we obtain after combining (1.7) and (1.8) that

$$2[v, Av]_0 + \|Bv\|^2_{\mathfrak{L}_2(\mathbb{R}^{d_1},\mathbb{L}_2)} \leq - \delta'\|v\|_{1,2}^2 + N\|v\|_0^2.$$

Due to equivalence of the norms $\|\cdot\|_{1,2}$ and $\|\cdot\|$, the latter inequality proves the coercivity of the LSES which is equivalent to problem (1.1), (1.2) in the normal triple $(\mathbb{H}^1,\mathbb{L}_2,\mathbb{H}^{-1})$.

1.2. In this item it is assumed that the number $m\in\mathbb{N}\cup\{0\}$.

Theorem 2. Suppose that the superparabolic condition (0.3) *holds true as well as the following:*

(*i*) *The functions* a^{ij}, b^i, c, σ^{il}, $h^l(i_1, j=1,2,\dots,d, l=1,2,\dots,d_1)$ *are differentiable in* x *up to the order of* m *for all* t, x, *and* ω, *and together with the derivatives mentioned above are uniformly bounded (with respect to* t, x, *and* ω) *by a constant* K.

(*ii*) $\varphi\in\mathbb{L}_2(\Omega;H^m)$, $f\in\mathbb{L}_2([T_0,T]\times\Omega;\mathbb{H}^{m-1})$, $g^l\in\mathbb{L}_2([T_0,T])\times\Omega;\mathbb{H}^m$, *where* $l=1,2,..,d_1$.

Then the generalized solution u *of problem* (1.1), (1.2) *belongs to the class* $\mathbb{L}_2([T_0,T],\mathcal{P};\mathbb{H}^{m+1}) \cap C([T_0,T];\mathcal{P};\mathbb{H}^m)$ *and satisfies equality* (1.3) *for all* $t\in[T_0,T]$ *and all* ω *from some subset of* Ω *of probability* 1.

There exists $N\in\mathbb{R}_+$ *depending only on* K, d, d_1, m, T_0, *and* T *such that*

$$\mathbb{E} \sup_{t\in[T_0,T]} \|u(t)\|_m^2 + \mathbb{E} \int_{[T_0,T]} \|u(t)\|_{m+1}^2 \, dt \leq N \,\mathbb{E}\Big(\|\varphi\|_m^2 +$$

$$+ \int_{[T_0,T]} \Big(\|f(t)\|_{m-1}^2 + \sum_{l=1}^{d_1} \|g^l(t)\|_m^2\Big)dt\Big). \quad \square$$

Proof. Since $\mathbb{H}^{m+1} \subset \mathbb{H}^1$, $\mathbb{H}^m \subset \mathbb{H}^0 = \mathbb{L}_2$, *and* $\mathbb{H}^{m-1} \subset \mathbb{H}^{-1}$, and all the imbeddings are normal, the hypotheses of Theorem 1 are fulfilled and thus problem (1.1), (1.2) has a unique generalized solution belonging to $\mathbb{L}_2([T_0,T];\mathcal{P};\mathbb{H}^1) \cap \mathbb{L}_2(\Omega; \mathbb{C}([T_0,T],\mathbb{L}_2))$. Moreover, as was shown in the previous item, this is the unique solution in the normal triple $(\mathbb{H}^1,\mathbb{L}_2,\mathbb{H}^{-1})$ of the LSES (3.0.1) with the operators A and B defined by equalities (1.5), (1.6), and

$$\mathbb{M}(t):= \sum_{i=1}^{d_1} \int_{[T_0,t]} g^l(s)\, dw^l(s).$$

We now show that Theorem 3.1.8, where $\mathbb{V}: = \mathbb{H}^{m+1}$, $\mathbb{U}:= \mathbb{H}^m$ and $\mathbb{V}:= \mathbb{H}^{m+1}$, is applicable to this LSES. For the case of $m=0$ it was shown in the previous theorem. So we can confine ourselves to the case of $m\geq 1$.

If $v\in\mathbb{H}^{m+1}$, $y\in\mathbb{H}^m$, then via integration by parts (see §3.4.6), it is easy to obtain from (1.5) that $[y, Av]_0 = (y, \mathcal{L}_{div} v)_0$, where $\mathcal{L}_{div}v:= (a^{ij} v_i)_j + f^i v_i + cv$.

Thus, the restriction of $A(t,\omega)$ to $\mathbb{H}^{m+1}$ coincides with $\mathcal{L}_{div}(t,\omega)$ and the following inequality holds:

$$\|A(t,\omega)v\|_{m-1} \leq N\|v\|_{m+1}, \quad \forall v\in\mathbb{H}^{m+1}. \tag{1.9}$$

From the same Proposition 3.4.6 (i), it follows from $B(t,\omega)$ is a uniformly

continuous operator from $\mathbb{H}^{m+1}$ into $\mathfrak{L}_2(\mathbb{R}^{d_1},\mathbb{H}^m)$.

We show that condition (A_1) of Theorem 3.1.8 is fulfilled. Let $v\in C_0^\infty(\mathbb{R}^d)$ and $[\cdot,\cdot]_m$ be a CBF of the normal triple $\mathbb{H}^{m+1}$, $\mathbb{H}^m$, $\mathbb{H}^{m-1}$). From the second property of CBF (see §2.4.1), it follows that $[v, Av]_m = (v, \mathcal{L}_{div}v)_m$. On the other hand, in view of (3.2.2) and self-adjointness of the operator $\Lambda = (\mathbb{I}-\Delta)^{1/2}$, we have that $(v, \mathcal{L}_{div} v)_m = (\mathbb{I}-\Delta)^m v, \mathcal{L}_{div} v)$

Clearly,

$$(\mathbb{I}-\Delta)^m v = \sum_{|\alpha|\leq m} (-1)^{|\alpha|} C^\alpha v_{2\alpha}, \tag{1.10}$$

where $\alpha = (\alpha^1, \quad ,\alpha^d)$ is a multiindex, $v_{2\alpha}:= \partial^{2|\alpha|}/\partial(x^1)^{2\alpha^1} ...\partial(x^d)^{2\alpha^d}$ and C^α is a positive constant.

Thus

$$2[v, Av]_m = -\sum_{|\alpha|=m} C^{\alpha}(a^{ij}(v_i)_{\alpha}, (v_j)_{\alpha})_0$$

$$+\sum_{|\alpha|=m} C^{\alpha}(b^i(v_i)_{\alpha}, v_{\alpha})_0 - \tag{1.11}$$

$$-\sum\nolimits^{\alpha} C^{\alpha}((a^{ij})_{\beta}(v_i)_{\gamma}, (v_i)_{\alpha})_0 + U_1(v).$$

where $\sum^{\alpha}$ denotes summation over all $\alpha = \beta+\gamma$ such that $|\alpha| = m$, $|\beta| \geq 1$ and $|U_1(v)| \leq N\|v\|_m^2$ ($U(v)$ is a finite sum of scalar products in $\mathbf{L}_2$ which includes derivatives of v up to the m-th order).

Obviously,

$$\|Bv\|^2_{\mathfrak{L}_2(\mathbb{R}^{d_1},\mathbb{H}^m)} = \sum_{l=1}^{d_1} \|\mathcal{M}^l v\|_m^2 = \sum_{l=1}^{d_1} \|\sigma^{ij} v_i\|_m^2 +$$

$$+ 2\sum_{l=1}^{d_1} (\sigma^{ij} v_i, h^l v)_m + \sum_{l=1}^{d_1} \|h^l v\|_m^2. \tag{1.12}$$

From relation (1.10) it follows that

$$\sum_{l=1}^{d_1} \|\sigma^{ij} v_i\|_m^2 = \sum_{l=1}^{d_1} ((\mathrm{I}-\Delta)^m \sigma^{il} v_i, \sigma^{il} v_i)_0$$

$$= \sum_{l=1}^{d_1} \sum_{|\alpha|=m} C^{\alpha}(\sigma^{ij}(v_i)_{\alpha}, \sigma^{ij}(v_i)_{\alpha})_0 + \tag{1.13}$$

$$+ 2\sum_{l=1}^{d_1} \sum\nolimits^{\alpha} C^{\alpha}(\sigma^{il}(v_i)_{\alpha}, \sigma^{il}(v_i)_{\alpha}) + U_2(v),$$

where $|U_2(v)| \leq N\|v\|_m^2$.

In the same way, it is easy to see that

$$(1.14)\qquad 2\sum_{l=1}^{d_1}(\sigma^{il}v_i,\ h^l v)_m = \sum_{l=1}^{d_1}\sum_{|\alpha|=m}(\sigma^{il}(v_i)_\alpha,\ h^l v_\alpha)_0 + U_3(v),$$

where $|U_3(v)| < N\|v\|_m^2$.

Hence from the expressions (1.11) to (1.14) and (0.3) we obtain,

$$(1.15)\qquad \begin{aligned} &2[v,\ Av]_m + \|Bv\|^2_{\mathfrak{L}_2(\mathbb{R}^d,\mathbb{H}^m)} = \\ &= -\delta\sum_{|\alpha|=m} C^\alpha \sum_{i=1}^{d_1}\|(v_i)_\alpha\|_0^2 + \mathbf{V}(v) + U(v), \end{aligned}$$

where $|U(v)| < N\|v\|_m^2$, and

$$V(v) = \sum_{|\alpha|=m} C^\alpha(b^i(v_i)_\alpha,\ v_\alpha)_0 - \Sigma^\alpha\ C^\alpha((a^{ij})_\beta\ (v_i)_\gamma,\ (v_i)_\alpha)_0 +$$

$$+\ 2\sum_{l=1}^{d_1}\ \Sigma^\alpha\ C^\alpha(\sigma^{ij}_\beta(v_i)_\gamma,\ \sigma^{il}(v_i)_\alpha)_0 +$$

$$+\ 2\sum_{l=1}^{d_1}\sum_{|\alpha|=m}(\sigma^{il}(v_i)_\alpha,\ h^l v_\alpha)_0.$$

Clearly, there exist $\delta'>0$ and $N\in\mathbf{R}_+$ such that

$$(1.16)\qquad -\delta\sum_{|\alpha|=m} C^\alpha\sum_{i=1}^{d}\|(v_i)_\alpha\|_0^2 \le -\delta'\|v\|_{m+1} + N\|v\|_m^2$$

On the other hand, in view of the Schwarz inequality and (3.1.10)

$$(1.17)\qquad |V(v)| \le N\|v\|_{m+1}\ \|v\|_m \le \varepsilon 2^{-1}N\|v\|^2_{m+1} + \varepsilon^{-1}2^{-2}N\|v\|_m^2.$$

Taking ε to be sufficiently small, we obtain from (1.15) - (1.17) that there exist constants $\sigma''\geq 0$ and $K\in\mathbf{R}_+$ such that for all (t,ω) and every $v\in C_0^\infty(\mathbf{R}^d)$ the following inequality holds:

$$(1.18)\quad 2[v, A(t,\omega)v]_m + \|B(t,\omega)v\|^2_{\mathfrak{L}_2(\mathbf{R}^d,\mathbf{H}^m)} \leq \delta''\|v\|^2_{m+1} + K'\|v\|^2_m.$$

Since $C_0^\infty(\mathbf{R}^d)$ is dense in $\mathbf{H}^{m+1}$ (see Proposition 3.4.2 (ii)) this inequality can be extended to all $v\in\mathbf{H}^{m+1}$. Taking into account (1.9) and (1.18) we find that the conditions (A_1), (B_1) of Theorem 3.1.8 are satisfied. The validity of the other conditions of this theorem follows from the assumptions made above and the properties of the Hilbert scale $\{\mathbf{H}^s\}$. Thus the statement of the theorem of this paragraph follows from Theorem 3.1.8 and Proposition 3.1.3.

Suppose that the assumptions of the theorem are fulfilled and there is a collection $\{a_n^{ij}, b_n^i, c_n, \sigma_n^{il}, h_n^l, f_n, g_n^l, \varphi_n\}$ where $i, j=1,2,\dots,d$, $l=1,2,\dots,d_1$ and $n\in\mathbf{N}$, satisfying the same assumptions as the functions a^{ij}, b^i, c, σ^{il}, h^l, f, g^l, φ, respectively (it is assumed that constants in the assumptions do not depend on n). Suppose then that u^n is a solution of problem (1.1), (1.2), where a^{ij} is replaced by a_n^{ij}, b^i by b_n^i etc.

Corollary 2. Let $a_n^{ij}\to a^{ij}$, $b_n^i\to b$, $c_n\to c$, $\sigma_n^{il}\to\sigma^{il}$, *and* $h_n^l\to h^l$ *as* $n\to\infty$ *for all* i, j, *and* l, $(l\times l_d\times\mathbf{P}\text{-a.s.})$ *and*

$$\lim_{n\to\infty}\mathbf{E}\Bigg(\int_{[T_0,T]}\Big(\|f_n(t)-f(t)\|^2_m + \sum_{l=1}^{d_1}\|g_n^l(t)-g^l(t)\|^2_m\Big)\,dt +$$

$$+\|\varphi_n-\varphi\|^2_m = 0.$$

Then

$$\lim_{n\to\infty}\mathbf{E}\sup_{t\in[T_0,T]}\|u^n(t)-u(t)\|^2_m + \int_{[T_0,T]}\|u^n(t)-u^2(t)\|^2_{m+1}dt = 0.\ \square$$

Since (as it was shown in the proof of the theorem of this item) the generalized solution of problem (1.1), (1.2), and consequently the generalized solution of the similar problems, where a^{ij}, b^i, c^i, σ^{il}, h^l, f, g^l and φ are replaced by a_n^{ij}, b_n^i, c_n^i, σ_n^{il}, h_n^l, f_n, g_n^l and φ_n, are solutions of the corresponding LSES in the triple

$(\mathbb{H}^{m+1},\mathbb{H}^{m},\mathbb{H}^{m-1})$, the statement of the corollary follows easily from Theorem 3.3.2.

1.3. Theorems of the type proved in the previous item are usually called increasing of smoothness theorems because they determine the dependence between the differential properties of the coefficients, the initial value and the external forces of a problem and the differential properties of its generalized solution.

In this item we shall derive from Theorem 2 an important corollary which shows that if the coefficients, the initial values, and the external forces of the problem are sufficiently smooth, the generalized solution of this problem possesses a "classical" version, that is, the solution of (1.1) is in $\mathbf{C}^2(\mathbf{R}^d)$. The proof will be based on the following simple but important result.

Proposition 3. *Let (S, Σ) be a measurable space and ξ be a Σ-measurable mapping of S into $\mathbb{W}_p^m$, where $p\geq 1$, and for some $n\in\mathbb{N}\cup\{0\}$, $m>n + d/p$. Then there exists a function $\tilde{\xi}$ $S\times\mathbb{R}^d\to\mathbb{R}^1$ with the properties:*

(a) $\tilde{\xi}$ is $\Sigma\otimes\mathfrak{B}(\mathbb{R}^d$-measurable,

(b) $\tilde{\xi}(s,\cdot)\in \mathbf{C}_b^n(\mathbb{R}^d)\cap\mathbb{W}_p^m$ and $\|\tilde{\xi}(s,\cdot)\|_{\mathbf{C}_b^n(\mathbb{R}^d)}\leq N\|\xi(s,\cdot)\|_{m,p}$

for every $s\in S$, where N depends only on m,p, d and n;

(c) $\|\tilde{\xi}(s,\cdot)-\xi(s,\cdot)\|_{m,p}=0$ for every $s\in S$. □

Proof. Let $\{x_k\}$, $k\in\mathbb{N}$, be a countable dense subset of $\mathbf{R}^d$, R be a positive number, $B_k:=\{x\in\mathbb{R}^d, |x-x_k|_d<R\}$ and $\xi_k(\Delta)$ be a restriction of $\xi(s)$ on B_k.

Evidently, ξ_k is a Σ-measurable mapping of S into $\mathbf{W}_p^m(B_k)$. By Theorem 3.4.4 $\mathbb{W}_p^m(B_k)$ is continuously imbedded into $\mathbf{C}_b^n(B_k)$. Therefore, there exists a function $\tilde{\xi}$: $S\to \mathbf{C}_b^n(B_k)$ with the properties:

(i) $\|\tilde{\xi}_k(s,\cdot)\|_{\mathbf{C}_b^n(B_k)}\leq N\|\xi_k(s,\cdot)\|_{B_k,m,p}\leq N\|\xi(s,\cdot)\|_{m,p}$;

(ii) $\tilde{\xi}_k$ *is a Σ-measurable function in* $\mathbf{C}_b^n(B_k)$;

(iii) $\|\tilde{\xi}_k(s,\cdot)-\xi(s,\cdot)\|_{B_k,m,p}=0,\quad \forall\in S.$

(Note that item (i) follows from the continuity of the imbedding operator).

From (ii) it follows that $\tilde{\xi}_k(s,x)$ is a Σ-measurable function for every $x\in B_k$. This, together with the continuity of $\tilde{\xi}_k(s,x)$ with respect to x, yields that $\tilde{\xi}_k(s,x)$ is a $\Sigma\otimes\mathfrak{B}(B_k)$-measurable mapping of $S\times B_k$ into $\mathbf{R}^1$.

Write

$$\tilde{\xi}(s,x) := \sum_k \tilde{\xi}_k(s,x)\, 1_{\{x \in B'_k\}}.$$

where $B'_k = B_k \setminus \bigcup_{i=1}^{k-1} B_i$.

It is easy to see that this function possesses the properties (a) to (c).

Theorem 3. Let $\mathbb{N} \ni n \geq 2$ *and the assumptions of Theorem 2 be satisfied for* $m > n + d/2$. *Then a generalized solution of problem* (1.1), (1.2) *has a version* v *(with respect to* x*), with the following properties:*

(*a*) $v(t,x,\omega)$ *is a predictable stochastic process for every* $x \in \mathbb{R}^d$;

(*b*) $v(t,x,\omega) \in \mathbf{C}^n_b(\mathbb{R}^d)$ *for* $(t,\omega) \in [T_0,T] \times \Omega$;

(*c*) $v(t,x,\omega)$ *as the generalized solution possesses properties enumerated in Theorem* 2;

(*d*) $\mathbb{E} \sup_{t \in [0,T]} \|v(t)\|^2_{\mathbf{C}^n_f(\mathbb{R}^d)} < \infty$;

(*e*) *If* α *is a d-dimensional multiindex with the length* $|\alpha| \leq n\text{-}2$, $\tilde{f}$ *and* $\tilde{g}^l$ *are the versions of* f *and* g^l *whose existence is ensured by the proposition (for* $(S,\Sigma) = (T_0,T] \times \Omega, \mathcal{P})$*), then for every* $x \in \mathbb{R}^d$ *and for all* $(t,\omega) \in [T_0,T] \times \Omega_{x,t}$, *where* $\Omega_{x,t} \subset \Omega$ *and* $\mathbf{P}(\Omega_{x,t}) = 1$ *the following equality holds**.

$$v_\alpha(t) = \varphi_\alpha + \int_{[T_0,T]} ((a^{ij} v_i(s))_i + b^i v_i(s) +$$

(1.19)

$$+ cv(s) + \tilde{f}(s))_\alpha ds + \int_{[T_0,T]} (\sigma^{il} v_i(s) + h^l v(s) + \tilde{g}^l(s))_\alpha dw^l(s);$$

(*f*) *if* $v_1(t,x,\omega)$ *and* $v_2(t,x,\omega)$ *are generalized solutions of problem* (1.1), (1.2) *having the properties* (*a*), (*b*), *then*

---

*The stochastic integral in (1.19) has a continuous in (x,t) version. For this version equality (1.19) holds on the set Ω' of probability 1 which does not depend on x and t.

$$\mathbb{P}\Bigg(\sup_{\substack{t\in[T_0,T]\\ x\in\mathbb{R}^d}} |v_1(t,x,\omega) - v_2(t,x,\omega)| > 0\Bigg) = 0. \ \square$$

Proof. Since by Theorem 2 the generalized solution u of problem (1.1), (1.2) belongs to $\mathbb{H}^{m+1}$ for all $(t,\omega) \in [T_0,T]\times\Omega$, we obtain, in view of the proposition, that there exists a version v of the generalized solution having properties mentioned in the item (a) to (c) of the theorem. The validity of the item (d) follows easily from item (b) of the proposition and inequality of Theorem 2.

In the proof of item (d) we shall use the averaging operator defined in §1.4.10.

In Lemma 3 below, we formulate the properties of the operator which are used later in the book. The proofs of these properties can be found for example in [105], § 1.4, § 1.5.

Lemma 3. *(i)* T_ε *is a continuous linear operator in* $\mathbb{W}_p^m$ *and* $\mathbb{R}(T_\varepsilon) = \mathbb{C}^\infty(\mathbb{R}^d)\cap\mathbb{W}_p^m$, *for every* $m\in\mathbb{N}\cup\{0\}$.

(ii) If $f\in\mathbb{W}_p^m$, *the* $\lim_{\varepsilon\to 0}\|T_\varepsilon f - f\|_{m,p} = 0$.

(iii) If $f\in\mathbb{W}_p^m$, *then for every multiindex* α *such that* $|\alpha|\le m$, $(T_\varepsilon f)_\alpha = T_\varepsilon f_\alpha$.

(iv) For $f\in\mathbb{W}_p^m$, $\|T_\varepsilon f\|_{m,p} \le \|f\|_{m,p}$. $\square$

Let α be a d-dimensional multiindex whose length $|\alpha|$ does not exceed $n+1$. For every $x\in\mathbb{R}^d$, $T_\varepsilon v_\alpha(t,x)$ is a scalar product $(v_\alpha(t), \zeta_{\varepsilon,x})_0$, where $\zeta_{\varepsilon,x}(y):= \varepsilon^{-d}\zeta(x-y)/\varepsilon)$ (see § 1.4.10). Therefore, for all $(t,\omega) \in [T_0,T]\times\Omega'$, where $\Omega'\subset\Omega$ and $\mathbb{P}(\Omega') = 1$, the following equality holds:

$$T_\varepsilon v_\alpha(t,x) = T_\varepsilon\varphi_\alpha(x) + \int_{[T_0,t]} T_\varepsilon[\mathcal{L}_{div} v + \tilde{f}]_\alpha(s,x)ds +$$

$$(1.20)\qquad + \sum_{l=1}^{d_1} \int_{[T_0,t]} T_\varepsilon[\mathcal{M}^l v + \tilde{g}^l]_\alpha(s,x)\, dw^l(s).$$

From the lemma and Theorem 3.4.4 it follows that for every t,ω, $T_\varepsilon v_\alpha(t,x) \to v_\alpha(t,x)$, $T_\varepsilon\varphi_\alpha(x) \to \varphi_\alpha(x)$, $T_\varepsilon[\mathcal{L}_{div} v(t,x) \to f(t,x)]_\alpha \to [\mathcal{L}_{div}\, v(t,x) + f(x,x)]_\alpha$

and $T_\varepsilon[\mathcal{M}^l v(t,x) + \tilde{g}(tx,)]_\alpha \to \mathcal{M}^l v(t,x) + \tilde{g}^l(t,x)$ *as* $\varepsilon \to 0$ in the norm of $\mathbf{C}_b^0(\mathbf{R}^d)$.

Thus, by the dominated convergence theorem

$$\int_{[T_0,t]} T_\varepsilon[\mathcal{L}_{div} v + \tilde{f}]_\alpha(s,x)ds \to \int_{[T_0,t]} [\mathcal{L}_{div} v + \tilde{f}]_\alpha(s,x)ds \ \forall (t,x,\omega);$$

$$\mathbf{E}\left|\sum_{l=1}^{d_1} \int_{[T_0,t]} [T_\varepsilon(\mathcal{M}^l v + \tilde{g}^l)_\alpha(s,x) - (\mathcal{M}^l v + \tilde{g}^l)_\alpha(s,x)]dw^l(s)\right|^2 \le$$

$$\le \sum_{l=1}^{d_1} \int_{[T_0,T]} |T_\varepsilon[\mathcal{M}^l v + \tilde{g}^l]_\alpha(s,x) - |\mathcal{M}^l v + \tilde{g}^l]_\alpha(s,x)|^2 ds \to 0 \ \forall x.$$

Making use of this relations and passing in (1.19) to the limit over some subsequence $\varepsilon_i \to 0$, we find easily that $v_\alpha(t,x)$ satisfies (1.19) for every $(t,x) \in [T_0,T]\times\mathbf{R}^d$ on the set $\Omega_{t,x}$ such that $\Omega_{t,x} \subset \Omega$ and $\mathbf{P}(\Omega_{t,x}) = 1$.

The proof of the item (f) is clear. ☐

Definition 3. *A function* v: $[T_0,T]\times\Omega\times\mathbf{R}^d \to \mathbf{R}^1$ *belonging to* $\mathbf{C}^{0,2}([T_0,T]\times\mathbf{R}^d)$, ($\mathbf{P}$-*a.s.*), *and possessing properties* (*a*) *and* (*d*) (*the last one for* $\alpha=0$) *from Theorem* 3, *will be called a classical solution of problem* (1.1), (1.2). ☐

Corollary 3. *If the conditions of Theorem* 2 *are fulfilled for all* $m\in\mathbf{N}$ (*it may be with different constants for different* m) *then the classic solution of problem* (1.1), (1.2) *is infinitely differentiable in* x ($\mathbf{P}$-*a.s.*). ☐

The corollary follows from the theorem of this item.

1.4. In this item and in the one to follow we consider the problem of solvability of system (1.1), (1.2) in $\mathbf{L}_p([T_0,T]\times\Omega;\ \mathbf{W}_p^m)$. The numbers $m\in\mathbf{N}\cup\{0\}$, $p\in[2,\infty[$ and $q = p/(p-1)$ are assumed to be fixed. It is also assumed that $f(t,\cdot,\omega) \in \mathbf{H}^{m-1}$ for all t,ω, and consequently in view of Proposition 3.4.6 (ii) $f(t,x,\omega) = f^0(t,x,\omega) + \sum_{i=1}^{d} f_i^i(t,x,\omega)$ where $f^i(t,x,\omega) \in \mathbf{H}^m$ for all t,ω and $i=1,2,\ldots,d$. Below we state assumptions of f as assumptions on f^i.

Theorem 4. Suppose that the assumptions of Theorem 2 are fulfilled and in addition $\varphi \in \mathbb{L}_p(\Omega; \mathbb{W}^m_p)$, f^j *and* $g^l \in \mathbb{L}_p(T_0, T] \times \Omega;\ \mathbb{W}^m_p)$ *for* $j=0,1,2,\dots d$, $l=1,2,\dots,d_1$. *Then the generalized solution* u *of the problem* (1.1), (1.2) *belongs to* $\mathbb{L}_p([T_0,T] \times \Omega,\ C_w\ \mathbb{W}^m_p)$ *and*

$$\mathbb{E} \sup_{t\in[T_0,T]} \|u(t)\|^p_{m,p} + \frac{\delta}{2}\, p(p-1) \sum_{|\alpha|\le m} \sum_{i=1}^{d_1} \mathbb{E} \int_{[T_0,T]} |u_\alpha|^{p-2} |(u_i)_\alpha dx ds \le$$

(1.21)

$$\le N\,\mathbb{E}\left(\|\varphi\|^p_{m,p} + \int_{[T_0,T]} \left(\sum_{i=0}^{d} \|f^i(t)\|^p_{m,p} + \sum_{l=1}^{d_1} \|g^l(t)\|^p_{m,p} \right) dt \right),$$

where the constant N *depends only on* m, p, δ, K, T_0, T, d, *and* d_1. □

Corollary 4. Suppose that assumptions of the theorem are fulfilled for $n \in \mathbb{N} \cup \{0\}$ *and* $(m-n)p > d$, *then a generalized solution of problem* (1.1), (1.2) *has a version* v *that belongs to* $\mathbb{C}^{0,n}_b([T_0,T] \times \mathbb{R}^d$, ($\mathbb{P}$-*a.s.*), *and* $\mathbb{E} \sup_{t\in[T_0,T]} \|v(t)\|^p_{\mathbb{C}^n_b(\mathbb{R}^d)}$ *is dominated by the right-hand side of the inequality* (1.21). *If* $n \ge 2$ *then* v *is a classical solution of problem* (1.1), (1.2). □

Proof. Let B be a sphere in $\mathbb{R}$ of finite radius. By Theorem 3.4.4 $\mathbb{W}^m_p(B)$ is continuously imbedded in $\mathbb{C}^n_b(B)$. Thus the theorem yields that the restriction of u on B has a version v_B belonging to $\mathbb{C}^n_b(B)$ for every t such that $\mathbb{E} \sup_{t\in[T_0,T]} \|v_B(t)\|^p_{\mathbb{C}^n_f(B)}$ is dominated by the right-hand side of (1.21).

It is well-known (see e.g. [87], Ch. II, §2) that the imbedding operator of $\mathbb{W}^m_p(B)$ into $\mathbb{C}^n_b(B)$ is compact. From this fact and the weak continuity of v_B with respect to t in $\mathbb{W}^m_p$ follows the continuity of $v_B(int)$ in $\mathbb{C}^n_b$. Thus we have proved that v_B belongs to $\mathbb{C}^{0,n}(B)$. Making use of this statement and arguing in the same way as in the proof of the items (e) and (c) of Theorem 3 the reader will easily finish the proof of the corollary. □

For the proof of the theorem we need the following lemma.

Lemma 4. Suppose that the following conditions hold:

(*a*) *the coefficients* a^{ij}, b^i, c, σ^{il} *and* h^l, *where* j, $i=1,2,\dots,d$ *and* $l=1,2,\dots,d_1$, *of equation (1.1) satisfy condition (0.3) and the assumption (i) of Theorem 2.*

(*b*) *The functions* $u\in W_p^{m+1}$, $h^l\in W_p^{m+1}$, *and* $f^j\in W_p^m$, *where* $l=1,2,\dots,d_1$ *and* $j=0,1,\dots,d$, *are given.*

Then for every $(t,\omega)\in[T_0,T]\times\Omega$,

$$U(m,p,u,f,g,t):=\int_{\mathbf{R}^d}\sum_{|\alpha|\le m}|u_\alpha|^{p-2}\Big(-p(p-1)\,(u_j)_\alpha\,(a^{ij}u_i+$$

$$+f^j)_\alpha+pu_\alpha(b^iu_i+cu+f^0)_\alpha+$$

$$+\frac{1}{2}\,p(p-1)\sum_{l=1}^{d_1}|\sigma^{il}u_i+h^lu+g^l)_\alpha|^2\Big)\,dx\le$$

$$\le-\frac{\delta}{2}\,p(p-1)\sum_{|\alpha|\le m}\sum_{i=1}^{d}\int_{\mathbf{R}^d}|u_\alpha|^{p-2}\,|(u_i)_\alpha|^2\,dx+$$

$$+N_0(\|u\|_{m,p}^p+\sum_{i=0}^{d}\|f^i\|_{m,p}^p+\sum_{l=1}^{d_1}\|g^l\|_{m,p}^p),$$

where the constant N_0 *depends only on* p, d, d_1, δ, m, K.

Proof. For $s\in\mathbf{R}^1$ set $G(s)=|s|^p$. For real functions ξ, η on $\mathbf{R}^d$ we write $\xi\prec\eta$ if $\xi=\eta+0$ and there exists a constant N such that for all x

$$\theta\le N\sum_{|\alpha|\le m}\Big(|u_\alpha|^p+\sum_{i=0}^{d}|f^i_\alpha|^p+\sum_{l=1}^{d_1}|g^l_\alpha|^p\Big).$$

As a rule the argument u_α of the function $G(u_\alpha)$ and its derivatives $G'(u_\alpha)=p|u_\alpha|^{p-2}u_\alpha$ and $G''(u_\alpha)=p(p-1)\,|u_\alpha|^{p-2}$ will be dropped. Throughout the proof it is supposed that $|\alpha|\le m$.

First we note that by the inequality

$$(1.22)\qquad |ab|\le p^{-1}b^p+q^{-1}a^q,\ a,\ b\ge\mathbf{R}_+,$$

we have

$$\sum_{|\alpha|\le m} G'[cu+f^0]_\alpha \prec 0$$

and

$$\sum_{|\alpha|\le m} G'' \sum_{l=1}^{d_1} |(h^l u + g^l)_\alpha|^2 \prec 0.$$

Next

$$\sum_{|\alpha|\le m} G'\cdot(b^i u_i)_\alpha \prec \sum_{|\alpha|\le m} G''\cdot(u_\alpha)\, b^i(u_i)_\alpha.$$

From the inequality (3.1.10) it follows that

$$\sum_{|\alpha|\le m} G''\cdot(u_\alpha)\, b^i(u_i)_\alpha \prec \frac{\varepsilon}{2} N \sum_{|\alpha|\le m} \sum_{i=1}^{d} |(u_i)_\alpha|^2\, G''.$$

By the same inequality,

$$G''\cdot(u_i)_\alpha\, f^j_\alpha \le G''\left(\frac{\varepsilon}{2} \sum_{j=1}^{d} |(u_j)_\alpha|^2 + \frac{1}{2\varepsilon} \sum_{j=1}^{d} |f^j_\alpha|^2\right).$$

Thus with the help of (3.1.10), we obtain

$$\sum_{|\alpha|\le m} G''\cdot(u_j)_\alpha\, f^j_\alpha \prec \frac{\varepsilon}{2} \sum_{|\alpha|\le m} G'' \sum_{j=1}^{d} |(u_j)_\alpha|^2.$$

Making use of (1.22) and (3.1.10), we find

$$\sum_{|\alpha|\le m} G'' \sum_{l=1}^{d_1} (\sigma^{il} u_i)_\alpha\, (h^l u + g^l)_\alpha \prec \sum_{|\alpha|\le m} G'' \sum_{l=1}^{d_1} \sigma^{il}(u_i)_\alpha (h^l u_\alpha + g^l_\alpha) \le$$

$$\le \frac{\varepsilon}{2} N \sum_{|\alpha|\le m} G \sum_{i=1}^{d} |(u_i)_\alpha|^2 + \frac{N}{2\varepsilon} \sum_{|\alpha|\le m} G''\cdot\Big(|u_\alpha|^2 +$$

$$+ \sum_{l=1}^{d_1} |g^l_\alpha|^2\Big) \prec \frac{\varepsilon}{2} N \sum_{|\alpha|\le m} G'' \sum_{i=1}^{d} |(u_i)_\alpha|^2.$$

At last from the superparabolic condition (0.3) it follows that

$$\sum_{|\alpha|\le m} |u_\alpha|^{p-2} \Big(-p(p-1)\, (u_j)_\alpha\, (a^{ij} u_i)_\alpha +$$

$$+ \frac{1}{2}\, p(p-1) \sum_{i=1}^{d_1} |\sigma^{il} u_i)_\alpha|^2\Big) =$$

$$= \sum_{|\alpha|\le m} |u_\alpha|^{p-2} \Big(-p(p-1)\, (u_j)_\alpha a^{ij} (u_i)_\alpha +$$

$$+ \frac{1}{2}\, p(p-1) \sum_{l=1}^{d_1} |\sigma^{il} (u_i)_\alpha|^2\Big) +$$

$$+ \Sigma^\alpha |u_\alpha|^{p-2} \Big(-p(p-1)\, (u_j)_\alpha\, a^{ij}_\beta (u_i)_\gamma +$$

$$+ \frac{1}{2}\, p(p-1) \sum_{l=1}^{d_1} |\sigma^{ij}_\beta (u_i)_\gamma|^2\Big)^3 \le -\delta \sum_{|\alpha|\le m} G'' \sum_{i=1}^{d} |(u_i)_\alpha|^2 +$$

$$+ \Sigma^\alpha\, |u_\alpha|^{p-2} \Big[-p(p-1)(u_j)_\alpha a^{ij}_\beta (u_i)_\gamma + \frac{1}{2}\, p(p-1) \sum_{l=1}^{d_1} |\sigma^{il}_\beta (u_i)_\gamma|^2\Big].$$

Here and in the sequel we use the notation,

$$\Sigma^\alpha := \sum_{\substack{\alpha := \beta+\gamma \\ |\alpha|\le m,\ |\beta|\ge 1}} .$$

By (3.1.10) and (1.22) we obtain that

$$\sum^{\alpha} (|u_\alpha|^{p-2} [-p(p-1) (u_j)_\alpha a^{ij}_{\beta}(u_i)_\gamma] \prec \frac{\varepsilon}{2} \sum_{|\alpha|\le m} G'' \sum_{j=1}^{d} |(u_j)_\alpha|^2$$

and

$$\sum^{\alpha} |u_\alpha|^{p-2} \left[\frac{1}{2} p(p-1) \sum_{l=1}^{d_1} |\sigma^{il}_{\beta}(u_i)_\gamma|^2\right] \prec 0.$$

Collecting all the inequalities obtained above and choosing ε to be sufficiently small, it is easy to find that

$$U(m,p,u,f,g,t) \prec -\frac{\delta}{2} \sum_{|\alpha|\le m} G'' \sum_{i=1}^{d} |(u_i)_\alpha|^2$$

So, we have proved the lemma. □

1.5. Proof of Theorem 4.1. Let T_ε be Sobolev's averaging operator (see 1.4.10).

In this item we shall use the following reduced notation: $\xi_{(\varepsilon)} := T_\varepsilon \xi$. Consider the problem

$$du(t) = \left[(a^{ij}_{(\varepsilon)} u_i(t))_i + b^i_{(\varepsilon)} u_i(t) + c_{(\varepsilon)} u(t) + \sum_{j=0}^{d} (f^j_{(\varepsilon)}(t))_j\right] dt +$$

$$+ [\sigma^{il}_{(\varepsilon)} u_i(t) + h^l_{(\varepsilon)} u(t) + g^l_{(\varepsilon)}(t)]\, dw^l(t),$$

(1.23) $\qquad (t,x,\omega) \in]\,T_0, T] \times \mathbb{R}^d \times \Omega,$

(1.24) $\qquad u(T_0) = \varphi_{(\varepsilon)}(x,\omega) \in \mathbb{R}^d \times \Omega.$

The coefficients of the problem as well as $\varphi_{(\varepsilon)}$, $g^l_{(\varepsilon)}$ and $f^j_{(\varepsilon)}$ satisfy the assumptions of the theorem for all m.

Indeed, we first observe that the initial value $\varphi_{(\varepsilon)}$ and the external forces $f^j_{(\varepsilon)}$, $g^l_{(\varepsilon)}$ of problem (1.23), (1.24) are infinitely differentiable in x. Next, obviously the coefficients of the problem and their derivatives are bounded by the same constant as the coefficients of the original equation (1.1) and their derivatives (Lemma 1.4.10), and the initial value and the external forces belong to $\mathbb{L}_{p'}([T_0,T]\times\Omega;\mathbb{W}_p^m)$ for $p'=p$ and 2, and $2m\in\mathbb{N}\cup\{0\}$ (Lemma 3 (i)). Making use of the obvious inequality $|\xi_{(\varepsilon)}|^2 \le (\xi^2)_{(\varepsilon)}$, it is easy to see that problem (1.23). (1.24) satisfies the superparabolic condition (0.3).

Let $u^{(\varepsilon)}$ be a classical solution of problem (1.23), (1.24). The existence of this solution is ensured by Theorem 3. From Corollary 3 it follows that the solution is infinitely differentiable in x. Thus, there exists a subset $\Omega'\subset\Omega$ such that $\mathbf{P}(\Omega')=1$ and for all $(t,x,\omega) \in [T_0,T]\times\mathbb{R}^d\times\Omega$ and every d-dimensional multiindex α the following equality is satisfied (see Theorem 3, (e)):

$$u_\alpha^\varepsilon(t) = (\varphi_\varepsilon)_\alpha + \int_{[T_0,t]} [a^{ij}_{(\varepsilon)}(s)]_i + b^i_{(\varepsilon)}\, u_i^\varepsilon(s) + c_{(\varepsilon)}u^\varepsilon(s) +$$

$$+ \sum_{j=0}^d (f^j_{(\varepsilon)}(s))_j]_\alpha\, ds + \int_{[T_0,t]} (\sigma^{il}_{(\varepsilon)}u_i^\varepsilon(s) + h^l_{(\varepsilon)}u^\varepsilon(s) +$$

$$+ g^l_{(\varepsilon)}(s))_\alpha\, dw^l(s).$$

Applying Ito's formula (§1.4.2) to u_α^ε we obtain from the latter equality that for every $x\in\mathbb{R}^d$, $N\in\mathbb{R}_+$, and all $(t,\omega) \in [T_0,T]\times\Omega_x$, where $\mathbf{P}(\Omega_x) = 1$, the following equality holds:

$$e^{-Nt}|u_\alpha^\varepsilon(t)|^p = e^{-NT_0}|(\varphi_{(\varepsilon)})_\alpha|^p + \int_{[T_0,t]} e^{-Ns}\Big\{G'(u_\alpha^\varepsilon(s))_j +$$

$$+ b^i_{(\varepsilon)}u_i^\varepsilon(s) + c_{(\varepsilon)}u^\varepsilon(s) + \sum_{j=0}^d (f^j_{(\varepsilon)})_j(s))_\alpha +$$

(1.25)

$$+ \frac{1}{2}\, G''(u_\alpha^\varepsilon(s)) \sum_{l=1}^{d_1} |\sigma^{il}_{(\varepsilon)}u_i^\varepsilon(s) + h^l_{(\varepsilon)}u^\varepsilon(s) + g^l_{(\varepsilon)}(s))_\alpha|^2\Big\}ds +$$

$$+ \int_{[T_0,t]} e^{-Ns} G'(u_\alpha^\varepsilon(s))\, (\sigma^{il}_{(\varepsilon)}u_i^\varepsilon(s) + h^l_{(\varepsilon)}u^\varepsilon(s) +$$

$$+ g^l_{(\varepsilon)}(s))_\alpha \, dw^l(s) - N \int_{[T_0,t]} e^{-Ns} |u^\varepsilon_\alpha(s)|^p \, ds,$$

where G' and G'' are the functions introduced in the proof of Lemma 4. In the sequel the argument of the functions G' and G'' will be omitted.

Note that the stochastic integral in (1.25) has a version continuous in (t,x). Thus, considering just this version, we may suppose that equality (1.25) is valid on a ω-set of probability 1 for all t,x.

Observe also that

$$\int_{\mathbf{R}^d} G'\Big((a^{ij}_{(\varepsilon)} u^\varepsilon_i(s,x))_j + \sum_{j=1}^d (f^j_{(\varepsilon)})_j(s,x)\Big)_\alpha \, dx =$$

(1.26)

$$= - \int_{\mathbf{R}^d} G''(u^\varepsilon_j)_\alpha \, (a^{ij}_{(\varepsilon)} u^\varepsilon_j(s,x) + \sum_{j=1}^d (f^j_\varepsilon)(s,x))_\alpha \, dx.$$

Integrating (1.25) over $\mathbf{R}^d$ and changing the order of integration (see §1.4.8) it is easy to obtain with the help of (1.26) that

$$e^{-Nt} \|u^\varepsilon(t)\|^p_{m,p} = e^{-NT_0} \|\varphi_{(\varepsilon)}\|^p_{m,p} + \int_{[T_0,t]} e^{-Ns} \, U(m,p,u^\varepsilon, f_{(\varepsilon)}, g_{(\varepsilon)}, s) ds +$$

(1.27)

$$+ \int_{[T_0,t]} e^{-Ns} \int_{\mathbf{R}^d} \sum_{|\alpha| \le m} G'[\sigma^{il}_{(\varepsilon)} u^\varepsilon_i(t) + h^l_{(s)} u^\varepsilon(s) +$$

$$+ g^l_{(\varepsilon)}(s)]_\alpha \, dx dw^l(s) - N \int_{[T_0,t]} e^{-Ns} \|u^{(\varepsilon)}(s)\|^p_{m,p} \, ds,$$

where U is the function introduced in Lemma 4. Write for $r \in \mathbf{N}$ $\Gamma_r(\omega) := \{t: \|u^\varepsilon(t,\omega)\|_{\mathbf{C}^m_b(\mathbf{R}^d)} > r, \; t \in [T_0,T]\}$, and

$$\tau_r(w) := \begin{cases} \inf_t \Gamma_r(\omega), & if \;\; \Gamma_r(\omega) \neq \oslash; \\ T, & if \;\; \Gamma_r(\omega) = \oslash. \end{cases}$$

Evidently, $\{\tau_r\}$, where $r \in \mathbb{N}$, is a sequence of stopping times localizing the stochastic integral in (1.27). Thus for every $r \in \mathbb{N}$,

$$\mathbb{E}\|u^{\varepsilon}(t\wedge\tau_r)\|^p_{m,p}\, e^{-N(t\wedge\tau_r)} = \mathbb{E}\|\varphi_{(\varepsilon)}\|^p_{m,p}\, e^{-NT_0} +$$

$$\mathbb{E} \int_{[T_0, t\wedge\tau_r]} e^{-Ns}\, U(m,p,u^{\varepsilon}, f_{(\varepsilon)}, g_{(\varepsilon)}, s)\, ds -$$

$$- N \int_{[T_0, t\wedge\tau_r]} e^{-Ns}\|u^{\varepsilon}(s)\|^p_{m,p}\, ds.$$

From this equality and Lemma 3, in view of continuity of $u^{\varepsilon}(t)$ in $\mathbf{W}^m_p$ with respect to t, it follows that

$$\mathbb{E}\|u^{\varepsilon}(t)\|^p_{m,p}\, e^{-Nt} \leq \mathbb{E}\|\varphi_{(\varepsilon)}\|^p_{m,p}\, e^{-NT_0} -$$

$$- \frac{\delta}{2}\, p(p-1) \sum_{|\alpha|\leq m} \sum_{i=1}^{d} \int_{[T_0,t]} \mathbb{E} \int_{\mathbb{R}^d} |u^{\varepsilon}_{\alpha}(s)|^{p-2} \times$$

$$\times |(u^{\varepsilon}_i(s))_{\alpha}|^2\, dxds + N_0 \int_{[T_0,t]} e^{-Ns}\Big(\|u^{\varepsilon}(s)\|^p_{m,p} + \sum_{j=0}^{d} \|f^j_{(\varepsilon)}(s)\|^p_{m,p} +$$

$$+ \sum_{l=1}^{d_1} \|g^l_{(\varepsilon)}(s)\|^p_{m,p}\Big)\, ds - N \int_{[T_0,t]} e^{-Ns}\|u^{\varepsilon}(s)\|^p_{m,p}\, ds.$$

Taking N to be sufficiently large, we obtain from the latter inequality that

$$\sup_{t\in[T_0,T]} \mathbb{E}\|u^{\varepsilon}(t)\|^p_{m,p} +$$

$$+ \sum_{|\alpha| \le m} \sum_{i=1}^{d} \int_{[T_0,T]} \mathbf{E} \int_{\mathbf{R}^d} |u^{\varepsilon}_{\alpha}(s)|^{p-2} \, |(u^{\varepsilon}_{i}(s))_{\alpha}|^2 \, dxds \le$$

$$\le N_1 \mathbf{E}\Bigg(\|\varphi_{(\varepsilon)}\|^p_{m,p} + \int_{[T_0,T]} \Bigg(\sum_{j=0}^{d} \|f^j_{(\varepsilon)}(s)\|^p_{m,p} +$$

(1.28)

$$+ \sum_{l=1}^{d_1} \|g^l_{(\varepsilon)}(s)\|^p_{m,p} \Bigg) ds \Bigg) \le N_1 \mathbf{E}\Bigg(\|\varphi\|^p_{m,p} +$$

$$+ \int_{[T_0,T]} \Bigg(\sum_{j=0}^{d} \|f^j(s)\|^p_{m,p} + \sum_{l=1}^{d_1} \|g^l(s)\|^p_{m,p} \Bigg) ds \Bigg).$$

We now show that $\mathbf{E} \sup_{t \in [T_0,T]} \|u^{\varepsilon}(t)\|^p_{m,p}$ is also bounded (provided N_1 is large enough) by the right-hand side of the inequality (1.28). For this purpose, take $N=0$ in equality (1.27). From the obtained equality, inequality (1.28), and Lemma 4 it follows that

$$\mathbf{E} \sup_{t \in [T_0,T]} \|u^{\varepsilon}(t)\|^p_{m,p} \le N_2 \mathbf{E}\Bigg(\|\varphi\|^p_{m,p} +$$

$$+ \int_{[T_0,T]} \Bigg(\sum_{j=0}^{d} \|f^j(s)\|^p_{m,p} + \sum_{l=1}^{d_1} \|g^l(s)\|^p_{m,p} \Bigg) ds \Bigg) +$$

(1.29)

$$+ \mathbf{E} \sup_{t \in [T_0,T]} \Bigg| \int_{[T_0,t]} \int_{\mathbf{R}^d} \sum_{|\alpha| \le m} G^l[\sigma^{il}_{(\varepsilon)} u^{\varepsilon}_{i}(s) +$$

$$+ h^l_{(\varepsilon)} u^{\varepsilon}(s) + g^l_{(\varepsilon)}(s)]_{\alpha} \, dxdw^l(s) \Bigg|.$$

From the Burkholder-Davis inequality (§2.1.7), it follows that the last term in the right-hand side of (1.29) is dominated by

$$\mathbf{E}\int_{[T_0,T]}\left[\sum_{l=1}^{d_1}\left(\int_{\mathbf{R}^d}\sum_{|\alpha|\le m} G^l[\sigma^{il}_{(\varepsilon)}u^{\varepsilon}_i(s) + h^l_{(\varepsilon)}(s) + g^l_{(\varepsilon)}(s)]_\alpha\right)^2 ds\right]^{\frac{1}{2}}$$

By the Schwarz inequality, we find that this value does not exceed

$$N\mathbf{E}\left(\int_{[T_0,T]}\|u^{\varepsilon}(s)\|^p_{m,p}\sum_{l=1}^{d_1}\sum_{|\alpha|\le m}\int_{\mathbf{R}^d} G''|\sigma^{il}_{(\varepsilon)}u^{\varepsilon}_i(s) + h^l_{(\varepsilon)}u^{\varepsilon}(s) +\right.$$

$$\text{(1.30)}\qquad \left. + g^l_{(\varepsilon)}(s))_\alpha|^2\, dxds\right)^{\frac{1}{2}} \le \frac{\lambda N}{2}\,\mathbf{E}\sup_{t\in[T_0,T]}\|u^{\varepsilon}(s)\|^p_{m,p} +$$

$$+\frac{N}{2\lambda}\sum_{l=1}^{d_1}\sum_{|\alpha|\le m}\int_{[T_0,T]}\int_{\mathbf{R}^d} G''|\sigma^{il}_{(\varepsilon)}u^{\varepsilon}_i(s)+h^l_{(\varepsilon)}u^{\varepsilon}(s)+g^l_{(\varepsilon)}(s))_\alpha|^2 dxds.$$

The last inequality in (1.30) follows from (3.1.10) and is valied for every $\lambda>0$.

At last applying inequality (1.28) to the right-hand side of (1.30) and making use of (1.22), it is easy to see that this expression is majorized by

$$\frac{\lambda N}{2}\,\mathbf{E}\sup_{t\in[T_0,T]}\|u^{\varepsilon}(s)\|^p_{m,p} + N(\lambda)\,\mathbf{E}\left(\|\varphi\|^p_{m,p} +\right.$$

$$\left. + \int_{[T_0,T]}\left(\sum_{j=0}^{d}\|f^j(s)\|^p_{m,p} + \sum_{l=1}^{d_1}\|g^l(s)\|^p_{m,p}\right)ds\right).$$

From (1.28), (1.29) and the estimates following inequality (1.29) taking λ to be sufficiently small, we obtain

$$\mathbf{E}\sup_{t\in[T_0,T]}\|u^{\varepsilon}(t)\|^p_{m,p} +$$

$$(1.31)\qquad +\mathbb{E}\sum_{i=1}^{d}\sum_{|\alpha|\le m}\int_{[T_0,T]}\int_{\mathbf{R}^d}|u_i^{\varepsilon}(s))_\alpha|^2\,|u_\alpha^{\varepsilon}(s,x)|^{p-2}\,dxds\le$$

$$\le N\,\mathbb{E}\left[\|\varphi\|_{m,p}^p+\int_{[T_0,T]}\left(\sum_{i=0}^{d}\|f^i(s)\|_{m,p}^p+\sum_{l=1}^{d_1}\|g^l(s)\|_{m,p}^p\right)ds\right],$$

where the constant N does not depend on ε (the reader will easily check that, as a matter of fact, N depends only on m, p, δ, k, T_0, T.

Thus it has been proved that $u^{\varepsilon}\in\mathbb{L}_p(\Omega;\ \mathbb{C}([T_0,T];\ \mathbf{W}_p^m)$.

The next step of the proof is the passage to the limit as $\varepsilon\to 0$ in problem (1.23), (1.24) and in equality (1.31).

From Lemma 3 we obtain that

$$\lim_{\varepsilon\to 0}\mathbb{E}\left\{\|\varphi_{(\varepsilon)}-\varphi\|_{m,p}^p+\int_{[T_0,T]}\left(\sum_{j=0}^{d}\|f^i_{(\varepsilon)}(t)-f^i(t)\|_{m,p}^p+\right.\right.$$

$$\left.\left.+\sum_{l=1}^{d_1}\|g^l_{(\varepsilon)}(t)-g(t)\|_{m,p}^p\right)dt\right\}=0.$$

In particular this relation holds in the case of $p=2$. Hence by the arguments given in the beginning of the item we find that Corollary 2 is applicable to problems (1.1), (1.2) and (1.23), (1.24).

From the corollary it follows that there exists a sequence $\{\varepsilon_n\}$, where $\varepsilon_n\to 0$ as $n\to\infty$, such that ($l\times l\times\mathbb{P}$-a.s.),

$$(1.32)\qquad \lim_{n\to\infty}\sum_{|\alpha|\le m}|u_\alpha^{\varepsilon_n}(t,x,\omega)-u_\alpha(t,x,\omega)|=0.$$

Here u is the generalized solution of problem (1.1), (1.2), belonging, in view of Theorem 2, to $\mathbb{L}_2(\Omega;\ \mathbb{C}([T_0,T];\ \mathbb{H}^m))$.

If we drop the first term in (1.31) and pass to the limit in the obtained inequality over the sequence mentioned above, then by Fatou's lemma it follows that

$$\sum_{i=1}^{d} \sum_{|\alpha|\le m} \int_{[T_0,T]} \int_{\mathbf{R}^d} |(u_i(s,x))_\alpha|^2 \, |u_\alpha(s,x)|^{p-2} dx \le$$

$$\le N\mathbf{E}\left\{\|\varphi\|_{m,p}^p + \int_{[T_0,T]} \left(\sum_{j=0}^{d} \|f^j(s)\|_{m,p}^p + \sum_{l=1}^{d_1} \|g^l(s)\|_{m,p}^p \right) ds\right\}.$$

Next, we have to check that $\mathbf{E}\sup_{t\in[T_0,T]} \|u(t)\|_{m,p}^p$ is also majorized by the right-hand side of the latter inequality.

Observe that $\|u(t,\omega)\|_{m,p}^p$ is a predictable function, so that the expectation $\mathbf{E}\sup_{t\in[T_0,T]} \|u(t)\|_{m,p}$ is well-defined. To prove that $\|u(t,\omega)\|_{m,p}^p$ is predictable, note that $\|u(t,\omega)\|_{m,p} = \sup_{v\in Z} |(u,t,\omega),v)_0|$, where Z is a dense set of function from $\mathbf{C}_0^\infty(\mathbf{R}^d)$ in the unit ball of $\mathbf{W}_p^m$, and $u(t,\omega)$ is a predictable function in $\mathbf{L}_2$.

From (1.32), it follows that there exist a countable set $\{t^i\}$ which is a dense subset of $[T_0,T]$ and a countable, dense in $\mathbf{L}_2$ set $\{y^i\}$, where $y^i\in\mathbf{C}_0^\infty(\mathbf{R}^d)$ for every i, such that on some ω-set of probability 1 for all $|\alpha|\le m$, $t^i\in\{t^i\}$ and $y^i\in\{y^i\}$,

$$|(u_\alpha(t^i),\, y^i)_0| = \lim_{\varepsilon_n\to 0} |u_\alpha^{\varepsilon_n}(t^i),\, y^i)|_0 \le \lim_{\varepsilon_n\to 0} \|u^{\varepsilon_n}(t^i)\|_{m,p}\, \|y^i\|_{0,q}.$$

As it has been mentioned above, $u\in\mathbf{L}_2(\Omega;\, \mathbf{C}([T_0,T];\, \mathbf{H}^m)$, which implies that there exists $\tilde\Omega\subset\Omega$ such that $\mathbf{P}(\tilde\Omega)=1$ and for all $\omega\in\tilde\Omega$, $y^i\in\{y^i\}$, and $|\alpha|\le m$, $(u_\alpha(t),\, y^i)$ is a continuous function on $[T_0,T]$.

Thus

$$\sup_{t\in[T_0,T]} \|u_\alpha(t)\|_{0,p} = \sup_{t\in[T_0,T]} \sup_j |u_\alpha(t),\, y^i)_0| \le$$

$$\le \sup_j \sup_i |u_\alpha(t^i),\, y^i)_0| \le \sup_i \lim_{\varepsilon_n\to 0} \|u_\alpha^{\varepsilon_n}(t^i)\|_{m,p} \le$$

$$\le \lim_{\varepsilon_n\to 0} \sup_{t\in[T_0,T]} \|u_\alpha^{\varepsilon_n}(t)\|_{m,p}.$$

From this inequality, (1.31), and Fatou's lemma we can easily obtain the following inequality

$$\mathbb{E}\sup_{t\in[T_0,T]}\|u(t)\|^p_{m,p} \le N\mathbb{E}\Big(\|\varphi\|^p_{m,p} +$$

$$+\int_{[T_0,T]}\Big(\sum_{j=0}^{d}\|f^j(s)\|^p_{m,p} + \sum_{l=1}^{d_1}\|g^l(s)\|^p_{m,p}\Big)ds\Big).$$

where the constant N depends on the same parameters as the constant N in (1.31).

Since, by Theorem 2, $(u(t),y)$ is a continuous ($\mathbb{P}$-a.s.) function of t for every $y\in\mathbb{C}_0^\infty(\mathbb{R}^d)$, and $\mathbb{C}_0^\infty(\mathbb{R}^d)$ is dense in $\mathbb{L}_q$, $u(t)$ is weakly continuous in $\mathbb{W}_p^m$.

Thus the theorem is proved. □

1.6. *Remark* 6. *Sometimes it is useful to consider an equation a slightly more general than* (1.1), *namely: we can add to* (1.1) *the term* $\tilde{c}(t,x,\omega)(\tilde{b}^i(t,x,\omega)u(t,x,\omega))_i$, *where* $\tilde{b}^i$, $\tilde{c}$, *and* $\tilde{c}_i$ *are to satisfy the same assumptions as* c *in Theorem* 4. *In equality* (1.3) *this term gives the contribution* $(\tilde{b}^i u, (\tilde{c}y)_i)_0$.

It is easy to see that the statements of Theorem 4 *and Corollary* 4 *remain valid for the equation* (1.1) *modified in such a way.* □

4.2. The Cauchy Problem for Ito's Second Order Equations.

2.0. In this section we consider the Cauchy problem (0.1), (1.2). The assumptions on the coefficients, on the initial value, and on the external forces made in §10 are assumed to be in force. It is also assumed that the coefficients a^{ij} $(i,j=1,2,\ldots,d)$ are differentiable in x for all $(t,\omega)\in[T_0,T]\times\Omega$ and their derivatives are bounded (uniformly in t,x and ω), and $f\in\mathbb{L}_2^\omega([T_0,T];\ \mathbb{L}_2)$.

We prove for problem (0.1), (1.2) results similar to the ones proved for problem (1.1), (1.2) in the previous section.

2.1. *Definition* 1. *A function* $u\in\mathbb{L}_2^\omega(T_0,T],\ \mathcal{P};\ \mathbb{H}^1)$ *is said to be a generalized solution of problem* (0.1), (1.2) *if for every* $y\in\mathbb{C}_0^\infty(\mathbb{R}^d)$, *it satisfies* ($l\times\mathbf{P}$-*a.s.*) *the following equation*

$$(u(t),y)_0 = (\varphi,y)_0 + \int_{[T_0,t]}[-(a^{ij}u_i(s),y_i)_0 +$$

$$(2.1) \qquad + ((b^i - a^{ij}_j)u_i(s) + cu(s) + f(s),y)_0]ds +$$

$$+ \int_{[T_0,t]} (\sigma^{il}u_i(s) + h^l u(s) + g^l(s),y)_0 dw^l(s). \quad \square$$

Evidently, under the assumptions that were made, all the integrals in (2.1) are well-defined.

Warning 1. *As before, (see §1.1) we may suppose that the generalized solution of problem* (0.1), (1.2) *belongs to* $\mathbf{C}([T_0,T];\mathcal{P};\mathbf{L}_2)$ *and satisfies equality* (2.1) *for all* $t\in[T_0,T]$ *and* $\omega\in\Omega'\subset\Omega$, *where* $\mathbf{P}(\Omega') = 1$. $\square$

It is assumed that the numbers $m\in\mathbf{N}$, $p\in[2,\infty[$, and $q = p/(p-1)$ are fixed.

Theorem 1. *Suppose that for all* $(t,\omega)\in[T_0,T]\times\Omega$, *the coefficients* a^{ij} *are differentiable in* x *up to the order of* $2\vee m$, σ^{il} *and* h^l *up to the order of* $m+1$, *and* b^i *and* c *up to the order of* m, *where* $i,j=1\ d$ *and* $l=1\ d$. *It is also assumed that all the above functions, as well as their derivatives, are bounded by the constant* K. *Let us also suppose that the parabolic condition* (0.2) *is fulfilled. Then a generalized solution of problem* (0.1), (1.2) *(provided one does exist) is unique.*

If in addition to the assumption made above, we suppose that for $p'=p$ *and* 2, $i=1,2,\dots,d$, *and* $l=1,2,\dots,d$, f, g^l, *and* $g^l_i\in\mathbf{L}_{p'}([T_0,T]\times\Omega;\ \mathbf{W}^m_{p'})$ *and* $\varphi\in\mathbf{L}_{p'}(\Omega;\ \mathbf{W}^m_{p'})$, *then problem* (0.1), (1.2) *has a generalized solution* u *in the class* $\mathbf{L}_2(\Omega;\ \mathbf{C}([T_0,T];\ \mathbf{H}^{m-1})) \cap \mathbf{L}_2(T_0,T]\times\Omega;\ \mathbf{C}_w\mathbf{H}^m) \cap \mathbf{L}_p([T_0,T]\times\Omega;\ \mathbf{C}_w\ \mathbf{W}^m_p)$. *Apart from that, for* $p'=p$ *and* 2 *the following inequality holds*

$$\mathbf{E} \sup_{t\in[T_0,T]} \|u(t)\|^{p'}_{m,p} \le N\mathbf{E}\Bigg(\|\varphi\|^{p'}_{m,p'} +$$

$$+ \int_{[T_0,T]} \Big(\|f(t)\|^{p'}_{m,p'} + \sum_{l=1}^{d_1} \|g^l(t)\|^{p'}_{m+1,p'}\Big)dt\Bigg),$$

where the constant N *depends only on* p, d, d_1, K, m, T_0, *and* T. $\square$

Arguing in the same way as in the proof of Corollary 1.4 and Theorem 1.3, we derive from the theorem stated above the following result.

Corollary 1. *Suppose that the assumptions of the theorem are fulfilled, the number* $n\in\mathbb{N}\cup\{0\}$ *is given and* $(m-n)p>d$.

Then the generalized solution of problem (0.1), (1.2) *has a version* v *possessing the following properties:*

(*a*) *For every* $x\in\mathbb{R}^d$, $v(t,x,\omega)$ *is a predictable real stochastic process.*

(*b*) *For all* ω, $v(t,x,\omega)\in\mathbf{C}_f^{0,n}([T_0,T])\times\mathbb{R}^d)$.

(*c*) *As a generalized solution of problem* (0.1), (0.2) *it possesses the properties mentioned in the theorem.*

(*d*) $\mathbb{E}\sup_{t\in[T_0,T]}\|v(t)\|^p_{\mathbf{C}_b^n(\mathbb{R}^d)}$ *is dominated by the right-hand the inequality of the theorem.*

(*e*) *If* $n\geq 2$ *and* α *is a d-dimensional multiindex with the length* $|\alpha|\leq n-2$, $\tilde{f}$ *and* $\tilde{g}$ *are the versions of* f *and* g^l *whose existence is ensured by Proposition* 1.3 (*for* $(S;\Sigma) = ([T_0,T]\times\Omega;\mathcal{P})$, *then for every* $(t,\omega)\in[T_0,T]\times\Omega_{x,t}$, *where* $\Omega_{x,t}\subset\Omega$ *and* $\mathbb{P}(\Omega_{x,t})=1$ *the following equality holds.**

$$v_\alpha(t) = \varphi_\alpha + \int_{[T_0,t]} (a^{ij}v_{ij}(t) + b^i v_i(s) + c(s) + \tilde{f}(s))_\alpha ds +$$

$$+ \int_{[T_0,t]} (\sigma^{il}v_i(s) + h^l v(s) + \tilde{g}^l(s))_\alpha dw^l(s);$$

(*f*) *If* $v_1(t,x)$ *and* $v_2(t,x)$ *are generalized solutions of problem* (0.1), (1.2) *possessing properties* (*a*), (*b*), *then*

$$\mathbb{P}\Big(\sup_{\substack{t\in[T_0,T]\\ x\in\mathbb{R}^d}} |v_1(t,x,\omega) - v_2(t,x,\omega)| \geq 0\Big) = 0. \quad \square$$

2.2. The notion of a classical solution of a problem (0.1), (1.2) is defined similarly to the notion of a classical solution of problem (1.1), (1.2) (see Definition 1.3).

---

*The stochastic integral in this equality has a continuous in (x,t) version. For this version equality (1.19) holds on the set Ω' of probability 1 which does not depend on x and t.

Corollary 2. If the assumptions of Theorem 1 are fulfilled for all $m\in\mathbb{N}$ *(there may be with different constants for different* m*), the a classical solution of problem* (0.1), (1.2) *is infinitely differentiable in* x *and all the derivatives are continuous in* (t,x) ($\mathbb{P}$*-a.s.*).

2.3. The proof of Theorem 1 is based on the following important estimate.

Lemma 3. Given $m\in\mathbb{N}\cup\{0\}$ *and* $p\in[2,\infty[$ *we assume that*

a) The coefficients of equation (0.1) *satisfy the conditions of Theorem* 1 *and for* $i,j=1,2,\dots,d$, b^i_j *is uniformly bounded.*

b) The functions $u\in\mathbb{W}^{m+2}_p$, $g^l\in\mathbb{W}^{m+1}_p$, *where* $l=1,2,\dots,d$, $f\in\mathbb{W}^m_p$ *and* $r^\alpha\in\mathbb{R}_+\setminus\{0\}$, *where* $|\alpha|\le m$, *are given.*

Then for every $(T,\omega)\in[T_0,T]\times\Omega$,

$$V(m,p,u,f,g,\{r^\alpha\}):=\int_{\mathbb{R}^d}\sum_{|\alpha|\le m} r^\alpha\Big(G'(u_\alpha)\,(a^{ij}u_{ij}+b^iu_i+$$

$$+\,cu+f)_\alpha+\tfrac{1}{2}\,G''(u_\alpha)\sum_{j=1}^{d_1}|(\sigma^{ij}u_i+h^ju+g^j)_\alpha|^2\Big)dx\le \tag{2.2}$$

$$\le N\Big(\|u\|^p_{m,p}+\|f\|^p_{m,p}+\sum_{j=1}^{d_1}\|g^j\|^p_{m+1,p}\Big),$$

where the constant N *depends only on* r^α, p, d, d_1, K *and* m, $G'(s):=p|s|^{p-2}s$, *and* $G''=p(p-1)\,|s|^{p-2}$ *for every* $s\in\mathbb{R}^1$. □

Proof. For ξ, $\eta\in\mathbb{L}_1$ we write $\xi\sim\eta$ if ξ and η have the same integrals over $\mathbb{R}^d$ and we write $\xi\ll\eta$ if $\xi\sim\eta+\theta$, where

$$\theta\le N\Big(\sum_{|\alpha|\le m}(|u_\alpha|^p+|f_\alpha|^p+\sum_{|\alpha|\le m+1}\sum_{j=1}^{d_1}|g^j_\alpha|^p\Big).$$

In the course of the proof it will be assumed that $|\alpha|\le m$. As a rule we shall drop the argument u_α of the functions $G(u_\alpha)=|u_\alpha|^p$, $G'(u_\alpha)$, and $G''(u_\alpha)$.

From inequality (1.22) it follows that $G'f_\alpha\ll 0$ and $G'(Cu)_\alpha\ll 0$. It is also evident that

$$G'(b^i u_i)_\alpha \ll (u_i)_\alpha G' b^i = G_i b^i = (Gb^i)_i - Gb^i_i \sim Gb^i_i \ll 0.$$

By inequality (1.22),

$$G'' \sum_{j=1}^{d_1} |g^j_\alpha|^2 \ll 0, \; G'' \sum_{j=1}^{d_1} |h^j u)_\alpha|^2 \ll 0$$

and

$$G''(h^j u)_\alpha g^j_\alpha \leq \frac{1}{2} G'' \sum_{j=1}^{d_1} [(h^j u^2_\alpha) + (g^j_\alpha)^2] \ll 0,$$

$$G''(\sigma^{ij} u_i)_\alpha g^j_\alpha \ll (u_i)_\alpha \, G'' \sigma^{ij} g^j_\alpha = G'_i \sigma^{ij} g^j_\alpha.$$

Integrating by parts, we obtain, with the help of inequality (1.22), that

$$G'_j \sigma^{ij} g^j_\alpha \sim - G'(\sigma^{ij} g^j_\alpha)_i \ll 0,$$

$$G''(\sigma^{ij} u_i)_\alpha (h^j u)_\alpha \ll (u_i)_\alpha \, G' \sigma^{ij} (h^j u)_\alpha = G'_j \sigma^{ij} (h^j u)_\alpha \ll$$

$$\ll - G' \sigma^{ij} (h^i u_i)_\alpha \ll - (u_i)_\alpha G' \sigma^{ij} h^j =$$

$$= - G_i \sigma^{ij} h^j \sim G(\sigma^{ij} h^j)_i \ll 0.$$

It remains to prove the inequality,

$$\sum_{|\alpha| \leq m} r^\alpha \left((a^{ij} u_{ij})_\alpha \, G' + \frac{1}{2} G'' \sum_{j=1}^{d_1} |(\sigma^{ij} u_i)_\alpha|^2 \ll 0. \right. \tag{2.3}$$

For $m=0$, (2.3) follows from the parabolic condition (0.2). Actually, by integration by parts we obtain

$$(a^{ij} u_{ij}) \, G' = (a^{ij} u_i)_j G' - a^{ij}_i u_i G' \sim \tag{2.4}$$

$$- a^{ij}u_iG''u_j - a^{ij}_j(G)_i \sim - G''a^{ij}u_iu_j + a^{ij}_{ij}G \ll - G'a^{ij}u_iu_j$$

Suppose now that $|\alpha|\geq 1$. Evidently,

$$(a^{ij}u_{ij})_\alpha G' \ll \Sigma_\alpha (a^{ij})_\beta (u_{ij})_\gamma G' + a^{ij}(u_{ij})_\alpha G',$$

where

$$\Sigma_\alpha := \sum_{\substack{\beta+\gamma=\alpha \\ |\beta|=1}} .$$

In the same way as in (2.4), we obtain

$$a^{ij}(u_{ij})_\alpha G' = (a^{ij}(u_i)_\alpha)_j G' - a^{ij}_j(u_i)_\alpha G' \sim -a^{ij}(u_i)_\alpha G''(u_j)_\alpha -$$

$$- a^{ij}_j(G)_i \sim -G''a^{ij}(u_i)_\alpha (u_j)_\alpha + a^{ij}_{ij}G \ll - G''a^{ij}(u_i)_\alpha (u_j)_\alpha.$$

Thus it has been proved that for $|\alpha|\geq 1$,

$$(a^{ij}u_{ij})_\alpha G' \ll - G''a^{ij}(u_i)_\alpha (u_j)_\alpha + G'\Sigma_\alpha (a^{ij})_\beta (u_{ij})_\gamma. \tag{2.5}$$

Obviously,

$$G'' \sum_{l=1}^{d_1} |(\sigma^{il}u_i)_\alpha|^2 - G'' \sum_{l=1}^{d_1} |(\sigma^{il}u_i)_\alpha - \sigma^{il}(u_i)_\alpha + \sigma^{il}(u_i)_\alpha|^2 \ll$$

$$\ll 2(u_j)_\alpha G''\sigma^{jl}[(\sigma^{il}u_i)_\alpha - \sigma^{il}(u_i)_\alpha] + G'' \sum_{l=1}^{d_1} |\sigma^{il}(u_i)_\alpha|^2 =$$

$$= 2G'_j\sigma^{jl}[(\sigma^{il}u_i)_\alpha - \sigma^{il}(u_i)_\alpha] + G'' \sum_{l=1}^{d_1} |\sigma^{il}(u_i)_\alpha|^2.$$

In the sequel we use the following notation

$$\sum_{|\beta|\neq 1}^{\alpha} := \sum_{\substack{\beta+\gamma=\alpha \\ |\beta|>1}} .$$

It is clear that

$$2G'_j\sigma^{jl}[(\sigma^{il}u_i)\alpha - \sigma^{il}(u_i)\alpha] =$$

$$= 2G'_j\sigma^{jl} \sum_{|\beta|\neq 1}^{\alpha} \sigma^{il}_{\beta}(u_i)_\gamma + 2G'_j\sigma^{jl} \sum\alpha\, \sigma^{il}_{\beta}(u_i)_\gamma .$$

If $|\beta|>1$, then $|\gamma|\leq m$-2 and integration by parts yields,

$$2G'_j\sigma^{jl} \sum_{|\beta|\neq 1}^{\alpha} \sigma^{il}_{\beta}(u_i)_\gamma \sim 2G'\Big(\sigma^{il} \sum_{|\beta|\neq 1}^{\alpha} \sigma^{il}_{\beta}(u_i)_\gamma\Big)_j \ll 0.$$

Moreover, it is clear that

$$2G'_j\sigma^{jl} \sum\alpha\, \sigma^{il}_{\beta}(u_i)_\gamma \ll -\, 2G'\sigma^{jl} \sum\alpha\, \sigma^{il}_{\beta}(u_{ij})_\alpha .$$

Thus we have proved the relation,

$$(2.6) \quad G'' \sum_{l=1}^{d_1} |(\sigma^{il}u_i)\alpha|^2 \ll G'' \sum_{l=1}^{d_1} |\sigma^{il}(u_i)\alpha|^2 - 2G' \sum\alpha\, \sigma^{jl}\sigma^{il}_{\beta}(u_{ij})_\gamma$$

From relations (2.5), (2.6) it follows that

$$\eta := \sum_{1\leq|\alpha|\leq m} r^{\alpha}\left[(a^{ij}u_{ij})\alpha G' + \tfrac{1}{2}\, G'' \sum_{l=1}^{d_1} |(\sigma^{il}u_i)\alpha|^2 \right] \ll$$

$$(2.7) \qquad \ll \sum_{1\le\alpha\le m} r^\alpha\Big(G''\big(-a^{ij}(u_i)_\alpha\,(u_j)_\alpha + \tfrac{1}{2}\sum_{l=1}^{d_1}|\sigma^{il}(u_i)_\alpha|^2\big) +$$

$$+\; G'\sum_\alpha\,[(a^{ij})_\beta\,(u_{ij})_\gamma - \sigma^{jl}\sigma^{il}_\beta(u_{ij})_\gamma] =$$

$$= \sum_{1\le|\alpha|\le m} r^\alpha\left[\;G'\sum_\alpha A^{ij}_\beta(u_{ij})_\gamma - \sum_{1\le|\alpha|\le m} G''A^{ij}(u_i)_\alpha(u_j)_\alpha\right],$$

where $A^{ij} := a^{ij} - \frac{1}{2}\,\sigma^{il}\sigma^{jl}$.

We will need for further use, the following well-known result (see e.g. [104], Ch. 1, §7).

Proposition 3. Suppose that for i, $j=1,2,\dots,d$, $b^{ij}\in C^2_b(\mathbf{R}^d)$, *and* $b^{ij}\xi^i\xi^j\ge 0$ *for every* $x\in\mathbf{R}^d$, *where* ξ^i, ξ^j *belonging to* $\mathbf{R}^1$.

Then for every function $v\in C^2(\mathbf{R}^d)$,

$$(b^{ij}_l(x)v_{ij}(x))^2 \le Nb^{ij}(x)\;v_{il}(x)\;v_{jl}(x),$$

where the constant N depends only on the second derivatives of b^{ij}. □

In view of the proposition

$$(2.8) \qquad \sum_\alpha|\,A^{ij}_\beta(u_{ij})_\gamma|^2 \le N\sum_{|\gamma|\le m-1} A^{ij}(u_{il})_\gamma(u_{jl})_\gamma \le$$

$$\le N_1\sum_{1\le|\alpha|\le m} r^\alpha A^{ij}(u_i)_\alpha\,(u_j)_\alpha.$$

From the inequality (3.1.10) it follows that

$$(2.9) \quad \sum_{1\le|\alpha|\le m} r^\alpha G'\sum_\alpha A^{ij}_\beta(u_{ij})_\gamma \le 2\varepsilon^{-1}G + \varepsilon 2^{-1}N_2G''\sum_\alpha|A^{ij}_\beta(u_{ij})_\gamma|^2.$$

Making ε sufficiently small, we obtain from inequalities (2.7) to (2.9) that $\eta \ll 0$.

Thus inequality (2.2) is proved. By a simple analysis of the proof, the reader will find that this constant depends only on p, d, d_1, r^α, K and m. □

2.4. In the next step of the proof we shall apply Theorems 3.2.2 and 3.3.1. We shall use the Hilbert scale of Sobolev's spaces. Put $\lambda = m$.

As was mentioned in §4.1.0, we can transform equation (0.1) to the form (1.1). Thus, arguing as in the proof of Theorem 1.1, it is easy to demonstrate that a generalized solution of problem (0.1), (1.2) is also a solution of the LSES of the type (3.0.1) in $(\mathbb{H}^1, \mathbb{L}_2, \mathbb{H}^{-1})$ whose the operators A, B and the martingale $M(t)$ are given by the formulas,

$$[y,\ A(t,\omega)v]_0 := -(a^{ij}(t,\omega)v_{i}, y_{i})_0 +$$

$$+ ((b^i(t,\omega) - a^{ij}_j(t,\omega))v_i + c(t,\omega)v, y)_0, \quad \forall v,\ y \in \mathbb{H}^1,$$

$$B(t,\omega)v := (\mathcal{M}^1(t,\omega)v, \ldots, \mathcal{M}^{d_1}(t,\omega)v), \quad \forall v \in \mathbb{H}^1, \quad (\textit{see } §.1),$$

$$M(t) := \int_{[0,t]} g^l(s)\,dw^l(s).$$

From the arguments given in the beginning of the proof of Theorem 1.1, it follows that the operator $A(t,\omega)$ is uniformly (relative to (t,ω)) continuous from $\mathbb{H}^1$ into $\mathbb{H}^{-1}$, that is, there exists a constant K such that for all t,ω,

$$\|A(t,\omega)v\|_{-1} \le K\|v\|_1, \quad \forall v \in \mathbb{H}^1. \tag{2.10}$$

For reasons similar to the ones presented in the beginning of the proof of Theorem 1.2, it follows that the operator A acting on functions from $\mathbb{H}^{m+1}$ coincides with $\mathcal{L}(\cdot) = a^{ij} \times (\cdot)_{ij} + b^i \times (\cdot)_i + c \times (\cdot)$ and the following inequality holds:

$$\|A(t,\omega)v\|_{m-1} \le K\|v\|_{m+1}, \quad \forall v \in \mathbb{H}^{m+1},\ (t,\omega) \in [T_0, T] \times \Omega.$$

Since the differentiation operator is bounded as an operator from $\mathbb{H}^{m+1}$ into $\mathbb{H}^m$ and from $\mathbb{H}^1$ into $\mathbb{H}^0$ (see Proposition 3.4.6), we have that for all $(t,\omega)) \in [T_0, T] \times \Omega$,

$$\|B(t,\omega)v\|^2_{\mathfrak{L}_2(\mathbb{R}^d,\mathbb{H}^n)} \le K\|v\|_{n+1} \;\forall v\in\mathbb{H}^{n+1},\; n=0,\, m.$$

Thus condition (B) of Theorem 3.2.2 is fulfilled. The validity of condition (C) of the same theorem follows immediately from the assumptions of our theorem. We show that condition (A) of Theorem 3.2.2 is also satisfied.

Let $u\in\mathbb{H}^{m+2}$. It is clear that $A\cdot\mathbb{H}^{m+2}\to\mathbb{H}^m$. Thus $[u,Au]_m = (u,\, \mathcal{L}u)_m$ for every $u\in\mathbb{H}^{m+2}$. From (3.2.2), in view of self-adjointness of the operator $\Lambda := (\mathbb{I}-\Delta)^{1/2}$, it follows that

$$(u,\, \mathcal{L}u)_m = (\mathbb{I}-\Delta)^m u,\, \mathcal{L}u)_0.$$

Thus representation (1.10) yields that

$$[u,\, A(t,\omega)u]_m + \tfrac{1}{2}\,\|B(t,\omega)u\|^2_{\mathfrak{L}_2(\mathbb{R}^d,\mathbb{H}^m)} =$$

$$= \sum_{|\alpha|\le m} C^\alpha\Big[((\mathcal{L}(t,\omega)u)_\alpha, u_\alpha)_0 + \tfrac{1}{2}\,\|\mathcal{M}^l(t,\omega)u)_\alpha\|_0^2\Big] =$$

$$= NV(m,\, 2,\, u,\, 0,\, 0,\, \{c^\alpha\},\, t),$$

where V is the function introduced in the lemma.

From the last inequality and (2.2) it follows that condition (A) holds for $u\in\mathbb{H}^{m+2}$. Since $\mathbb{H}^{m+2}$ is dense in $\mathbb{H}^{m+1}$ and $A(t,\omega)$ is a uniformly continuous operator from $\mathbb{H}^{m+1}$ into $\mathbb{H}^m$, the last statement can be carried over to the case of $u\in\mathbb{H}^{m+1}$ by passing to the limit.

Thus the LSES discussed above has a solution in $(\mathbf{H}^1,\mathbf{L}_2,\mathbf{H}^{-1})$ belonging to $\mathbf{L}_2([T_0,T];\, \mathcal{P};\, \mathbb{H}^m)$ which is simultaneously a generalized solution of problem (0.1), (1.2).

From (2.10), (2.11) it follows that the latter LSES satisfies the assumptions of Theorem 3.3.1 and thus problem (0.1), (1.2) has a unique generalized solution.

The rest of the statement of the theorem can be proved in complete analogy with the proof of corresponding assertion in Theorem 1.4.

In particular, the following auxiliary result holds, which by the way is of independent.

Corollary 4. Suppose that the assumptions of Theorem 1 are satisfied. Let $\{a_n^{ij}, b_n^i, c_n, \sigma_n^{il}, h_n^l, f_n, g_n^l, \varphi_n\}$, *where* $i,j=1,2,\ldots,d$, $l=1,2,\ldots,d_1$, *and* $n\in\mathbb{N}$, *be a collection of functions satisfying the same assumptions (with constants that do not depend on* n*) as the functions* $a^{ij}, b^i, c, \sigma^{il}, h^l, f, g^l, \varphi$. *Suppose, as well, that* u^n *is the generalized solution of problem (0.1), (1.2), where* a^{ij} *is replaced by* a_n^{ij}, b^i *by* b_n^i *etc.*

If for all $i,j,l, a_n^{ij} \to a^{ij}$, $b_n^i \to b^i$, $c_n \to c$, $\sigma_n^{il} \to \sigma^{il}$ *and* $h_n^l \to h^l$, $(l\times l\times\mathbb{P}$-*a.s.), as* $n\to\infty$, *and*

$$\lim_{n\to\infty} \mathbb{E}\left(\int_{[T_0,T]} \left(\|f_n(t)-f(t)\|_m^2 + \sum_{l=1}^{d_1} \|g_n^l(t)-g(t)\|_{m+1}^2\right)dt + \right.$$

$$\left. + \|\varphi_n-\varphi\|_m^2 \right) = 0,$$

then

$$\lim_{n\to\infty} \mathbb{E} \sup_{t\in[T_0,T]} \|u^n(t)-u(t)\|_m^2 = 0. \quad \square$$

The proof of this assertion and of the rest of the statements of the theorem can easily be provided by the reader.

4.3. The Forward Cauchy Problem and the Backward One in Weighted Sobolev Spaces.

3.0. In this section we consider the Cauchy problem for equations (0.1) and (1.1) with the initial value φ and the external forces $f(t)$, $g^l(t)$, where $l=1,2,\ldots,d$, taking values in weighted Sobolev spaces $\mathbf{W}_p^m(r)$.

In addition, we consider the following versions of the backward Cauchy problem (see §1.4.12):

(3.1) $$-dv(t,x,\omega) = [\mathcal{L}_{div}\, v(t,x,\omega) + f(t,x,\omega)]\,dt +$$
$$+ [\mathcal{M}^l v(t,x,\omega) + g^l(t,x,\omega)]*dw^l(t),$$
$$(t,x,\omega)\in[T_0,T[\times\mathbb{R}^d\times\Omega;$$

(3.2) $$v(T,x,\omega) = \varphi(x,\omega),\ (x,\omega)\in\mathbb{R}^d\times\Omega,$$

and

(3.3) $$-dv(t,x,\omega) = [\mathcal{L}v(t,x,\omega) + f(t,x,\omega)]dt + $$
$$+ [\mathcal{M}^l v(t,x,\omega) + g^l(t,x,\omega)]*dw^l(t),$$
$$(t,x,\omega)\in[T_0,T[\times\mathbf{R}^d\times\Omega,$$

(3.4) $$v(T,x,\omega) = \varphi(x,\omega),\ (x,\omega)\in\mathbf{R}^d\times\Omega.$$

Here, just as above,

$$\mathcal{L}_{div}\, v(t,x,\omega):= (a^{ij}(t,x,\omega)v_i(t,x,\omega))_j +$$
$$+ b^i(t,x,\omega)v_i(t,x,\omega) + c(t,x,\omega)v(t,x,\omega);\ \mathcal{L}v(t,x,\omega):= a^{ij}(t,x,\omega)\ \ v_{ij}(t,x,\omega)$$
$$+ b^i(t,x,\omega)v_i(t,x,\omega) + c(t,x,\omega)v(t,x,\omega);\ \mathcal{M}^l v(t,x,\omega):=$$
$$= \sigma^{il}(t,x,\omega)v_i(t,x,\omega) + h(t,x,\omega)v(t,x,\omega);\ i,j=1,2,...,d,\ l=1,2,...,d_1.$$

The coefficients, the initial value and the external forces in problems (3.1), (3.2) and (3.3), (3.4) depend on the "the future" of the Wiener process $w(t)$.

In the above problems, the most important case is the one when $r\leq 0$ because it includes the case of bounded or mildly growing initial values and external forces.

The core of the sections is a collection of existence and uniqueness theorems for problems ((0.1), (1.2)), (1.1), (1.2)), ((3.), (3.2)), and ((3.3), (3.4)) in the space $\mathbf{L}^p([T_0,T]\times\Omega;\ \mathbf{W}_p^m(r))$ and in the space of differentiable functions.

Let us fix $r\in\mathbf{R}^1$, $K\in\mathbf{R}_+$, $p\in[2,\infty[$, *and* $m\in\mathbf{N}\cup\{0\}$.

Throughout this section it is supposed that the coefficients a^{ij}, b^i,c,σ^{il} *and* h^l, where $i,j=1,2,...,d$ and $l=1,2,...,d_1$, are $\mathcal{B}([T_0,T]\times\mathbf{R}^d)\otimes\mathcal{F}$-measurable functions. Considering problem (0.1), (1.2), we shall suppose that the coefficients a^{ij}, where $i,j=1,2,...,d$, possess bounded derivatives of first order.

3.1. In this item it is assumed that φ is an $\mathcal{F}_{T_0}$-measurable function taking values in $\mathbf{L}_2(r)$, $f(t):= f^0(t) + \sum_{i=1}^{d} f_i^i(t)$, where f^0 and $f_i^i\in\mathbf{L}_2^\omega([T_0,T];\ \mathcal{P};\ \mathbf{L}_2(r))$, $i=1,2,...,d$ and $g^l\in\mathbf{L}_2^\omega([T_0,T];\ \mathcal{P};\ \mathbf{L}_2(r))$, where $l=1,2,...,d_1$. The coefficients of equation (1.1) are assumed to be predictable (for every x).

Definition 1. *A function* $u\in\mathbf{L}_2^\omega([T_0,T];\ \mathcal{P};\ \mathbf{H}^1(r))$ *is called an r-generalized solution of problem* (1.1), (1.2) *if it satisfies equality* (1.3) *for every* $y\in\mathbf{C}_0^\infty(\mathbf{R}^d)$, $(l\times\mathbf{P}$*-a.s.*). □

Obviously, the definitions of an r-generalized solution and a generalized one coincide.

Define $S(x) = (1+|x|_d^2)^{r/2}$ *and* $S(i,x):= rx^i(1+|x|_d^2)^{-1}$.

It is clear that for every $i=1,2,\dots,d$, $(S^{-1}(x))_i\, S(x) = - S(i,x)$. In the sequal the argument x of the functions $S(x)$ and $S(i,x)$ will be omitted.

Theorem 1. *Suppose that the following conditions are fulfilled.*

(*i*) *The functions* a^{ij}, b^i, c, σ^{il}, *and* h^l, *where* $i,j=1,2,\dots,d$ *and* $l=1,2,\dots,d_1$ *are differentiable in* x *up to order* m *and satisfy condition* (0.3). *The functions and their derivatives mentioned above are bounded by a constant* K *which does not depend on* t,x,ω.

(*ii*) *For* p' *equal to* p *and* 2, $\varphi\in \mathbb{L}_{p'}(\Omega,\mathbb{W}^m_{p'}(r))$, f^0, $f^i\in\mathbb{L}_{p'}([T_0,T]\times\Omega;$ $\mathbb{W}^m_{p'}(r))$, *where* $l=1,2,\dots,d_1$, *and* $g^l\in\mathbb{L}_{p'}([T_0,T]\times\Omega;\ \mathbb{W}^m_{p'}(r))$, *where* $l=1,2,\dots,d_1$.

Then problem (1.1), (1.2) *has* r-*generalized solution belonging to* $\mathbb{L}_2(\Omega;\mathbb{C}([T_0,T],\ \mathbb{H}^{m-1}(r))\ \cap\ \mathbb{L}_2([T_0,T]\times\Omega;\ \mathbb{C}_w\mathbb{H}^m(r))\ \cap\ \mathbb{L}_p([T_0,T]\times\Omega;$ $\mathbb{C}_w\mathbb{W}^m_p(r))$ *and for* $p'=p$ *and* 2 *satisfies the following inequality*

$$
\mathbb{E}\sup_{t\in[T_0,T]}\|u(t)\|^{p'}_{m,p',r} + \frac{\delta}{2}\,p(p-1)\,\mathbb{E}\int_{[T,T_0]}\sum_{|\alpha|\ge m}\sum_{i=1}^{d}\int_{\mathbb{R}^d}|(Su)_\alpha|^{p-2}\times
$$

$$
(3.5)\qquad \times\,|((Su)_i)_\alpha|^2\,dx\,dt \le N\mathbb{E}\Bigg(\|\varphi\|^{p'}_{m,p',r} + \int_{[T_0,T]}\Big(\sum_{j=0}^{d}\|f^j(t)\|^{p'}_{m,p',r} +
$$

$$
\sum_{l=1}^{d_1}\|g^l(t)\|^{p'}_{m,p',r}\Big).\ \square
$$

Remark 1. *Condition* (*i*) *of the theorem is sufficient for the uniqueness of* r-*generalized solutions because the difference of two* r-*generalized solutions is also an* r-*generalized solution of the problem with zero initial value and external forces and satifies condition* (*ii*) *automatically.*

Proof. If $r=0$ the statement of the theorem coincides with the one of Theorem 4.1.4. We now show that the case of $r\neq 0$ can be easily reduced to the case of $r=0$.

From Lemma 3.4.7 (ii) it follows that $S\varphi\in\mathbb{L}_{p'}(\Omega;\ \mathbb{W}^m_{p'})$ Sf^0 and Sf^n_n belong to $\mathbb{L}_{p'}([T_0,T];\ \mathcal{P};\ \mathbb{W}^m_{p'})$, and $g^l\in\mathbb{L}_{p'}([T_0,T];\ \mathcal{P};\ \mathbb{W}^m_{p'})$ for $n=1,2,\dots,d$, $l=1,2,\dots,d_1$

and p' equal to p and 2.

Consider the problem

$$dv(t) = [a^{ij}v_i(t))_j + \tilde{b}^i v_i(t) - S(i)(a^{ij}v(t))_j + \tilde{c}v(t) + \tilde{f}^0(t) -$$

$$- \sum_{i=1}^{d} (Sf^i(t)_i]dt + [\sigma^{il}v_i(t) + \tilde{h}^l(t) + Sg^l(t)]dw^l(t), \tag{3.6}$$

$$(t;\ x,\ \omega)\in[T_0,T]\times\mathbf{R}^d\times\Omega,$$

$$v(T_0) = S\varphi,\ (x,\omega)\in\mathbf{R}^d\times\Omega, \tag{3.7}$$

where

$$\tilde{b}^i := b^i - a^{ij}S(j),\ \tilde{c} := c - S(i)b^i + S(S^{-1})_{ij}a^{ij},\ \tilde{f}^0 := Sf^0 -$$

$$- \sum_{i=1}^{d} S(i)\ Sf^i,\ \tilde{h}^l = h^l - S(i)\sigma^{il}.$$

Note that problem (3.6), (3.7) could informally be obtained by termwise multiplying of (1.1), (1.2) and S and subsequent change of variable $v := Su$.

The coefficients, the initial value and the external forces in problem (3.6), (3.7) satisfy the assumptions of Theorem 1.4. Thus this problem has a generalized solution $v\in\mathbf{L}_2(\Omega;\mathbf{C}([T_0,T];\ \mathbf{H}^{m-1}))\cap\mathbf{L}_2([T_0,T]\times\Omega;\ \mathbf{C}_w\mathbf{H}^m)\cap\mathbf{L}_p([T_0,T]\times\Omega;\ \mathbf{C}_w\mathbf{W}_p^m)$, which satisfies inequality (1.21).

Let $\eta\in\mathbf{C}_0^\infty(\mathbf{R}^d)$. If we replace y by $S^{-1}\eta$ in the integral equality (of the form (1.3)) for the generalized solution of problem (3.6), (3.7) and define $u := S^{-1}v$ it can easily be seen that for all $\eta\in\mathbf{C}_0^\infty(\mathbf{R}^d)$ and $(t,\omega)\in[T_0,T]\times\Omega$, where $\Omega'\subset\Omega$ and $\mathbf{P}(\Omega') = 1$, the following holds

$$(u(t),\eta)_0 = (\varphi,\eta) + \int_{[T_0,t]} [-(a^{ij}u_i(s);\eta_j)_0 + (b^iu_i(s) + cu(s) +$$

$$+ f^0(s),\eta)_0 -)f^i(s),\eta_i)_0]ds +$$

$$+ \int_{[T_0,t]} (\sigma^{il}u_i(s) + h^lu(s) + g^l(s),\eta)_0dw^l(s).$$

From this equality and Lemma 3.4.7 (ii), it follows that u is an r-generalized solution of problem (1.1), (1.2).

From the properties of v (see Theorem 1.4.) and Lemma 3.4.7 (ii) it follows that the above solution u belongs to

$$\mathbb{L}_2(\Omega;\ \mathbb{C}([T_0,T];\ \mathbb{H}^{m-1}(r))\cap\mathbb{L}_2([T_0,T]\times\Omega;\ \mathbb{C}_w\mathbb{H}^m(r))\cap\mathbb{L}_p([T_0,T]\times\Omega;\ \mathbb{C}^w\mathbb{W}_p^m(r))$$

and satisfies inequality 3.4.7 (ii).

It remains to verify the uniqueness of r-generalized solutions of problem (1.1), (1.2).

Let u_1 and u_2 be two r-generalized solutions of problem (1.1), (1.2).

Write $\bar{u}:= u_1-u_2$. Obviously $\bar{u}$ satisfies equality

$$(\bar{u}(t),\eta) = \int_{[T_0,t]} [-(a^{ij}\bar{u}_i(s),\eta_j)_0 + (b^i u_i(s) + cu(s),\eta)_0]ds +$$

$$+ \int_{[T_0,t]} (\sigma^{il}u_i(s) + h^l u(s),\eta)_0 dw^l(s),$$

($l\times\mathbb{P}$-a.s.) for every $\eta\in\mathbb{C}_0^\infty(\mathbb{R}^d)$. By substituting in the equality Sy instead of η, where $y\in\mathbb{C}_0^\infty(\mathbb{R}^d)$, we easily obtain (see Lemma 3.4.7 (ii)) that $\bar{v}:= S\bar{u}$ belongs to $\mathbb{L}_2^\omega([T_0,T];\ \mathfrak{P};\ \mathbb{H}^1)$ and satisfies the following equality

$$(\bar{v}(t),y)_0 = \int_{[T_0,t]} [-(a^{ij}\bar{v}_i(s) - S(i)a^{ij}\bar{v}(s),y_i)_0 +$$

$$+ (\tilde{b}^i\bar{v}_i(s) + \tilde{c}\bar{v}(s),y)_0]ds + \int_{[T_0,t]} (\sigma^{il}\bar{v}_i(s) + \tilde{h}^l\bar{v}(s),y)_0 dw^l(s),$$

($l\times\mathbb{P}$-a.s.), for every $\eta\in\mathbb{C}_0^\infty(\mathbb{R}^d)$.

Theorem 1.1 and Remark 1.6 yield that $\|\bar{v}(t)\|_m = 0$, ($l\times\mathbf{P}$-a.s.). From this and Lemma 3.4.7 (ii), it follows that $\|\bar{u}(t)\|_{m,2,r} = 0$, ($l\times\mathbf{P}$-a.s.).

From Proposition 1.3 we readily obtain the following result.

Proposition 1. *Let* (U,Σ) *be a measurable space and* ξ *be a* Σ*-measurable mapping of* U *into* $\mathbb{W}_p^m(r)$, *where* $p\geq 1$ *and* $m>n+d/p$ *for some* $n\in\mathbb{N}\cup\{0\}$. *Then there exists a function* $\tilde{\xi}$: $U\times\mathbb{R}^d \to \mathbb{R}$ *with the following properties.*

(*a*) $\tilde{\xi}$ *is* $\Sigma\otimes\mathfrak{B}(\mathbb{R}^d)$*-measurable.*

(*b*) *For every* $u\in U$ *and* $\omega\in\Omega$, $\tilde{\xi}(u,\cdot)\in\mathbb{W}_p^m(r)$, $S(\cdot)\tilde{\xi}(u,\cdot)\in\mathbb{C}_b^n(\mathbb{E}^d$ *and*

$\|S(\cdot)\tilde{\xi}(u,\cdot)\|_{C_b^n(\mathbf{R}^d)} \leq N\|\xi(u,\cdot)\|_{m,p}$, *where N depends only on m,p,d,r and n.*

(c) $\|\tilde{\xi}(u,\cdot) - \xi(u,\cdot)\|_{m,p,r} = 0$ *for every* u,ω. □

Making use of the theorem of this item and arguing in the same way as in the proof of Corollary 1.4, we easily obtain the following statement.

Corollary 1. *Suppose that the assumptions of Theorem* 1 *are fulfilled and for some* $n\in\mathbf{N}\cup\{0\}$, $(m-n)p>d$. *Then the r-generalized solution of problem* (1.1), (1.2) *has a version (in x)* $v(t,x,\omega)$ *with the following properties.*

(a) *For every* $x\in\mathbf{R}^d$, *the function* $v(t,x,\omega)$ *is a predictable real-valued stochastic process.*

(b) *For every* $w\in\Omega$ *the function* $S(x)\ v(t,x,\omega)\in C_b^{0,n}([T_0,T]\times\mathbf{R}^d)$.

(c) *The process* $v(t,x,\omega)$ *is an r-generalized solution of problem* (1.1), (1.2).

(d) $\mathbf{E} \sup_{t\in[T_0,T]} \|Sv(t)\|^p_{C_b^n(\mathbf{R}^d)} < \infty$.

(e) *If* $n\geq 2$ *and* $|\alpha|\leq n-2$, $\tilde{f}$ *and* $\tilde{g}^l$ *are the versions of f and* g^l *whose existence is ensured by the proposition (for* $(S;\Sigma):= ([T_0,T]\times\Omega;\mathcal{P})$), *then for every* $(t,\omega)\in[T_0,T]\times\Omega_{x,t}$, *where* $\Omega_{x,t}\subset\Omega$ *and* $\mathbf{P}(\Omega_{x,t}) = 1$ *the following equality holds*

$$v_\alpha(t) = \varphi_\alpha + \int_{[T_0,t]} (\mathcal{L}_{div}v(s) + \tilde{f}(s))_\alpha ds +$$

$$+ \int_{[T_0,t]} (\mathcal{M}^l v(s) + g^l(s))_\alpha dw^l(s);$$

(f) *If* $v_1(t,x)$ *and* $v_2(t,x)$ *are r-generalized solutions of problem* (1.1), (1.2) *possessing properties* (a), (b), *then*

$$\mathbf{P}\Big(\sup_{\substack{t\in[T_0,t]\\ x\in\mathbf{R}_d}} |v_1(t,x,\omega) - v_2(t,x,\omega)| > 0\Big) = 0. \quad \square$$

3.2. In this item it is supposed that φ is an $\mathcal{F}_{T_0}$-measurable random variable taking values in $\mathbf{L}_2(r)$, $f(t)$ and $g^l(t)$ then belong to $\mathbf{L}_2^\omega([T_0,T];\ \mathcal{P};\ \mathbf{L}_2(t))$ and the coefficients of equation (0.1) are predictable for every x.

Definition 2. *A function* $u \in \mathbb{L}_2^{\omega}([T_0,T];\ \mathcal{P};\ \mathbb{H}^1(r))$ *is called an r-generalized solution of problem* (0.1), (1.2) *if it satisfies equality* (2.1) *for every* $y \in C_0^{\infty}(\mathbb{R}^d)$, $(l \times \mathbb{P}$-*a.s.*).

Theorem 2. Given $m \in \mathbb{N}$ *we suppose that the following assumptions hold.*

(*i*) *The condition* (0.2) *is fulfilled.*

(*ii*) *The coefficients* a^{ij} *are differentiable up to the order of* $2 \vee m$, σ^{il} *and* h^l *up to the order of* $m+1$, *and* f^i *and* c *up the order of* $m (i,j=1,2,\dots,d,\ l=1,2,\dots,d_1)$. *The absolute values of these coefficients and their derivatives just mentioned above are bounded by a constant K.*

(*iii*) *Given* $p'=p$ *and* 2, $i=1,2,\dots,d$, *and* $l=1,2,\dots,d_1$, f, g^l, *and* $g_i^l \in \mathbb{L}_{p'}([T_0,T] \times \Omega;\ \mathbb{W}_{p'}^m(r))$ *and* $\varphi \in \mathbb{L}_{p'}(\Omega;\ \mathbb{W}_{p'}^m(r))$.

Then problem (0.1), (1.2) *has a unique r-generalized solution u. This solution belongs to*

$$\mathbb{L}_2(\Omega;\mathbb{C}([T_0.T];\mathbb{H}^{m-1}(r)) \cap \mathbb{L}_2([T_0,T]\times\Omega;\mathbb{C}_w\mathbb{H}^m(r)) \cap \mathbb{L}_p([T_0,T]\times\Omega;\mathbb{C}_w\mathbb{W}_p^m(r)).$$

Moreover, for $p'=2$ *and* p, *the following inequality holds*

$$\mathbb{E} \sup_{t\in[T_0,T]} \|u(t)\|_{m,p',r}^{p'} \le N\mathbb{E}\Big(\|\varphi\|_{m,p',r}^{p'} + \tag{3.8}$$

$$+ \int_{[T_0,T]} \Big(\|f(t)\|_{m,p',r}^{p'} + \sum_{l=1}^{d_1} \|g^l(t)\|_{m+1,p',r}^{p'} \Big) dt \Big),$$

where the constant N depends only on p,d,d_1,K,m,T_0,T,r. □

Remark 2. Conditions (*i*), (*ii*) *suffice to ensure uniqueness of an r-generalized solution (see Remark* 1). □

Corollary 2. Suppose that the assumptions of the theorem are fulfilled and for some $n \in \mathbb{N} \cup \{0\}$, $(m-n)p > d$. *Then the r-generalized solution of problem* (0.1), (1.2) *has a version (in x)* $v(t,x,\omega)$ *which possess the properties* (*a*) *to* (*d*) *stated in Corollary* 1 *as well as the following ones:*

(*e*) *If the assumptions of item* (*e*) *of Corollary* 1 *are satisfied, then for every* $(t,\omega) \in [T_0,T] \times \Omega_{x,t}$, *where* $\Omega_{x,t} \subset \Omega$ *and* $\mathbb{P}(\Omega_{x,t}) = 1$ *the following equality holds*

$$v_\alpha(t) = \varphi_\alpha + \int_{[T_0,t]} (\mathcal{L}v(s) + \tilde{f}(s))_\alpha ds +$$

$$+ \int_{[T_0,t]} (\mathcal{M}^l v(s) + \tilde{g}^l(s))_\alpha dw^l(s);$$

(*f*) *If* $v_1(t,x)$ *and* $v_2(t,x)$ *are r-generalized solutions of problem* (0.1), (1.2) *possessing properties* (*a*), (*b*), *then*

$$\mathbf{P}(\sup_{\substack{t\in[T_0,T]\\ x\in\mathbf{R}^d}} |v_1(t,x,\omega) - v_2(t,x,\omega)|>0) = 0. \ \square$$

The above theorem and corollary can be derived from Theorem 2.1 *and Corollary* 2.1 *by the same method used in the proof of Theorem* 1 *and Corollary* 1.

3.3. *Corollary* 3. *If the assumptions of Theorem* 1 (*Theorem* 2) *are fulfilled for all* $m\in\mathbf{N}$ (*possibly with different constants for different* m), *then the solutions of problem* (1.1), (1.2) ((0,1), (1.2)) *are infinitely differentiable in* x *and all the derivatives are continuous in* t,x, ($\mathbf{P}$-*a.s.*). □

3.4. Here and in the following item, we consider problems (3.1), (3.2) and (3.3), (3.4).

Given a family $\{\mathcal{F}^t_T\}$, $t\in[0,T]$ of a sub-σ-algebras $\mathcal{F}$. We suppose that $\mathcal{F}^{t_1}_T \supset \mathcal{F}^{t_2}_T$ for $t_1\le t_2$ and $\bigcap_{\varepsilon>0} \mathcal{F}^{t-\varepsilon}_T = \mathcal{F}^t_T$ for $t\le T$. It is assumed also that the σ-algebra $\mathcal{F}^T_T$ is completed with respect to the measure $\mathbf{P}$. Denote $w_T(t):= w(T) - w(T-t)$ and assume that $w_T(t)$ is a Wiener martingale with respect to the family $\{\mathcal{F}^{T-t}_T\}$, $t\in[0,T]$ (cf. §1.4.1 and §1.4.12).

Let a^{ij}, f^i, c, σ^{il}, *and* h^l for i,j=1-d and l=1-d_1 be bounded,

---

$\mathcal{B}([T_0,T]\times\mathbf{R}^d)\otimes\mathcal{F}$-measurable functions on $[T_0,T]\times\Omega\times\mathbf{R}^d$. These functions are also assumed to be backward predictable for every $x\in\mathbf{R}^d$ with respect to the family $\{\mathcal{F}^t_T\}$. When (3.3), (3.4) is considered we will suppose in addition that the functions $a^{ij}(t,x,\omega)$ have bounded derivatives of the first order in x.

Warning 4. *Further in this chapter we consider backward predictable functions exclusively with respect to the family* $\{\mathcal{F}^t_T\}$. *Thus, later when discussing backward predictablility we will omit references to the related family of* σ-*algebras.* □

The following additional hypothesis are suppose to hold throughout this chapter:

a) φ is $\mathcal{F}^T_T$-measurable random variable taking values in $\mathbf{L}_2(r)$;

b) For $l=1\text{-}d$, $g^l(t)$ is a $\mathbb{L}_2(r)$-valued backward predictable stochastic process and $g^l \in \mathbb{L}_2^{\omega}([T_0,T];\ \mathbb{L}_2(r))$;

c) In problem (3.1), (3.3), $f(t) := f^0(t) + \sum_{i=1}^{d} f_i^i(t)$, where for $i=0\text{-}d$, $f^i(t)$ are $\mathbb{L}_2(r)$-valued backward predictable stochastic processes which belong to $\mathbb{L}_2^{\omega}([T_0,T];\ \mathbb{L}_2(r))$. In problem (3.3), (3.4) $f(t)$ is assumed to be a $\mathbb{L}_2(r)$-valued backward predictable stochastic process belonging to $\mathbb{L}_2^{\omega}([T_0,T];\ \mathbb{L}_2(r))$.

Definition 4. A backward predictable stochastic process $u \in \mathbb{L}_2^{\omega}([T_0,T];\ \mathbb{H}^1(r))$ *is called an r-generalized solution of problem* (3.1) *(respectively* (3.4)*) if for every* $y \in C_0^{\infty}(\mathbb{R}^d)$ *it satisfies* ($l \times \mathbb{P}$-*a.s.) the equality*

$$
\begin{aligned}
(u(t),y_0) = (\varphi,y)_0 &+ \int_{[t,T]} [-(a^{ij}u_i(s) + f^j(s),y_j)_0 + \\
&+ (b^i u_i(s) + cu(s) + f^0(s),y)_0]ds + \\
&+ \int_{[t,T]} (\sigma^{il}u_i(s) + h^l u(s) + g^l(s),y)_0 * dw^l(s),
\end{aligned}
\tag{3.9}
$$

$$
\begin{aligned}
(u(t),y)_0 = (\varphi,y)_0 &+ \int_{[t,T]} [-(a^{ij}u_i(s),y_j)_0 + ((b^i - a_j^{ij})u_i + \\
&+ cu(s) + f(s),y)_0]ds + \\
&+ \int_{[t,T]} (\sigma^{il}u_i(s) + h^l u(s) + g^l(s),y)_0 * dw^l(s),
\end{aligned}
\tag{3.10}
$$

respectively). □

Consider the problem

$$
\begin{aligned}
dv(t,x,\omega) = [&a^{ij}(T-t,x,\omega)v_{ij}(t,x,\omega) + \\
&+ b^i(T-t,x,\omega)v_i(t,x,\omega) + c(T-t,x,\omega)v(t,x,\omega) +
\end{aligned}
\tag{3.11}
$$

$$+ f(T\text{-}t,x,\omega)]dt + [\sigma^{il}(T\text{-}t,x,\omega)v_i(t,x,\omega) +$$

$$+ h^l(T\text{-}t,x,\omega)v(t,x,\omega) + g^l(T\text{-}t,x,\omega)]\, dw^l_T(t),$$

$$(t,x,\omega)\in]0,\ T\text{-}T_0]\times\mathbf{R}^d\times\Omega,\ v(0,x,\omega) = \varphi(x,\omega),\ (x,\omega)\in\mathbf{R}^d\times\Omega. \tag{3.12}$$

By the definition of a backward predictable stochastic process, the coefficients and the external forces in equation (3.11) are predictable relative to the family of σ-algebras $\{\mathcal{F}_T^{T\text{-}t}\}$, where $t\in[0,T\text{-}T_0]$. Note that the process w_T is a Wiener martingale with respect to the same family and the initial condition φ is measurable (as a random variable in $\mathbf{L}_2(r)$) with respect to the minimal σ-algebra of this family. Thus (3.11), (3.12) is a problem of the type (0.1), (1.2) considered on the probability space $\mathbb{F}_T := (\Omega,\ \mathcal{F},\ \{\mathcal{F}_T^{T\text{-}t}\}_{t\in[0,T\text{-}T_0]},\ \mathbf{P})$.

Let $v(t)$ be an r-generalized solution of problem (3.11), (3.12). Then by the definition of r-generalized solution, for every $y\in\mathbf{C}_0^\infty(\mathbf{R}^d)$, the following equality holds on $[T_0,T]\times\Omega$, ($\mathbf{P}$-*a.s.*)

$$(v(t\text{-}s),y)_0 = (\varphi,y) + \int_{[0,T\text{-}s]} [-(a^{ij}(T\text{-}t)v_i(t),y_j)_0 +$$

$$+ ((b^i(T\text{-}t)\text{-}a^{ij}_j(T\text{-}t)v_i(t) +$$

$$+ c(T\text{-}t)v(t) + f(T\text{-}t),y)_0]ds + \int_{[0,T\text{-}s]} (\sigma^{il}(T\text{-}t)v_i(t) +$$

$$+ h^l(T\text{-}t)v(t) + g^l(T\text{-}t),y_0)dw^l_T(t). \tag{3.13}$$

Write $u(s) := v(T\text{-}s)$. Since $v(t)\in\mathbf{L}_2^\omega([0,T\text{-}T_0];\ \overleftarrow{\mathcal{P}};\ \mathbf{L}_2(r))$, then $u(t)\in\mathbf{L}_2^\omega;\ \mathcal{P}(\mathcal{F}^{T\text{-}});\ \mathbf{L}_2(r))$. Changing the variables $T\text{-}t := \tau$ and $v(T\text{-}t) := u(t)$ in equality (3.13), we find that $u(t)$ satisfies, ($l\times\mathbf{P}$-a.s.), equality (3.10) on $[T_0,T]\times\Omega$ for every $y\in\mathbf{C}_0^\infty(\mathbf{R}^d)$. Thus u is an r-generalized solution of problem (3.3), (3.4).

Moreover, changing the variables in equality (3.10), we easily obtain that $v(t)$ is an r-generalized solution of problem (3.11), (3.12).

Thus we have proved that problems (3.3). (3.4) and (3.11), (3.12) are equivalent (i.e. the first one could be obtained from the latter by passage to the inverse time).

We see that a problem of the type (1.1), (1.2) is equivalent (in the same sense as above) to problem (3.1), (3.2). Therefore all the results obtained in this chapter for problems (1.1), (1.2) and (0.1), (1.2) are naturally carried over to problem (3.1), (3.2) and (3.3), (3.4), respectively.

3.5. *Warning* 5. *In the sequel dealing with problem* (3.1), (3.2) *and* (3.3), (3.4) *we shall refer (possibly with no special reservations) to the corresponding results for problems* (1.1), (1.2) *and* (0.1), (1.2). □

Remark 5. *Real-valuedness of the coefficients, the initial values, and the external forces which were assumed above are of no importance. Statements similar to the ones proved above are still valid in the case of complex-valued data.*

In this case, conditions (0.2), (0.3) *have to be replaced by*

$$(0.2') \quad 2\,Re\, a^{ij}\xi^i\xi^j - \sum_{l=1}^{d_1} |\sigma^{il}\xi^i|^2 \geq 0,$$

$$(0.3') \quad 2\,Re\, a^{ij}\xi^i\xi^j - \sum_{l=1}^{d_1} |\sigma^{il}\xi^i|^2 \geq \delta|\xi|_d^2, \; \forall (t,x,\omega)\in[T_0,T]\times\Omega\times\mathbf{R}^d,$$

respectively. □

Chapter 5

ITO'S PARTIAL DIFFERENITAL EQUATIONS AND DIFFUSION PROCESSES

5.0. Introduction

0.1. In this chapter we proceed with the study of the Cauchy problem for parabolic Ito's equations of second order. However, in contrast to the previous section where the problem was considered from the analytical viewpoint, we now concern ourselves with qualitative aspects of the problem.

We are going to demonstrate that the Ito's parabolic equations of second order are connected with diffusion processes as closely as deterministic parabolic equations of second order are.

0.2. Fix $T_0 \leq T \in \mathbf{R}_+$ and a pair of positive integers, d and d_1. Let $\mathsf{F} := (\Omega, \mathcal{F}, \{\mathcal{F}_t\}_{t\in[0,T]}, \mathbf{P})$ be a standard probability space and $w(t)$

be a standard Wiener process on F. Both F and w are assumed to be fixed throughout this chapter.

Let $\mathcal{F}_t^s$ be a σ-algebra generated by increments $w(\tau_1)$-$w(\tau_2)$ of the Wiener process, where τ_1 and τ_2 belong to $[s,t]$, and assume it is completed with respect to measure $\mathbf{P}$.

Warning 4.0.1 is still in force throughout this chapter. It is supposed in this chapter that for every $(t,x,\omega)\in[T_0,T]\times\mathbf{R}^d\times\Omega$, the matrix $a := (a^{ij}(t,x,\omega))$ can be represented as follows

$$(0.1) \qquad a(t,x,\omega) := \frac{1}{2}\left(\sigma\sigma^*(t,x,\omega) + \hat{\sigma}\hat{\sigma}^*(t,x,\omega)\right),$$

where $\sigma := (\sigma^{il})$, $\hat{\sigma} := (\hat{\sigma}^{ik})$, $i=1,2,\dots,d$, $l=1,2,\dots,d_1$, and $k=1,2,\dots,d_0$.

Remark 2. Obviously, condition (0.1) *implies the parabolic condition* (4.0.2). *Conversely, if the matrix a is symmetric and the parabolic condition* (4.0.2) *is fulfilled, then representation* (0.1) *follows automatically. To prove this, it suffices to choose $\hat{\sigma}$ as the square root of the matrix* $A := (2a^{ij} - \sigma^{il}\sigma^{jl})$. □

Denote by $\sum$ the matrix $(\sum^{il})$, where

$$\Sigma^{il} := \begin{cases} \hat{\sigma}^{il} & \text{if } l=1,2,\dots,d_0 \\ \sigma^{i(l-d)} & \text{if } l=d_0+1,\ d_0+2,\dots,d_0+d_1, \end{cases}$$

and $i=1,2,\dots,d$, and by B the vector (B^i), where $B^i := b^i - \sigma^{il}h^l$, $i=1,2,\dots,d$.

Also let $\hat{w}(t)$ be a d_0-dimensional standard Wiener process independent of $w(t)$, and $\nu(t)$ be a (d_0+d_1)-dimensional standard Wiener process whose first d_0 components coincide with $\hat{w}(t)$ and the last d_1 with $w(t)$.

Consider the diffusion process $\mathfrak{X}(t,x,s)$ given by the system of Ito ordinary equations,

$$\begin{aligned} \mathfrak{X}^i(t,x,s) = x^i + \int_{[s,t]} B^i(\tau,\mathfrak{X}(\tau,x,s))d\tau + \\ + \int_{[s,t]} \Sigma^{il}(\tau,\mathfrak{X}(\tau,x,s))d\nu^l(\tau), \end{aligned} \tag{0.2}$$

where $i=1,2,\dots,d$, $t\in[s,T]$, $s\in[T_0,T]$, and $x\in\mathbf{R}^d$ (x and s are assumed to be fixed).

Everywhere in this chapter except §2.2 it is assumed that the coefficients $B^i(t,x)$ and $\Sigma^{il}(t,x)$ are bounded and satisfy Lipschitz condition in x uniformly with respect to t. These assumptions ensure existence and uniqueness of a solution of system (0.2). If $\Sigma\equiv\sigma$ and $B\equiv b$ we shall denote the solution of system (0.2) by $X(t,x,s)$.

In §1.1 it will be shown that if the coefficients, the initial value φ, and the external forces f, g^l do not depend on "chance", then r-generalized solutions of the problems discussed in the previous chapter possess representation similar to the probability representation of a solution of the deterministic parabolic equation of the second order (1.4.7).

Specificially, if $f\equiv g^l\equiv 0$ this representation for the problem (4.3.3), (4.3.4) looks as follows

$$u(t,x) = \mathbf{E}[\varphi(\mathfrak{X}(T,x,t)\rho(T,t)|\mathfrak{F}^t_T], \tag{0.3}$$

where

$$\rho(s,t) := exp\Big\{ \int_{[t,s]} c(\tau,\mathfrak{X}(\tau,x,t))d\tau +$$

$$(0.4) \qquad + \int_{[t,s]} h^l(\tau,\mathfrak{X}(\tau,x,t))dw^l(\tau) - \frac{1}{2} \int_{[t,s]} h^l h^l(\tau,\mathfrak{X}(\tau,x,t))d\tau\Big\},\ t,\ s\in\mathbf{R}_+,\ t\le s.^*$$

Obviously, if $c\equiv\sigma^{il}\equiv h^l\equiv 0$, (in this case equation (4.3.3) is the backward Kolmogorov equation (1.4.8)) representation (0.3) coincides with (1.4.7).

From representation (0.3) it obviously follows that problem (4.3.3), (4.3.4) as well as the Kolmogorov backward equation can be solved by the method of random characteristic, the diffusion process $\mathfrak{X}(t,x,s)$ being used as the random characteristic.

In contrast to (1.4.7), representation (0.4) is a conditional averaging relative to the σ-algebra $\mathfrak{F}^t_T$ over the characteristic.

Representation of the type (0.3) for problem (4.0.1), (4.1.2) or (4.3.3), (4.3.4) will be called averaging over the characteristics (AOC) formula.

An important corollary of the AOC formula is the maximum principle for Ito's parabolic equation which will be proved in section 1.

0.3. The relationship between diffusion processes and Ito's parabolic equations discussed above has proved to be mutually beneficial. In particular, from AOC formula (0.3), where $\hat{\sigma}^{il}\equiv h^l\equiv c\equiv 0$ and $\varphi(x)\equiv x$ it follows that the i-th coordinate $X^i(t,x,s)$ of the process $X(t,x,s)$ as a function of x,s is an r-generalized solution of the problem,

$$(0.5) \qquad - du(s,x) = \mathcal{L}_0 u(s,x)ds + \mathcal{M}_0^l u(s,x)*dw^l(s),$$

$$(s,x,\omega)\in[T_0,t[\times\mathbf{R}^d\times\Omega,$$

$$(0.6) \qquad u(t,x) = x^i,\ (x,\omega)\in\mathbf{R}^d\times\Omega.$$

Here and below we denote by $\mathcal{L}_0$ the operator $\mathcal{L}$ with $a^{ij}=\frac{1}{2}\sigma^{il}\sigma^{jl}$ and $c\equiv 0$, and by $\mathcal{M}_0^l$ the operator $\mathcal{M}^l$ with $h^l\equiv 0$ (the operators $\mathcal{L}$ and $\mathcal{M}^l$ were defined in the previous chapter) i.e.

$$\mathcal{L}_0(t,x)(\cdot):= \frac{1}{2}\,\sigma^{il}\sigma^{jl}(t,x)(\cdot)_{ij} + b^i(t,x)(\cdot)_i,$$

$$\mathcal{M}_0^l(t,x)(\cdot):= \sigma^{il}(t,x)(\cdot)_i,\ l=1,2,\dots,d_1.$$

*Note that in our notation (see Warning 4.0.1) $h^l h^l$ is the same as $|h|^2_d$.

In the rest of the book problem (0.5), (0.6) is called the backward diffusion equation.

If we calculate the expectation of the both sides of the integral equality corresponding to problem (0.5), (0.6), then informally changing the order of integration and making the assumption that the stochastic integral is a martingale, we obtain from (0.5), (0.6) the backward Kolmogorov equation.

A slightly more general problem, namely equation (0.5) with the terminal condition

$$u(t,x) = \varphi(x),\ (x,\omega)\in\mathbf{R}^d\times\Omega, \tag{0.7}$$

will be called the backward Liouville equation for the diffusion process $X(t,x,s)$. The origin of this terminology is connected with the tradition to call Liouville's equation the equation for a first integral* of a solution of a differential equation. It will be shown in section 2 that for $\mathbf{P}$-a.s. ω, the solution of problem (0.5), (0.7) is the first integral for the diffusion process $X(s,x,0)$, where $s\in[0,t[$ and $x\in\mathbf{R}^d$ is fixed. It should be noted that in contrast to deterministic dynamical systems, the Ito ordinary equation has two different Liouville's equations, namely the forward one for the first integral depending on the "past" of the corresponding diffusion process (i.e. $\mathcal{F}^0_T$-adapted). The forward Liouville equation for the process $X(s,x,0)$ looks as follows

$$dv(s,x) = (\mathcal{M}^l_0\mathcal{M}^l_0 - \mathcal{L}_0)v(s,x)ds - \tag{0.8}$$

$$- \mathcal{M}^l_0 v(s,x)dw^l(s),\ (s,x,\omega)\in]0,t]\times\mathbf{R}^d\times\Omega,\ t\in]0,T];$$

$$v(0,x) = \varphi(x),\ (x,\omega)\in\mathbf{R}^d\times\Omega. \tag{0.9}$$

Since v, the solution of the above problem, where $\varphi(x)\equiv x^i$ for $i=1,2,\dots,d$, is a first integral of the $X(s,x,0)$, we find that $v(s,X(s,x,0)) = x^i$, ($\mathbf{P}$-a.s). Moreover, it will be shown in section 2 that the mapping $X(s,\cdot,0)$: $x\in\mathbf{R}^d \to X(s,x,0)\in\mathbf{R}^d$ under certain assumptions on the smoothness of the coefficients is a diffeomorphism of $\mathbf{R}^d$ on $\mathbf{R}^d$ and the ith coordinate of the inverse mapping $X^{-1}(s,x,0)$, call it $X^{-i}(s,x,0)$, is the unique solution of problem (0.8), (0.9) for $\varphi(x) = x^i$. This last variant of problem (0.8), (0.9) will be called the forward equation of the inverse diffusion.

---

*If y:$[T_0,T] \to \mathbf{R}^d$ and φ:$[T_0,T]\times\mathbf{R}^d\to\mathbf{R}^1$ are measurable functions and $\varphi(t,y(t)) = const$ for all $t\in[T_0,T]$, then φ is usually called the first integral of y. The notion of the first integral is of great importance in classical mechanics (see e.g. [1]).

From the AOC formula for problem (4.0.1), (4.1.2) we deduce easily that the backward equation of the inverse diffusion is a system of the backward ordinary Ito's equations for $\{X^{-i}(s,x,\cdot)\}$ (see §2.3).

Making use of the forward equation of the inverse diffusion we derive in section 2 a "formula of variation of constants" for an ordinary Ito's equation. To be more specific, for the system of Ito's equation

$$Y^i(s,x,0) = x^i + \int_{[0,s]} {}_Y b^i(\tau, Y(\tau,x,0))d\tau +$$

$$+ \int_{[0,s]} {}_Y\sigma^{il}(\tau, Y(\tau,x,0))dw^l(\tau),$$

where $i=1,2,...,d$ and $s\in[0,T]$, we define the diffusion process $Z(t,x,s)$ such that $Y^i(t,x,s) = X^i(t, Z(t,x,s),s)$ for every $i=1,2,...,d$.

Making use of the fact that the mapping $X(t,\cdot,s)$: $x\rightarrow X(t,x,s)$ is the diffeomorphism, we prove in §2.2 that an Ito's parabolic equation of the second order is equivalent in a sense to a parabolic equation of the second order with random coefficients of special form.

In section 3 we consider the problem conjugate to problem (0.1), (1.2) where the coefficients and free forces can depend on "chance". In this section we also prove another AOC formula for the functional $\int_{\mathbf{R}^d} f(t,x)u(t,x)dx$, where $u(t,x)$ is a solution of the problem mentioned above and f is a given functional.

5.1. The Method of Stochastic Characteristics.

1.0. Formulas of averaging over characteristics of the form (0.3) for problems (4.0.1), (4.1.2) and (4.3.3), (4.3.4) are obtained in this section. Making use of these formulas we deduce the maximum principle for the problems considered in the previous chapter.

Throughout this section it is supposed to that the coefficients a^{ij}, b^i, c, σ^{il}, and h^l as well as φ, b, and g^l for $i,j=1,2,...,d$ and $l=1,2,...,d_1$ do not depend on ω.

1.1. In addition to the system (0.2) of Ito's equations, we consdier the following similar system of Ito's backward equations:

$$\mathcal{Y}^i(t,x,s) = x^i + \int_{[s,t]} B^i(\tau, \mathcal{Y}(t,x,\tau))d\tau +$$

$$+ \int_{[s,t]} \Sigma^{il}(\tau, \mathcal{Y}(t,x,\tau)) * d\nu^l(\tau), \tag{1.1}$$

$$s\in[T_0,t],\ i=1,2,\ldots,d,\ t\in]T_0,T],\ x\in\mathbf{R}^d,$$

where t and x are assumed to be fixed.

The function B^i, Σ^{il} and the Wiener process ν are defined in §02. By the the assumptions made in §0.2, system (1.1) as well as system (0.2) has unique solutions which possess continuous versions in t,x,s (see §1.4.5 and §1.4.14). Throughout what follows, we consider these versions of the processes $\mathfrak{X}(t,x,s)$, $\mathfrak{Y}(t,x,s)$ denoting them in the same way.

Write

$$\gamma(t,s):=exp\Big\{\int_{[s,t]} c(\tau,\ \mathfrak{Y}(t,x,\tau))d\tau +$$

$$+\int_{[s,t]} h^l(\tau,\ \mathfrak{Y}(t,x,\tau))*dw^l(\tau) - \frac{1}{2}\int_{[s,t]} h^l h^l(\tau,\ \mathfrak{Y}(t,x,\tau))d\tau\Big\}.$$

Theorem 1. *Suppose that r is a fixed number and the following conditions are satisfied:*

(*i*) *The functions a^{ij}, b^i, c, σ^{il}, h^l, σ^{il}_j, and h^l_j for $i,j=1,2,\ldots,d$ and $l=1,2,\ldots,d_1$, and their derivatives of first order (in x) as well as the derivatives of second order (in x) for the functions a^{ij} are uniformly bounded by the constant K.*

(*ii*) $\varphi\in\mathbb{H}^1(r)$, $f\in\mathbb{L}_2([T_0,T];\ \mathbb{H}^1(r))$ *and* $g^l\in\mathbb{L}_2([T_0,T];\ \mathbb{H}^2(r))$.

Let $u(t,x)$ be an r-generalized solution of problem (4.0.1), (4.1.2) *and $v(t,x)$ be an r-generalized solution of problem* (4.3.3), (4.3.4). *Then the following formulas are valid for $l\times l_d$-a.a. t,x,* ($\mathbf{P}$-*a.s.*):

$$u(t,x) = \mathbf{E}\Big(\int_{[T_0,t]} f(s,\mathfrak{Y}(t,x,s))\gamma(t,s)ds +$$

(1.2) $$+\int_{[T_0,t]} g^l(s,\mathfrak{Y}(t,x,s))\gamma(t,s)*dw^l(s) +$$

$$+\ \varphi(\mathfrak{Y}(t,x,T_0))\gamma(t,T_0)|\ \mathcal{F}_t^{T_0}\Big),$$

and

$$v(t,x) = \mathbf{E}\Big(\int_{[t,T]} f(s,\mathfrak{X}(s,x,t))\rho(s,t)ds +$$

(1.3)
$$+ \int_{[t,T]} g^l(s,\mathfrak{X}(s,x,t))\rho(s,t)\,dw^l(s) +$$
$$+ \varphi(\mathfrak{X}(T,x,t))\rho(T,t)\Big|\ \mathcal{F}^t_T\Big). \quad \square$$

It is the above formulas that justifies calling the diffusion processess $\mathfrak{X}$(t,x,s) and $\mathcal{Y}(t,x,s)$ the stochastic characteristics of problems (4.3.3), (4.3.4) and (4.0.1), (4.1.2), respectively.

1.2. Before we proceed to the proof of the theorem we state some auxiliary results. Write

$$F(t,x) := \mathbf{E}\Big[\int_{[t,T]} f(s,X(s,x,t))\rho^0(s,t)\,ds + \varphi(X(T,x,t))\rho^0(T,t)\Big],$$

where

$$\rho^0(s,t) := exp\Big\{\int_{[t,s]} c(\tau,X(\tau,x,t))\,d\tau\Big\},$$

and X as it was mentioned above is the process X with $\hat{\sigma}^{il} \equiv 0$ and $h^l \equiv 0$.

Theorem 2. Let n be a given integer. Suppose

(i) For the coefficients σ^{il}, b^i and c the conditions of Theorem 1 are fulfilled.

(ii) $a^{ij} = \frac{1}{2}\sigma^{il}\sigma^{jl}$ for i,j=1,2,...,d.

(iii) There exists a constant $K \in \mathbf{R}_+$ such that for all $(t,x) \in [T_0,T] \times \mathbf{R}^d$

$$|\varphi(x)| + |f(t,x)| \leq K(1+|x|_d^2)^{n/2}$$

and for every $R \in \mathbf{R}_+$ for all $t \in [T_0,T]$ and z, $z' \in \mathbf{R}^d$ such that $|z|_d \leq R$ and $|z'|_d \leq R$,

$$|\varphi(z) - \varphi(z')| + |f(t,z) - f(t,z')| \leq K(1+R)^n\, |z-z'|_d.$$

Then the function $F(t,x)$ is continuous in (t,x) on $[T_0,T] \times \mathbf{R}^d$ and there exists a constant N depending only on K, n, T and T_0 such that for all $(t,x) \in [T_0,T] \times \mathbf{R}^d$, *the following inequalities hold**

$$(1.4)\qquad |F(t,x)| \leq N(1+|x|_d^2)^{n/2},$$

$$(1.5)\qquad \sum_{i=1}^{d} |F_i(t,x)| \leq N(1+|x|_d^2)^{n/2}.$$

Moreover, for every $y \in \mathbf{C}_0^\infty(\mathbf{R}^d)$ *and for all* $t \in [T_0, T]$, *the following equality holds true*

$$(F(t),y)_0 = (\varphi,y)_0 + \int_{[t,T]} [-(a^{ij}F_i(s),y_j)_0 +$$

$$(1.6)\qquad + ((b^i - a_j^{ij})F_i(s) + cF(s) + f(s),y)_0]\,ds. \quad \square$$

For the proof of this important theorem see [63], III 1.5, IV. 1.1, IV. 1.4.**

From Lemma 3.4.7 (i) inequalities (1.4), (1.5) it follows that $F \in \mathbf{L}_2([T_0,T]; \mathbf{H}^1(r))$, for

$$(1.7)\qquad r < -2^{-1}(d+2n).$$

From this and (1.6) we have for r satisfying (1.7), that $F(t,x)$ is an r-generalized solution of the backward Kolmogorov equation

$$(1.8)\qquad -\,dv(t,x) = [\mathcal{L}v(t,x) + f(t,x)]\,dt,\ (t,x) \in [T_0,T[\times \mathbf{R}^d,$$

$$(1.9)\qquad v(T,x) = \varphi(x),\quad x \in \mathbf{R}^d.$$

---

*The derivatives in inequality (1.5) are assumed to be in a generalized sense.

**In [63] it is assumed that $c \leq 0$. The purpose of this assumption is that in this monograph it is supposed that c may be an increasing function of x. If c is bounded this assumption is unnecessary, because we can always take "c" to be negative by changing the variable $\tilde{F}(t,x) := e^{-Nt}F(t,s)$, where $N > \sup_{t,x} c(t,x)$.

Note that problem (1.8), (1.9) is a special case of problem (4.3.3), (4.3.4) and an r-generalized solution of (1.8), (1.9) could be treated in the sense of Definition 4.3.4.

However, it will be helpful to state the spacial version of this definition.

Definition 2. A function $v \in \mathbf{L}_2([T_0,T];\ \mathbf{H}^1(r))$ *will be called an r-generalized solution of the problem* (1.8), (1.9) *if for every* $y \in \mathbf{C}_0^\infty(\mathbf{R}^d)$ *it satisfies the equation* (1.6), *where F is replaced by v and* F_i *by* v_i, *l-a.s. on* $[T_0,T]$. □

Remark 2. From the foregoing and Theorem 4.1.1 *it follows that under the assumptions of the theorem, problem* (1.8), 91.9) *has a unique r-generalized solution v for every* $r < -2^{-1}(d+2n)$. *This solution belongs to* $\mathbf{L}_2([T_0,T];\ \mathbf{H}^1(r)) \cap \mathbf{C}([T_0,T];\ \mathbf{L}_2(r))$ *and satisfies the equality* $v(t,x) = F(t,x)$ $l \times l_d$*-a.s.*. □

Denote by $B(0,R)$ the sphere in $\mathbf{R}^d$ with the center at the origin and finite radius R. The following result will be of use later in the book.

Corollary 2. Suppose that the assumptions of the theorem are fulfilled and, moreover, the functions $f(t,x)$, $\varphi(x)$, *and their first derivatives are bounded and vanish for* $x \notin B(0,R)$. *Then these statements hold*

(*i*) *Problem* (1.8), (1.9) *has the unique r-generalized solution v for every* $r \in \mathbf{R}^1$. *This solution belongs to* $\mathbf{L}_p([T_0,T];\ \mathbf{W}_p^1(r))$ *for every* $p \in [2,\infty[$ *and satisfies the equality* $v(t,x) = F(t,x)$ $l \times l_d$*-a.s.*

(*ii*) *For every* $n \in \mathbf{N}$

$$\overline{\lim_{|x|_d \to \infty}} \sup_{t \in [T_0,T]} (1 + |x|_d^2)^{n/2}\, |F(t,x)| = 0. \quad \Box$$

Proof. We begin with the second statement. Suppose for a moment that $f \equiv 0$. Since the drift and diffusion coefficients of the process $X(t,x,s)$ are bounded, it is easy to see that

$$\mathbf{E} \sup_{t \in [T_0,T]} |X(T,x,t) - x|_d^{n+1} \le N_0 < \infty, \tag{1.10}$$

where the constant N_0 depends only on T, T_0, n, and the constant that majorizes the coefficients of the equation for $X(T,x,t)$. From the boundedness of φ and c it follows that there exists a constant $\tilde{N}$ such that for $|x|_d \ge R$,

$$\sup_{t \in [T_0,T]} |F(t,s)| \le \tilde{N} \sup_{t \in [T_0,T]} \mathbf{P}(X(T,x,t) \in B(0,R)) \le$$

$$\leq \tilde{N} \sup_{t\in[T_0,T]} \mathbf{P}(|X(T,x,t)-x|_d \geq \rho(x,B(0,R))),$$

where $\rho(x,B(0,R))$ is the euclidean distance between the point x and the sphere $B(0,R)$. From this and Chedyshev's inequality it follows that

$$\text{(1.11)} \qquad \sup_{t\in[T_0,T]} |F(t,x)| \leq \tilde{N} \sup_{t\in[T_0,T]} \frac{\mathbf{E}|X(T,x,t)-x|^{n+1}}{|\rho(x,B(0,R))|^{n+1}}.$$

Statement (ii) readily follows from (1.10) and (1.11).

Now let $f\neq 0$. Evidently,

$$\left|\mathbf{E} \int_{[t,T]} f(s,X(s,x,t))\rho^0(s,t)\,ds\right| \leq$$

$$\leq N\mathbf{E} \sup_{s\in[T_0,T]} (|f(s,X(s,x,t))|\, \rho^0(s,t)).$$

Making use of the same arguments the reader will easily complete the proof of statement (ii).

In view of our assumptions, $\varphi\in\mathbf{W}^1_p(r)$ and $f\in\mathbf{L}_p([T_0,T];\ \mathbf{W}^1_p(r))$ for every $p\in[0,\infty[$ and $r\in\mathbf{R}^1$. Thus, we obtain from Theorem 4.3.2 that problem (1.8), (1.9) has a unique r-generalized solution v which belongs to $\mathbf{L}_p([T_0,T];\ \mathbf{W}^1_p(r))$ for every $p\in[2,\infty[$ and $r\in\mathbf{R}^1$.

On the other hand, from (ii) and the theorem of this paragraph we find that $F(t,x)$ is also an r-generalized solution of problem (1.8), (1.9) for every $r<-2^{-1}(d+2n)$. In view of the uniqueness of r-generalized solutions of the problem we obtain that $v(t,x) = F(t,x)$, $l\times l_d$-a.s., which completes the proof of the corollary. □

1.3. Fix $r\in\mathbf{R}^1$ as in Theorem 1 and define

$$S(x):= (1+|x|^2_d)^{r/2},$$

$$S(i,x):= rx^i(1+|x|^2_d)^{-1},$$

$$S(i,j,x):= [S(i,x)S(j,x) + r\delta_{ij}(1+|x|^2_d)^{-1} -$$

$$- 2rx^i x^j (1+|x|_d^2)^{-2}].$$

Clearly $S_i(x) = S(i,x,)S(x)$ and $S_{ij}(x) = S(i,j,x)S(x)$ for $i,j=1,2,\ldots,d$. Write

$$c^*(s,x) := - a^{il}_{ij}(s,x) - 2a^{ij}_i(s,x)\, S(j,x) -$$

$$- a^{ij} S(i,j,x) + b^i(s,x)S(i,x) + b^i_i(s,x) - c(s,x),$$

$$H_t := [T_0 t]\times\mathbf{R}^d.$$

$$\lambda := \sup_{(s,x)\in H_t} (-c^*(s,x)) \quad and \quad \tilde{\lambda} := \sup_{(s,x)\in H_t} (\rho\text{-}1)\, |c(s,x)|.$$

Evidently, under the assumptions of Theorem 2, $|\lambda|<\infty$.

Theorem 3 (The Krylov-Fichera inequality). Suppose that the conditions of Theorem 2 are satisfied and assume that numbers $s,t\in[T_0,T]$, $p\in[1,\infty[$, *and* $r\in\mathbf{R}^1$ *are fixed. Then for every* $\varphi\in\mathbf{L}_p(r)$, $\psi\in\mathbf{L}_p(H_t,r) := \mathbf{L}_p([T_0,t];\, \mathbf{L}_p(r))$ *the following inequalities hold.*

(1.12) $$\|\mathbb{E}\varphi(X(t,\cdot,s))\, \rho^0(t,s)\|^p_{\mathbf{L}_p(r)} \le e^{(\lambda+\tilde{\lambda})(t-s)}\, \|\varphi\|^p_{\mathbf{L}_p(r)},$$

$$\left\|\mathbb{E} \int_{[s,t]} \psi(r,X(r,\cdot,s))\, \rho^0(r,s)\,dr\right\|^p_{\mathbf{L}_p(H_t,r)} \le$$

(1.13) $$\le |t-T_0|^p\, e^{(\lambda\vee 0+2\tilde{\lambda})t}\, \|\psi\|^p_{\mathbf{L}_p(H_t,r)}. \quad \square$$

Proof. To begin with, we note that inequality (1.13) follows from (1.12) and Hölder's inequality, so it suffices to verify inequality (1.12). Moreover, it suffices to prove inequality (1.12) only for $p=1$ because the general case can be obtained from this one via the same Hölder's inequality. Clearly, it is sufficient to prove (1.12) for non-negative $\varphi\in\mathbf{C}_0^\infty(\mathbf{R}^d)$ which will be assumed in the sequel.

Let $F^0(t,x):=\mathbf{E}[\varphi(X(T,x,t))\,\rho^0(T,t)]$, $y(s,x):=\eta(s,x)S(x)$, and $\eta(s,x):=\eta_{(1)}(s)\,\eta_{(2)}(x/R)$, where $\eta_{(1)}(s)\in\mathbf{C}_0^\infty([T_0,t])$, $\eta_{(2)}(x)\in\mathbf{C}_0^\infty(\mathbf{R}^d)$ and in addition $\eta_{(2)}(0)=1$, $\eta_{(1)}, \eta_{(2)}\geq 0$, and $R\in\mathbf{R}_+$.

From Theorem 2 it follows that

$$\int_{H_t} F^0(s,x)\Big(-\frac{\partial y(s,x)}{\partial s}+(a^{ij}y(s,x))_{ij}-$$

$$-(b^i y(s,x))_i+cy(s,x)\Big)dxds=0,$$

which clearly implies

$$\int_{H_t} S(x)F^0(s,x)\Big(\frac{\partial\eta(s,x)}{\partial s}+c^*\eta(s,x)\Big)dxds=$$

$$=\int_{H_t} S(x)F^0(s,x)\,(2a_i^{ij}\eta_j(s,x)+a^{ij}\eta_{ij}(s,x)+$$

$$+2a^{ij}\eta_i(s,x)S(j,x)-b^i\eta_i(s,x))dxds.$$

Passing to the limit in the latter equality as $R\to\infty$ and taking into account that η_i, η_{ij} tends to zero as $R\to\infty$, we obtain by the dominated convergence theorem in view of Theorem 2 and Corollary 2 (ii) that

$$\int_{H_t} S(x)F^0(s,x)\Big(\frac{\partial\eta_1(s)}{\partial s}+c^*(s,x)\eta_1(s)\Big)dsdx=0,$$

and consequently

$$\int_{H_t} S(x)F^0(s,x)\Big(\frac{\partial\eta_1(s)}{\partial s}-\lambda\eta_1(s)\Big)dsdx\leq 0.$$

Denoting $\psi(s):= exp\{-\lambda s\}\ \eta_1(s)$ and $N(s):= \|F^0(s)\|_{\mathbf{L}_1(r)}$, we can rewrite the latter inequality as follows

$$\int_{[T_0,t]} N(s)\ e^{\lambda s} \frac{\partial}{\partial s}\ \psi(s)ds \leq 0. \tag{1.14}$$

Since, by the theorem and the corollary of paragraph 2, $N(s)$ is a continuous function of s, and $\psi(s)$ can be considered as arbitrary non-negative function belonging to $\mathbf{C}_0^\infty([T_0.t])$, inequality (1.14) yields that $N(s)exp\{\lambda s\}$ is an increasing function of s which implies inequality (1.12) for $p=1$. □

1.4. Let us formulate the last auxiliary result (see [44], §4.4, Lemma 5) we need to prove Theorem 1.

Write

$$q_T(t):= exp\Big\{ \int_{[t,T]} q^l(s)dw^l(s) - \frac{1}{2} \int_{[t,T]} q^l q^l(s)ds\Big\}, \tag{1.15}$$

where $q\in\mathbf{L}_\infty([T,T];\ \mathbf{R}^d):= \mathbf{L}_\infty([t,T];\ \overline{\mathfrak{B}([t,T]};\ l;\ \mathbf{R}^{d_1})$.

Lemma 4. *If* $\xi\in\mathbf{L}_2(\Omega;\ \mathcal{F}^t_T;\ \mathbf{R}^1)$ *and* $\mathbf{E}\xi q_T(t) = 0$ *for every* $q\in\mathbf{L}_\infty([t,T],\ \mathbf{R}^{d_1})$, *then* $\xi=0$, $\mathbf{P}$-*a.s.* □

1.5. Proof of Theorem 1. We shall only prove the part of the theorem concerning problem (4.3.3), (4.3.4). In fact, it is quite sufficient for the proof, because the forward Cauchy problem can be obtained from the backward one by the time inverse $s=T-t$ (see e.g. §4.3.4).

Suppose for the moment, that f, g^l, and φ as well as their first derivatives in x are uniformly bounded and vanish for $|x|_d\geq R$, where R is some fixed constant.

Let v be an r-generalized solution of problem (4.3.3), (4.3.4) and $y\in\mathbf{C}_0^\infty(\mathbf{R}^d)$. Applying Ito's formula (§1.4.2) to the product $(v(t),y)_0\ q_T(t)$, we obtain that $\tilde{v}(t,x):= v(t,x)\ q_T(t)$ is an r-generalized solution of the problem

$$-d\tilde{v}(t) = [a^{ij}\tilde{v}_{ij}(t)+\tilde{b}^i\tilde{v}_i(t)+\tilde{c}\tilde{v}(t)+\tilde{f}(t)]dt + [\sigma^{il}\tilde{v}_i(t)+\tilde{h}^l\tilde{v}(t) + \tag{1.16}$$

$$+ \tilde{g}^l(t)]*dw^l(t),\ (t,x,\omega)\in[T_0,T[\times\mathbf{R}^d\times\Omega,$$

$$\tilde{v}(T) = \varphi,\ (x,\omega)\in\mathbf{R}^d\times\Omega, \tag{1.17}$$

where

$$\tilde{b}^i := b^i + q^l\sigma^{il},\ \tilde{c} := c + q^l h^l,$$

$$\tilde{f} := q_T(f + q^l g^l),\ \tilde{h}^l := h^l + q^l,\ \tilde{g}^l := q_T g^l.$$

It is clear that under the assumption made above, $\varphi \in \mathbf{L}_p(\Omega;\ \mathbf{W}^1_p(r))$, $\tilde{f} \in \mathbf{L}_p([T_0,T]\times\Omega;\ \mathbf{W}^1{}_p(r))$, and $\tilde{g}^l \in \mathbf{L}_p([T_0,T]\times\Omega;\ \mathbf{W}^2_p(r))$ for every $p \geq 0$. Note also that $\tilde{f}$ and $\tilde{g}^l$ are backward predictable (relative to the family $\{\mathcal{F}^t_t\}$) functions. Thus from Theorem 4.3.2 and Corolllary 4.3.2 in view of the uniqueness of r-generalized solutions to a problem of the type (4.3.3), (4.3.4), it follows that $\tilde{v} \in L_p([T_0,T]\times\Omega;\ \mathbf{W}^1_p(r))$ for every $p \in [2,\infty[$ and it has a version (we shall denote it in the same way) which is continuous in t, x, ($\mathbf{P}$-$a.s.$), and satisfies the inequality

$$\mathbf{E} \sup_{\substack{t\in[T_0,T]\\ x\in\mathbf{R}^d}} |\tilde{v}(t,x)|^p < \infty.$$

(see Warning 4.3.5).

Denote $\mathbf{E}\tilde{v}$ by $\hat{v}$. From the arguments given above it follows in particular that $\hat{v} \in \mathbb{L}_2([T_0,T];\ \mathbb{H}^1(r))$. Let us calculate the expectation of both parts of the integral equality corresponding to problem (1.16), (1.17). Since $\tilde{v} \in \mathbf{L}_2([T_0,T];\ \mathcal{P};\ \mathbb{H}^1(r))$, the expectation of the stochastic integral is zero. Changing the order of integration in the obtained equality, we find that $\hat{v}$ is an r-generalized solution (in the sense of Definition 2) of the problem

$$-\,dF(t,x) = [a^{ij}F_{ij}(t,x) + \tilde{b}^i F_i(t,x) + \tag{1.18}$$
$$+\ \tilde{c}F(t,x) + \tilde{f}(t,x)]\,dt,\ (t,x)\in[T_0,T[\times\mathbf{R}^d,$$

$$F(T,x) = \varphi(x),\ x\in\mathbf{R}^d. \tag{1.19}$$

On the other hand, from the theorem and the corollary of paragraph 2 it follows that the following equality holds on $[T_0,T]\times\mathbf{R}^d$ ($l\times l_d$-a.s.)

$$\hat{v}(t,x) = \mathbf{E}\Bigg(\int_{[t,T]} \tilde{f}(s,\tilde{\mathfrak{X}}(s,x,t))\ exp\Big\{\int_{[t,s]} \tilde{c}(\tau,\tilde{\mathfrak{X}}(\tau,x,t)d\tau\Big\}ds +$$
$$+\ \varphi(\tilde{\mathfrak{X}}(T,x,t))\ exp\Big\{\int_{[t,T]} \tilde{c}(\tau,\tilde{\mathfrak{X}}(\tau,x,t))d\tau\Big\}\Bigg), \tag{1.20}$$

where $\tilde{\mathfrak{X}}(s,x,t)$ is a solution of the system of ordinary Ito's equations

$$\tilde{\mathfrak{X}}^i(s,x,t) = x^i + \int_{[t,s]} \tilde{b}^i(\tau,\tilde{\mathfrak{X}}(\tau,x,t))d\tau +$$

$$+ \int_{[t,s]} \Sigma^{il}(\tau,\tilde{\mathfrak{X}}(\tau,x,t))d\nu^l(\tau),$$

$$s\in[t,T],\ i=1,2,...,d.$$

Define

$$\tilde{\rho}(s,t) = exp\Big\{ \int_{[t,s]} \tilde{c}(\tau,\mathfrak{X}(\tau,x,t))d\tau +$$

$$+ \int_{[t,s]} (q^l+h^l)\ (\tau,\mathfrak{X}(\tau,x,t))dw^l(\tau) -$$

$$- \frac{1}{2} \int_{[t,s]} (q^l+h^l\ (q^l+h^l)\ (\tau,\mathfrak{X}(\tau,x,t))d\tau\Big\}.$$

By Girsanov's theorem (§1.4.6), equality (1.2) may be rewritten as follows

$$\hat{v}(t,x) = \mathbb{E}\left[\int_{[t,T]} \tilde{f}(s,\mathfrak{X}(s,x,t))\tilde{\rho}(s,t)ds + \varphi(\mathfrak{X}(T,x,t))\tilde{\rho}(T,t)\right] =$$

$$(1.21)\qquad = \mathbb{E}\Big(\int_{[t,T]} (f(s,\mathfrak{X}(s,x,t)) + q^l(s)g^l(s,\mathfrak{X}(s,x,t)))\rho(s,t)q_s(t)ds +$$

$$+ \varphi(\mathfrak{X}(T,x,t))\rho(T,t)q_T(t)].$$

Applying Ito's formula to the product

$$q_r(t)\Big(\int_{[t,r]} f(s,(\mathfrak{X}(s,x,t))\rho(s,t)ds +$$

$$+ \int_{[t,r]} g^l(s,\mathfrak{X}(s,x,t))\rho(s,t)\,dw^l(s)),$$

where t is assumed to be fixed, integrating over the interval $[t,T]$ and taking the expectation of the both sides of the obtained equality we find that

$$\mathbf{E}\Bigg(\int_{[t,T]} (f(s,\mathfrak{X}(s,x,t)) +$$

$$+ q^l(s)g^l(s,\mathfrak{X}(s,x,t)))\rho(s,t)q_s(t)\,ds\Bigg) =$$

(1.22)

$$= \mathbf{E}\ q_T(t)\Bigg(\int_{[t,T]} f(s,\mathfrak{X}(s,x,t))\rho(s,t)\,ds +$$

$$+ \int_{[t,T]} g^l(s,\mathfrak{X}(s,x,t))\rho(s,t)\,dw^l(s)\Bigg).$$

From (1.21), (1.22) it follows that for all $t\in[T_0,T]$, l_d-a.s., we have

$$\hat{v}(t,x) = \mathbf{E}\Big\{q_T(t)\ \mathbf{E}\Big(\int_{[t,T]} f(s,\mathfrak{X}(s,x,t))\rho(s,t)\,ds +$$

$$+ \int_{[t,T]} g^l(s,\mathfrak{X}(s,x,t))\rho(s,t)\,dw^l(s) + \varphi(\mathfrak{X}(T,x,,t))\rho(T,t)|\mathfrak{F}^t_T\Big)\Big\}.$$

This together with Lemma 4 yields equality (1.3).

We now show this equality holds without the additional assumptions on φ, f, and g^l made above. To prove this, we choose sequences $\{\varphi^n\}$, $\{f^n\}$ and $\{(g^n)^l\}$ whose elements are of the kind used on the first step of the proof and such that $(g^n)^l \to g^l$ in $\mathbf{L}_2([T_0,T];\ \mathbb{H}^2(r))$ for every $l=1,2,...,d_1$, $f^n \to f$ in $\mathbf{L}_2([T_0,T];\ \mathbb{H}^1(r))$, and $\varphi^n \to \varphi$ in $\mathbb{H}^1(r)$ as $n\to\infty$.

Denote by v^n, the r-generalized solution of problem (4.3.3), (4.3.4) corresponding to the terminal value φ^n and the external forces f^n, g^n. From inequality (4.3.8) it follows that for every $t\in[T_0,T]$, $v^n(t)$ converges in $\mathbf{L}_2(\mathbf{R}^d\times\Omega;\ \mathfrak{B}(\mathbf{R}^d)\otimes\mathfrak{F},\ l_d\times\mathbf{P},\mathbf{R}^1)$ to the solution of problem (4.3.3), (4.3.4), call it $v(t)$, corresponding to the terminal value φ and the external forces f and g^l, as $n\to\infty$. On the other hand, the right hand side of the equality of the form (1.3), where f, g^l, and φ are replaced by f^n, $(g^n)^l$, and φ^n, converges on some subsequence to the original expression for almost all t,x (**P**-a.s.). This follows from the Krylov-Fichera inequalities (Theorem 3).

Indeed, consider for example the expression

$$\mathbf{E}\left[\int_{[t,T]} (g^n)^l(s,\mathfrak{X}(s,x,t))\rho(s,t)dw^l(s)\Big|\ \mathfrak{F}_T^t\right].$$

By Girsanov's theorem, we obtain

$$U^n := \int_{H_T} S^2(x)\,\mathbf{E}\Big|\mathbf{E}\Big(\int_{[t,T]} ((g^n)^l -$$

$$(1.23)\qquad - g^l)(s,\mathfrak{X}(s,x,t))\rho(s,t)dw^l(s)\Big|\ \mathfrak{F}_T^t\Big]\Big|^2 dxdt \le$$

$$\le \int_{H_T} S^2(x)\,\mathbf{E}\int_{[t,T]}\sum_{l=1}^{d_1} |(g^n)^l - g^l|^2(s,\mathfrak{X}(s,x,t))\rho^2(s,t)dsdxdt =$$

$$= \int_{H_T} S^2(x)\,\mathbf{E}\int_{[t,T]}\sum_{l=1}^{d_1} |(g^n)^l - g^l|^2(s,\hat{\mathfrak{X}}(s,x,t,))\hat{\rho}(s,t)dsdxdt,$$

where $\hat{\mathfrak{X}}(s,x,t)$ is the solution of the system of Ito's equations,

$$\hat{\mathfrak{X}}^i(s,x,t) = x^i + \int_{[t,s]} (b^i - 2h^l\sigma^{il})(\tau,\hat{\mathfrak{X}}(\tau,x,t))d\tau +$$

$$+ \int_{[t,s]} \Sigma^{il}(\tau,x,t))dw^l(\tau),\ k=1,2,\dots,d,\quad s\in[t,T],$$

and

$$\hat{\rho}(s,t) := \exp\Big\{\int_{[t,s]} (2c + h^l h^l)(\tau,\hat{\mathfrak{X}}(\tau,x,t))d\tau\Big\}.$$

By inequality (1.13), we obtain that

$$U^n \le N \int_{H_T} S^2(x) \sum_{l=1}^{d_1} |(g^n)^l - g^l|^2(t,x)\,dt\,dx,$$

where the constant N does not depend on n. Consequently $\lim_{n\to\infty} U^n=0$.

We can pass to the limit in the remaining terms in the same way. So the theorem is proved. □

Some applications of the AOC formulas (1.2), (1.3) will be considered in the following sections of this chapter and also in Ch. 6. Here we confine ourselves to one simple case of exceptional importance, namely, we shall prove the maximum principle for problems (4.0.1), (4.1.2) and (4.3.3), (4.3.4).

From formulas (1.2), (1.3) we readily obtain

Corollary 5. (*The maximum principle*). *Suppose that conditions of Theorem* 1 *are satisfied.*

(*i*) *If* $g^l=0$ *for all* $l=1,2,\dots,d$, $f\ge 0$ ($l\times l_d$*-a.s.*) *and* $\varphi\ge 0$ (l_d*-a.s.*), *then the r-generalized solutions of problem* (4.0.1), (4.1.2) *and* (4.3.3), (4.3.4) *are non-negative,* ($l\times l_d\times \mathbb{P}$*-a.s.*).

(*ii*) *If* $h^l=g^l=0$ *for all* $l=1,2,\dots d$, $f\le 0$, $c\le 0$ ($l\times l_d$*-a.s.*), *and* $\varphi\le 1$ (l_d*-a.s.*), *then the r-generalized solutions of problems* (4.0.1), (4.1.2) *and* (4.3.3), (4.3.4) *do not exceed* 1, ($l\times l_d\times \mathbb{P}$*-a.s.*). □

In the proof of this corollary we can make use of Theorem 3, to show that a change of c, f, h^l, and g^l on a set of zero $l\times l_d$-measure and φ on a set of zero l_d-measure does not change the right hand side of formulas (1.2), (1.3) (up to $l\times l_d\times\mathbb{P}$-equivalence).

Remark 5. *If the superparabolic condition* (4.0.3) *is satisfied for the matrix* (a^{ij}), *then the conditions on the smoothness of the coefficients as well as* φ, f *and* g^l *can be made less restrictive than in Theorem* 1.

In fact it suffices to assume that all the coefficients are bounded, a^{ij} *are differentiable in* x *up to first order and all the derivatives are bounded,* f, $g^l\in \mathbb{L}_2([T_0,T];\ \mathbb{L}_2(r))$, *and* $\varphi\in\mathbb{L}_2(r)$. *From Theorem* 4.3.1 *it follows that problems* (4.1.1), (4.1.2) *and* (4.3.1), (4.3.2) *have unique r-generalized solutions.*

Of course, the above conditions are sufficient for the maximum principle to hold. □

1.6. *Remark* 6. *The maximum principle is still valid if the coefficients, the initial* (*terminal*) *conditions and the external forces of problems* (4.0.1), (4.1.2) *and* (4.3.3), (4.3.4) *are random* (*see* [72] *and* §5.3.1). □

5.2. Inverse Diffusion Processes, the Method of Variation of Constants and the Liouville Equations.

2.0. In this section we consider the diffusion process $X(t,x,s)$ which is the solution of the following system of ordinary Ito's equations

$$X^i(t,x,s) = x^i + \int_{[s,t]} b^i(r,\omega,X(r,x,s))dr +$$

$$+ \int_{[s,t]} \sigma^{il}(r,\omega,X(r,x,s))dw^l(r), \tag{2.1}$$

$$t\in[s,T],\ i=1,2,...,d,\ s\in[T_0,T[.$$

It is supposed that for $i=1,2,...,d$ and $l=1,2,...,d_1$, $b^i(t,x,\omega)$ and $\sigma^{il}(t,x,\omega)$ are continuous in (t,x), bounded, predictable (relative to the family $\{F_t^{T_0}\}$) for every x functions. It is also assumed that for every x, t, ω, i and l, f^i has derivatives in x up to first order and σ^{il} up to second order.

From the contents of §1.4.3 and §1.4.5 it follows that under our assumptions, system (2.1) has a unique predictable solution and this solution possesses a version continuous in t,x,s (**P**-a.s.). Only this version will be considered in the future.

It will be shown that for all t, s and (**P**-a.a.) ω, the mapping $X(t,\ ,s)$: $\mathbf{R}^d \to \mathbf{R}^d$ is a diffeomorphism of $\mathbf{R}^d$ onto $\mathbf{R}^d$. We shall derive equations for $X^{-1}(t,\cdot,s)$ with respect to t and to s and Liouville's equation for $X(t,x,T_0)$. We also discuss the method of variation of constants for a solution of equation (2.1).

Warning 0. Throughout what follows, where there is no danger of confusion, we shall denote $X(t,x,T_0)$ by $X(t,x)$. □

2.1. Let us fix the integer $m\geq 3$. We assume that for all $\omega\in\Omega$, $i=1,2,...,d$, $l=1,2,...,d_1$, $\sigma^{il}(\cdot,\cdot,\omega)\in\mathbf{C}_b^{0,k+1}([T_0,T]\times\mathbf{R}^d)$ and $f^i(\cdot,\cdot,\omega)\in\mathbf{C}_f^{0,m}([T_0,T]\times\mathbf{R}^d)$.

Definition 1. A family of mappings $f(t,\cdot,\omega)$: $\mathbf{R}^d\to\mathbf{R}^d$ for $t\in[T_0,T]$, $\omega\in\Omega$ will be called a stochastic flow of $C^{0,k}$-diffeomorphisms for $k\in\mathbf{N}\cup\{0\}$ of $\mathbf{R}^d$ in (onto) $\mathbf{R}^d$ if:

*(i) For **P**-a.a. ω it is one-to-one mapping of $\mathbf{R}^d$ in (onto) $\mathbf{R}^d$, $f^1(\cdot,\cdot,\omega)\in\mathbf{C}^{0,k}([T_0,T]\times\mathbf{R}^d)$, and $f^1(\cdot,\cdot,\omega)\in\mathbf{C}^{0,k}([T_0,T]\times\mathbf{R}(f))$.*

(ii) For every x, $f(t,x,\omega)$ and $f^1(t,x,\omega)$ are predictable stochastic processes.

*(iii) For **P**-a.a. ω, $f(t,\cdot,\omega)$ is a one-parameter group of mappings $\mathbf{R}^d$ in (onto)*

$\mathbf{R}^d$ *(see [1]) on* $[T_0,T]$ *(i.e.* $f(t,f(s,\cdot,\omega),\omega) = f(t+s,\cdot,\omega)$, $(\mathbf{P}$*-a.s.), for all* t, $s\in[T_0,T]$ *such that* $t+s\leq T$. □

Since, in the sequel, we shall consider only stochastic flows of diffeomorphisms the attribute "stochastic" will commonly be omitted.

Proposition 1. *The family of mappings* $X(t,\cdot)$: $x\in\mathbf{R}^d\rightarrow X(t,x)\in\mathbf{R}^d$ *is a flow of* $C^{0,m-1}$*-diffeomorphisms of* $\mathbf{R}^d$ *in* $\mathbf{R}^d$. *Moreover, for* $r<-(d/2+1)$ *the* i*-th coordinate* $(i=1,2,\ldots,d)$ *of the inverse mapping* $X^{-1}(t,x)$ *is the unique r-generalized solution of the problem*

$$du(t,x) = (\mathcal{M}_0^l\mathcal{M}_0^l-\mathcal{L}_0)\,(t,x)u(t,x)\,dt - \mathcal{M}_0^l(t,x)u(t,x)\,dw^l(t),\ (t,x,\omega)\in[T_0,T]\times\mathbf{R}^d\times\Omega; \tag{2.2}$$

$$u(T,x) = x^i,\ (x,\omega)\in\mathbf{R}^d\times\Omega \tag{2.3}$$

and belongs to the class $\mathbb{L}_2(\Omega;\ \mathbb{C}([T_0,T];\ \mathbb{H}^{m-1}(r)))\cap\mathbb{L}_p(\Omega\times[T_0,T];\ \mathbb{C}_w W_p^m(r))$ *for very* $p\in[2,\infty[$. □

Proof. From Lemma 3.4.7 it follows that $\varphi(x)\equiv x^i\in\mathbf{W}_p^m(t)$ for every $r<-(d/2+1)$ and $p\in[2,\infty[$. Thus from the theorem and the corollary of §4.3.2 it follows that for such p and r, problem (2.2), (2.3) has a unique r-generalized solution in the class $\mathbb{L}_2(\Omega;\ \mathbb{C}([T_0.T];\ \mathbb{H}^{m-1}(r))\cap\mathbb{L}_p(\Omega\times[T_0,T];\ \mathbb{C}_w\mathbf{W}_p^m(r))$ and this solution possesses a version belonging to $\mathbf{C}^{0,m-1}([T_0,T]\times\mathbf{R}^d)$ $(\mathbf{P}$-*a.s.*). Applying the Ito-Ventcel formula (§1.4.9) to the expression $u(t,X(t,x))$, we readily find that $du(t,X(t,x)) = 0$ which in turn implies that $u(t,X(t,x)) = x^i$ for all x,t, ($\mathbf{P}$-a.s.). In view of continuity of $u(t,x)$ and $X(t,x)$ in t,x we can choose an ω-set of probability 1, where the last equality holds for all x, t simultaneously. To complete the proof it suffices to observe that in view of the results of §1.4.4, $X(t,x)$ belongs to $\mathbf{C}^{0,m-1}([T_0,T]\times\mathbf{R}^d)$, ($\mathbf{P}$-a.s.), and is a one-parameter group of mappings of $\mathbf{R}^d$ in $\mathbf{R}^d$ on $[T_0,T]$.

Warning 1. *Throughout what follows we denote the* i*-th coordinate of the mapping* $X^{-1}(t,x,s)$ *by* $X^{-i}(t,x,s)$ *and for* $s = T_0$ *by* $X^{-i}(t,x)$. □

Now we are in a position to introduce the method of variation of constants (see [1]) for an ordinary Ito's differential equation.

Consider the system of ordinary Ito's equations

$$Y^i(t,x) = x^i + \int_{[T_0,t]} {}_Yb^i(s, Y(s,x))ds + \int_{[T_0,t]} {}_Y\sigma^{il}(s, Y(s,x))dw^l(s),$$

(2.4)

$$t\in[T_0,T], \ i=1,2,...,d.$$

It is assumed that ${}_Yb^i(s,x) := {}_Yb^i(s,x,\omega)$ and ${}_Y\sigma^{il}(s,x) := {}_Y\sigma^{il}(s,x,\omega)$ are $\mathcal{B}([T_0,T]\times\mathbf{R}^d)\otimes\mathcal{F}$-measurable, predictable (for every x) functions, and that the integrals in (2.4) are defined in the usual way.

Denote by ${}_Y\mathcal{L}$, ${}_Y\mathcal{M}^l$ the operators obtained from $\mathcal{L}_0$, $\mathcal{M}_0^l$, respectively, by replacing the coefficients in system (2.1) by those of system (2.4).

Theorem 1. (*i*) *System* (2.4) *has a solution if and only if there is a solution of the system*

$$Z^i(t,x) = x^i + \int_{[T_0,t]} {}_Zb^i(s,Z(s,x))ds + \int_{[T_0,t]} {}_Z\sigma^{il}(s,Z(s,x))dw^l(s),$$

(2.5)

$$t\in[T_0,T], \ i=1,2,...,d.$$

where

$${}_Zb^i(s,x) := [{}_Y\mathcal{L}-\mathcal{L}_0-({}_Y\mathcal{M}^l-\mathcal{M}_0^l)\mathcal{M}_0^l)X^{-i}]\ (s,X(s,x)),$$

$${}_Z\sigma^{il}(s,x) := [{}_Y\mathcal{M}^l-\mathcal{M}_0^l)X^{-i}]\ )(s,X(s,x));$$

moreover for all $(t,x,\omega)\in[T_0,T]\times\mathbf{R}^d\times\Omega'$, *where* $\Omega'\subset\Omega$ *and* $\mathbf{P}(\Omega') = 1$ *the following equality holds*

$$Y(t,x) = X(t, Z(t,x)). \tag{2.6}$$

(*ii*) *The solution of system* (2.4) *is unique if and only if the solution of system* (2.5) *is unique.* □

Proof. Applying the Ito-Ventcel formula to the right hand side of formula (2.6), we find that the obtained process is a solution of system (2.4). To prove this, one has to make some simple but cumbersome calculations based on the application of the formulas

$$\delta^{ij} = (X^{-1}{}_l^i(t,\ X(t,x))X_j^l(t,x) \tag{2.7}$$

where δ^{ij} is Kronecker's symbol, and

$$X^{-1}{}_{lk}^{i}(t,\ X(t,x))X_i^k(t,x))X_j^l(t,x) = -(X^{-1}{}_l^i(t,\ X(t,x))X_{ji}^l. \tag{2.8}$$

These formulas can easily be deduced from theorem on the derivative of the inverse function.

To prove the necessity part of (i) of the theorem we only have to apply the Ito-Ventcel formula to the expression $X^{-1}{}^i(t,\ Y(t,x))$, $i=1,2,\dots,d$, making use of the fact that $X^{-1}{}^i(t,x)$ is the solution of problem (2.2), (2.3).

Consider (ii). Suppose that a solution of system (2.4) is unique and let $Z(t,x)$ and $Z'(t,x)$ be two solutions of sytem (2.5). Then as it was proved above, $X(t,\ Z(t,x))$ and $X(t,\ Z'(t,x))$ are the solutions of system (2.4). The uniqueness of the solution of this system yields that for all t,x

$$1 = \mathbf{P}(X(t,\ Z(t,x)) = X(t,\ Z'(t,x))) = \mathbf{P}(X^{-1}(t,X(t,\ Z(t,x))) =$$

$$= X^{-1}(t,\ X(t,\ Z'(t,x)))) = \mathbf{P}(Z(t,x) = Z'(t,x)).$$

So the necessity is proved. The sufficiency can be proved in a similar way. □

We give below some applications of the formula of variation of constants (2.6).

The transformation killing a drift. Let ${}_Yb^i \equiv 0$ and ${}_Y\sigma^{il} \equiv \sigma^{il}$ then ${}_Z\sigma^{il} \equiv 0$ and ${}_Zb^i(t,x) = -b^j X^{-1}{}_j^i(t,\ X(t,x))$. The latter formula, in matrix notation, looks as ${}_Zb(t,x) = -(DX(t,x)/Dx)^{-1}\ b(t,\ X(t,x))$, where $(DX(t,x)/Dx)$ is the Jacoby matrix of the mapping $X(t,\cdot)$: $x \to X(t,x)$.

The transformation generating a drift and diffusion. Let $\sigma^{il} \equiv 0$ for $l \le d_2 \le d_1$, $b^i \equiv 0$, and ${}_Y\sigma^{il} \equiv \sigma^{il}$ for $l > d_2$. Then

$$Z^{b^{il}}(t,x)=\left[\frac{1}{2}\sum_{l=1}^{d_2}Y^{\sigma^{il}}Y^{\sigma^{i'l}}X^{-i}_{ji'}+Y^{b^i}X^{-i}_j\right](t,X(t,x));$$

$$Z^{\sigma^{il}}(t,x)=\begin{cases}\sigma^{il}X^{-i}_j(T,X(t,x)) & \text{for } l\le d_2,\\ 0 & \text{for } l>d_2.\end{cases}$$

Corollary 1. *The mapping* $X(t,\cdot)$: $x\in\mathbf{R}^d\to X(t,x)\in\mathbf{R}^d$ *is a flow of* $C^{0,m-1}$-*diffeomorphisms of* $\mathbf{R}^d$ *onto* $\mathbf{R}^d$.

Proof. In view of the proposition it suffices to determine that for all $\omega\in\Omega'$, where $\mathbf{P}(\Omega')=1$, and $t\in[T_0,T]$ the equality $X(t,X^{-1}(t,x))=x$ holds. On the other hand, from the theorem the existence of the process $Z(t,x)$ satisfying the equality $X(t, Z(t,x))=x$ for all t,x, ($\mathbf{P}$-a.s.) follows. Applying to both sides of this equality the mapping X^{-1}, we find that $Z(t,x) = X^{-1}(t,x)$ for all t,x, ($\mathbf{P}$-a.s.).

2.2. We give here an application of the theorem and the corollary of §.1 to the study of parabolic Ito's equation of second order. We show that if the coefficients of such an equation are sufficiently smooth, then it is equivalent in some sense to a parabolic differential equation of second order with random coefficients.

Consider problem (4.0.1), (4.1.2). It is assumed that the coefficients are

$\mathcal{B}(T_0,T]\times\mathbf{R}^d)\otimes\mathcal{F}$-measurable, predictable (for every x) functions, $f\equiv g^l\equiv0$, and

$\varphi(x)\in\mathbf{C}^{0,4}([T_0,T]\times\mathbf{R}^d)$ for all $\omega\in\Omega$, $i=1,2,\dots,d$, and $l=1,2,\dots,d_1$.

Let $\eta(t,x)$ be a diffusion process which is a solution of the system of ordinary Ito's equation

$$\eta^i(t,x)=x^i-\int_{[T_0,t]}\sigma^{il}(t,\eta(t,x))dw^l(t),\quad t\in[T_0,T],\ i=1,2,\dots,d.$$

From Corollary 1 we have that the process $\eta(t,x)$ has a version which is a flow of $C^{0,2}$-diffeomorphisms of $\mathbf{R}^d$ onto $\mathbf{R}^d$. We denote by $\eta^{-i}(t,x)$ the i-th coordinate of the inverse mapping $\eta^{-1}(t,x)$.

Proposition 1 and Corollary 4.3.2 yield that for every i, $\eta^{-i}(t,x)$ has a version which is the classical solution of problem (2.2), (2.3), where $b^i\equiv0$ and σ^{il} are replaced by σ^{il}.

In the sequel we consider only the versions of $\eta(t,x)$ and $\eta^{-i}(t,x)$ mentioned above, denoting them in the same way as the original processes.

Write

$$\psi(t,x) := exp\Big\{- \int_{[T_0,t]} h^l(t,\ \eta(t,x))dw^l(t) +$$

$$+ \frac{1}{2} \int_{[T_0,t]} h^l h^l(t,\ \eta(t,x))dt\Big\}.$$

In addition to problem (4.0.1), (4.1.2) consider the following

(2.9) $$dv(t,x) = \tilde{\mathfrak{L}} v(t,x)dt,\ (t,x,\omega)\in[T_0,T]\times\mathbf{R}^d\times\Omega$$

(2.10) $$v(T_0,x) = \varphi(x),\ (x,\omega)\in\mathbf{R}^d\times\Omega,$$

where

$$\tilde{\mathfrak{L}}v := \tilde{a}^{ij}v_{ij} + \tilde{b}^i v_i + \tilde{c}v,$$

$$\tilde{a}^{ij}(t,x) := \frac{1}{2}\,\hat{\sigma}^{li'}\hat{\sigma}^{ki}\eta^{-i}_l\eta^{-j}_k(t,\ \eta(t,x)),$$

$$\tilde{b}^i(t,x) := [(b^k - h^l\sigma^{kl} - \sigma^{kl}_j\sigma^{jl})\eta^{-i}_k + \frac{1}{2}\,\hat{\sigma}^{lj}\sigma^{kj}\eta^{-i}_{ik}]\ (t,\ \eta(t,x)) +$$

$$+ 2\psi(t,x)\psi^{-1}_j(t,x)\ \tilde{a}^{ij}(t,x),$$

$$\tilde{c}(t,x) := (c - h^l_i\sigma^{il})\ (t,\ \eta(t,x)) +$$

$$+ \psi\psi^{-1}_i(t,x)\ [b^k - h^l\sigma^{kl} - \sigma^{kl}_j\sigma^{jl})\ \overline{\eta}^i_k +$$

$$+ \frac{1}{2}\,\hat{\sigma}^{lj}\sigma^{kj}\eta^{-i}_{lk}]\ (t,\ \eta(t,x)) + \psi\psi^{-1}_{ij}\tilde{a}^{ij}(t,x).$$

Denote by $\mathfrak{R}$ the set of classical solutions of problem (4.0.1), (4.1.2) and by $\mathfrak{M}$

that of problem (2.9), (2.10).

Theorem 2. Suppose that for $\mathbf{P}$*-a.a.* ω, $\psi(t,x)\in \mathbf{C}^{0,2}([T_0,T]\times\mathbf{R}^d)$,

$\mathcal{L}$: $\mathbf{C}^{0,2}([T_0,T]\times\mathbf{R}^d) \to \mathbf{C}^{0,0}([T_0,T]\times\mathbf{R}^d)$, $\mathcal{M}^l$: $\mathbf{C}^{0,2}([T_0,T]\times\mathbf{R}^d) \to$

$\to \mathbf{C}^{0,1}([T_0,T]\times\mathbf{R}^d)$ $\forall l = 1,2,...,d$, $\mathcal{L}$: $\mathbf{C}^{0,2}([T_0,T)\times\mathbf{R}^d) \to$

$\to \mathbf{C}^{0,0}([T_0,T]\times\mathbf{R}^d)$.

Then the mapping ϕ: $\mathbf{C}^{0,2}([T_0,T]\times\mathbf{R}^d) \to \mathbf{C}^{0,2}([T_0,T]\times\mathbf{R}^d)$ *given by the formula*

(2.11) $$\Phi u(t,x) := \psi(t,x)\, u(t, \eta(t,x)),$$

for $\mathbf{P}$*-a.a.* ω *is a one-to-one mapping of* $\mathfrak{R}$ *onto* $\mathfrak{M}$ *and the inverse mapping is defined by the equality**

(2.12) $$\Phi^{-1} v(t,x) = \psi^{-1}(t, \eta^{-1}(t,x))\, v(t, \eta^{-1}(t,x)). \ \square$$

To prove the theorem we have to apply the Ito-Ventcel formula to the right hand side of equalities (2.11), (2.12), where u is a classical solution of problem (4.0.1), (4.1.2) and v is a classical solution of problem (2.9), (2.10). Simple but cumbersome calculations, which constitute the proof, are based on formulas (2.7), (2.8) and the equivalent to (2.7) formula

$$\delta^{ij} = X^l_j(t, X^{-1}(t,x))\, X^{-1}{}^i_l(t,x).$$

2.3. As it was mentioned in the introduction, we call problem (2.2), (2.3) the forward equation of inverse diffusion. This equations gives the dynamics of $X^{-1}(t,x,s)$ with respect to t for a fixed s. Of course, it is interesting to obtain the dynamics of $X^{-1}(t,x,s)$ with respect to s for a fixed t, i.e. the backward equation of inverse diffusion.

If the coeficients of system (2.1) do no depend on "chance", the latter equation can easily be deduced from the AOC formula (1.2) and the forward equation of inverse diffusion.

---

*We neither suppose nor assert that the sets $\mathfrak{R}$ and $\mathfrak{M}$ are non-empty.

Corollary 3. Suppose that the assumptions of §.1 are satisfied for m=3 and the functions b^i, σ^{il}, *where* $i=1,2,\dots,d$, $l=1,2,\dots,d_1$, *do not depend on* ω. *Denote by* $Y(t,x,s)$ *a continuous in* (t,x,s) *version of the solution of the Ito equation*

$$Y^i(t,x,s) = x^i + \int_{[s,t]} (\sigma^{il}_j \sigma^{jl} - b^i)\,(r,\ Y(t,x,r))\,dr$$

$$- \int_{[s,t]} \sigma^{il}(r,\ Y(t,x,r)) * dw^l(r),\ s\in[T_0,t],\ i=1,2,\dots,d. \tag{2.13}$$

Then for all $(t,x,\omega)\in[T_0,T]\times\mathbf{R}^d\times\Omega'$, *where* $\Omega'\subset\Omega$ *and* $\mathbf{P}(\Omega')=1$, *the following equality holds*

$$X^i(t,x) = Y^i(t,x,T_0). \quad \square \tag{2.14}$$

Proof. Clearly, problem (2.2), (2.3) is one of the type (0.1), (1.2), where $a^{ij} = 2^{-1}\sigma^{il}\sigma^{jl}$. From Theorem 1 it follows that the stochastic characteristic $\mathcal{Y}$ of this problem is $\mathcal{F}_t^{T_0}$-measurable for every t. Thus Proposition 1 and the AOC forumula (1.2) yields that equality (2.14) is valid for $l\times l_d$-a.a. t,x. Since both parts of equality (2.14) are continuous in t,x ($\mathbf{P}$-a.s.), we conclude that this equality holds for all t, x ($\mathbf{P}$-a.s.).

2.4. In this item we derive the backward diffusion equation. Here, as well as in the previous item, we suppose that the coefficients b^i, σ^{il} for all i, l do not depend on ω but we do not assume these coefficients are continuous in t.

Given $r < -(d/2+1)$ and $i=1,2,\dots,d$ consider the problem

$$-dv(s,x) = \mathcal{L}_0 v(s,x)\,ds + \mathcal{M}_0^l v(s,x) * dw^l(s), \tag{2.15}$$

$$(s,x,\omega)\in[T_0,T[\times\mathbf{R}^d\times\Omega,$$

$$v(T,x) = x^i,\ (x,\omega)\in\mathbf{R}^d\times\Omega. \tag{2.16}$$

By the theorem and the corollary of §4.2.1 this problem has a unique r-generalized solution, having a continuous in t,x ($\mathbf{P}$-a.s.) version, which we consider below.

Problem (2.15), (2.16) will be called the backward diffusion equation. The following result gives the reasons for this terminology.

Corollary 4. *Let* $X(T,x,s)$ *be a solution of problem* (2.1) *for* $t=T$ *and* $v(s,x)$ *be an* r*-generalized solution of problem* (2.15), (2.16). *Then for all* $(s,x,\omega)\in[T_0,T]\times\mathbf{R}^d\times\Omega'$, *where* $\Omega'\subset\Omega$ *and* $\mathbf{P}(\Omega')=1$,

$$X(t,x,s) = v(s,x). \quad \square$$

The assertion of the corollary readily follows from the AOC formula (1.3) and the continuity of $X(T,x,s)$ and $v(s,x)$ in x,s.

2.5. *Definition* 5. *The mapping* ξ: $[s,T]\times\mathbf{R}^d\times\Omega \to \bar{\mathbf{R}}^1$, $s\in[T_0,T[$, *is called a first integral of system* (2.1) *if for every* $t\in[s,t]$ *this mapping is* $\mathcal{B}(\mathbf{R}^d)\otimes\mathcal{F}$*-measurable and* $\xi(t, X(t,x,s)) = \xi(t+\tau; X(t+\tau,x,s))$, $\mathbf{P}$*-a.s. for all* t, $\tau\in[s,T]$ *such that* $t+\tau\le T$. $\square$

A first integral $\xi(t,x)$ *of system* (2.1) *will be called direct if it is predictable (for every* x*) and inverse if it is backward predictable (for every* x*).* $\square$

Proposition 5. *If for all* $\omega\in\Omega$, $i=1,2,\ldots,d$ *and* $l=1,2,\ldots,d_1$, $b^i(\cdot,\cdot,\omega)\in \mathbf{C}_b^{0,3}([T_0,T]\times\mathbf{R}^d)$ *and* $\sigma^{il}(\cdot,\cdot,\omega)\in\mathbf{C}_b^{0,4}([T_0,T]\times\mathbf{R}^d)$ *then for every function* $\varphi\in\mathbf{C}_b^2(\mathbf{R}^d)$, $\varphi(X^{-1}(t,x,T_0))$ *is the direct first integral of system* (2.1) *where* $s=T_0$, *and the unique classical solution of the problem*

$$du(t,x) = (\mathcal{M}_0^l\mathcal{M}_0^l - \mathcal{L}_0)\, u(t,x)dt - \mathcal{M}_0^l u(t,x)dw^l(t),$$

$$(t,x,\omega)\in]T_0,T]\times\mathbf{R}^d\times\Omega,$$

$$u(T_0,x) = \varphi(x),\ x\in\mathbf{R}^d. \quad \square$$

Evidently, the assertion of the corollary follows from the AOC (1.2), Proposition 1, and Corollary 4.2.1. $\square$

2.6. *Proposition* 6. *Suppose that the coefficients of system* (2.1) *do not depend on* ω *and for all* $t\in[T_0,T]$, $i=1,2,\ldots,d$ *and* $l=1,2,\ldots,d_1$, $b^i(t,\cdot)$ *and* $\sigma^{il}(t,\cdot)\in\mathbf{C}_b^3(\mathbf{R}^d)$. *Then for every function* $\varphi\in\mathbf{C}_b^2(\mathbf{R}^d)$, $\varphi(X(t,x,T_0))$ *is the inverse first integral of system* (2.1), *where* $s=T_0$, *and the unique classical solution of problem*

$$-dv(t,x) = \mathcal{L}_0 v(t,x,)dt + \mathcal{M}_0^l v(t,x)*dw^l(t), \tag{2.17}$$

$$(t,x,\omega)\in[T_0,T[\times\mathbf{R}^d\times\Omega,$$

(2.18) $$v(T_0,x) = \varphi(x),\ x\in\mathbf{R}^d.\ \square$$

The proposition is a simple corollary of the AOC (1.3) and Corollary 4.2.1. □

Remark 6. *As was mentioned in the introduction, the backward Liouville equation* (2.17), (2.18) *is the direct generalization of the backward Kolmogorov equation. The latter one can be obtained from the former by taking expectatin of the both sides of* (2.17).

5.3. A Representation of a Density-valued Solution.

3.0. Consider the problem

(3.1) $$du(t,x\omega) = \mathcal{L}^*u(t,x,\omega)dt + \mathcal{M}^{l*}u(t,x,\omega)dw^l(t),$$

$$(t,x,\omega)\in]o,T]\times\mathbf{R}^d\times\Omega,$$

(3.2) $$u(0,x,\omega) = \varphi(x,\omega),\ (x,\omega)\in\mathbf{R}^d\times\Omega,$$

where

$$\mathcal{L}^*u := \mathcal{L}^*(t,x,\omega)u := (a^{ij}(t,x,\omega)u)_{ij} - (b^i(t,x,\omega)u)_i + c(t,x,\omega)u,$$

$$\mathcal{M}^{l*}u := \mathcal{M}^{l*}(t,x,\omega)u := -(\sigma^{il}(t,x,\omega)u)_i + h^l(t,x,\omega)u.$$

Let $\mathfrak{F}_0$ be some sub-σ-algebra of $\mathcal{F}_0$, completed with respect to measure $\mathbf{P}$ and $\mathfrak{F}_t := \mathfrak{F}_0\vee\mathcal{F}_t^0$ for every $t\in[0,T]$. Write $\mathbf{F}(\mathfrak{F}) := (\Omega,\mathcal{F},\{\mathfrak{F}_t\}_{t\in[0,T]},\mathbf{P})$ and denote by $\mathcal{P}(\mathfrak{F})$ the σ-algebra of predictable sets on $\mathbf{F}(\mathfrak{F})$.

Note that the Wiener process $w(t)$ is a Wiener one on $\mathbf{F}(\mathfrak{F})$ also.

Definition 0. *A function* u: $[0,T]\times\Omega \rightarrow \mathbf{L}_2$ *will be called density-valued if for every* $t\in[0,T]$ *it belongs,* ($\mathbf{P}$-*a.s.*), *to the cone of non-negative functions from* $\mathbf{L}_1$. □

Throughout this section it is supposed that φ is the Radon-Nicodim derative of the regular conditional (relative to $\mathfrak{F}_0$) probability* $P_{\mathfrak{F}_0}(x_0\in)$ of some $\mathcal{F}_0$-measurable variable x_0 with respect to the Lebesgue measure on $\mathbf{R}^d$, that is

$$P_{\mathfrak{F}_0}(x_0\in\Gamma) = \int_\Gamma \varphi(x)ds, \quad \forall\Gamma\in\mathcal{B}(\mathbf{R}^d).$$

Apart from that it is assumed in the chapter that for all $(t,\omega)\in[0,T]\times\Omega$, $i,j=1,2,\ldots,d_1$, and $l=1,2,\ldots,d_1$, a^{ij} and σ^{il} are three times differentiable in x, b^i and h^l are two times differentiable in x, and c is one time differentiable in x. All the coefficients and their derivatives mentioned above are supposed to be uniformly bounded. It is also assumed that all the coefficients are $\mathcal{P}(\mathfrak{F})$-measurable for every $x\in\mathbb{R}^d$, and $\varphi\in\mathbb{L}_2(\Omega;\, \mathfrak{F}_0;\, \mathbb{H}^1)$.

Remark 0. *Evidently, under the assumptions made above* $\mathcal{L}^*u = a^{ij}u_{ij} - \tilde{b}^i u_i + \tilde{c}u$ *and* $\mathcal{M}^{l*}u = -\sigma^{il}u_i + \tilde{h}^l$, *where* $\tilde{b}^i := b_i + 2a^{ij}_i$, $\tilde{c} := c + a^{ij}_{ij} - b^i_i$, *and* $\tilde{h}^l := h^l - \sigma^{il}_i$. *Thus by Theorem* 4.2.1 *problem* (3.1), (3.2) *has a unique solution* $u\in\mathbb{L}_2([0,T];\, \mathcal{P}(\mathfrak{F});\, \mathbb{H}^1) \cap \mathbb{L}_2(\Omega;\, \mathbb{C}([0,T]);\, \mathbb{L}_2))$.

3.1. The main result of this section is as follows.

Theorem 1. *A generalized solution u of problem* (3.1), (3.2) *is density-valued and for every function* $\psi\in\mathbb{L}_\infty$ *for all* $t\in[0,T]$ *the following representation holds*

$$\int_{\mathbf{R}^d} \psi(x)\, u(t,x)dx = \mathbf{E}[\psi(\mathfrak{X}(t))\rho(t)|\, \mathfrak{F}_t], \quad (\mathbf{P}\text{-}a.s.) \tag{3.3}$$

where $\mathfrak{X}(t)$ *is a solution of the system of ordinary Ito's equations*

$$\mathfrak{X}^i(t) = x^i_0 + \int_{[0,t]} B^i(s,\, \mathfrak{X}(s))ds +$$

$$+ \int_{[0,t]} \Sigma^{il}(s,\, \mathfrak{X}(s))d\nu^l(s),\ i=1,2,\ldots,d,\ t\in[0,T] \tag{3.4}$$

---

* We recall that the function P_G: $\mathcal{B}(\mathbf{R}^d)\times\Omega\to\mathbf{R}^1$ such that: (a) $P_G(\Gamma,\omega)$ is an $\mathcal{F}$-measurable random variable; (b) $P_G(\Gamma,\omega) = \mathbf{P}(\xi\in\Gamma)|\mathcal{F})$, ($\mathbf{P}$-*a.s.*), for some random variable ξ; (c) for $\mathbf{P}$-a.a. ω $P_G(\cdot,\omega)$ is a probability measure on $\mathcal{B}(\mathbf{R}^d)$, is called the regular conditional probability distribution of ξ with respect to G.

(the coefficients B^i and Σ^{il} are defined in §.0.2) and

$$\rho(t):= exp\Big\{ \int_{[0,t]} c(s, \mathfrak{X}(s))ds + \int_{[0,t]} h^l(s, \mathfrak{X}(s))dw^l(s) - $$

$$- \frac{1}{2} \int_{[0,t]} h^l h^l(s, \mathfrak{X}(s))ds\Big\}. \quad \square$$

Remark 1. *Formula* (3.3) *is an AOC formula for the integral functional of the solution of system* (3.1), (3.2), *where the process* $\mathfrak{X}(t)$ *plays the part of a stochastic characteristic.*

The conjugate form of the operators in (3.1) *as well as the absence of the external forces are not crucial. The only purpose of such assumptions is the form of equation to be used in applications (see* §6.2).

Corollary 1. *There exists a* $\mathcal{P}(\mathfrak{F})$*-measruable function* $u\in\mathbb{L}_1([0,T]\times\Omega;\ \mathbb{C}_w\mathbb{L}_1)$ *which is the generalized solution of problem* (3.1), (3.2) *and belongs to the cone of all non-negative functions from* $\mathbb{L}_1$ *for all* $(t,\omega)\in[0,T]\times\Omega$. $\square$

3.2. Before passing to the proof of the theorem we note that it suffices to verify the representation (3.3) for $\psi\in\mathbb{C}_0^\infty(\mathbb{R}^d)$. For this case one can easily deduce that the generalized solution is density-valued and then by passing to the limit, that formula (3.3) is valid in the general situation (for $\psi\in\mathbb{L}_\infty$).

To prove the theorem we need the following auxiliary result.

Theorem 2. *For every* $\psi\in\mathbb{C}_b^{1,2}([0,T]\times\mathbb{R}^d)$ *the conditional expectation* $\mathbb{E}[\psi(t, \mathfrak{X}(t))\rho(t)|\ \mathfrak{F}_t]$ *has a version, call it* $\Phi_t[\psi]$, *which possesses the following stochastic differential*

$$(3.5) \qquad d\Phi_t[\psi] = \Phi_t\Big[\mathcal{L} + \frac{\partial}{\partial t}\ \psi\Big]dt + \Phi_t[\mathcal{M}^l\psi]dw^l(t),\ t\in[0,T]. \quad \square$$

Proof. We begin with the following result.

Lemma 2. *Let* $b\in\mathbb{L}_2([0,T];\ \mathcal{P};\ \mathbb{R}^{d_0})$, *where* $d_0\in\mathbb{N}$, *and* $\hat{w}(t)$ *be a Wiener process, relative to the family* $\mathcal{F}_t$, *independent of* w. *Then*

$$v:= \mathbb{E}\Big[\int_{[0,t]} b^l(s)d\hat{w}^l(s)|\ \mathfrak{F}_t\Big] = 0 \quad (\mathbf{P}\text{-}a.a.). \quad \square$$

Proof. Let us assume for the moment that for $s\in[0,t]$

$$b^l(s) = b_0^l 1_{\{0\}}(s) + \sum_{i=0}^{n-1} b_i^l 1_{\{[t_i,\, t_{i+1}]\}}(s),$$

where $0 = t_0<t_1<\ldots<t_n = t$, b_i^l is an $\mathcal{F}_{t_i}$-measurable random variable.

Since $\mathcal{F}_{t_i}$ is independent of $\mathcal{F}_T^{t_i}$, we have

$$U = \sum_{i=0}^{n-1} \mathsf{E}[b_i^l(\hat{w}^l(t_{i+1})-\hat{w}^l(t_i))|\, \mathfrak{F}_{t_i}] = \sum_{i=0}^{n-1} \mathsf{E}[b_i^l \mathsf{E}[\hat{w}^l(t_{i+1})-\hat{w}^l(t_i)|\, \mathfrak{F}_{t_i}]=0.$$

We complete the proof by passing to the limit. □

To prove equality (3.5), we have to differentiate by Ito's formula the product $\psi(t,\, \mathfrak{S}(t))\rho(t)$ and then take the conditional expectation (with respect to $\mathfrak{F}_t$) of both sides of the resulting integral equality. Making use of the above lemma, Theorem 1.4.7 and the equality

$$\mathsf{E}[\mathcal{L}\psi(s,\, \mathfrak{S}(s))\rho(s)|\, \mathfrak{F}_t] = \mathsf{E}[\mathcal{L}\psi(s,\, \mathfrak{S}(s))\rho(s)|\, \mathfrak{F}_s], \quad (\mathsf{P}\text{-}a.s.)$$

which easily follows from the independence of $\mathcal{F}_s$ and $\mathcal{F}_t^s$, we obtain equality (3.5).

3.3. It is clear that for any $f_1, f_2\in\mathbf{C}_b^2$ and $\alpha, \beta\in\mathbf{R}^1$

$$\Phi_t[\alpha f_1 + \beta f_2] = \alpha\Phi_t[f_1] + \beta\Phi_t[f_2]. \tag{3.6}$$

However, the ω-set for which equality (3.6) holds depends on f_1, f_2, α and β. Later on we shall need a version of $\Phi_t[\psi]$ possessing the stochastic differential (3.5) and which is also a linear functional on $\mathbf{C}_b^0(\mathbf{R}^d)$. The latter means that the equality (3.6) holds on an ω-set of probability one, not depending on f_1, f_2, α and β.

To prove the existence of such a version we need the following corollary of the Hahn-Banach theorem.

Proposition 3 (*see e.g.* [25] *Ch. II,* §5). *Let* $\mathbf{X}$ *be a Banach space and* $\mathbf{X}^*$ *be its conjugate space. Let* $\{x_n\}$, $n\in\mathsf{N}$ *be a sequence of elements of* $\mathbf{X}$ *and* $\{c_n\}$ *be a* *sequence of real numbers. The necessary and sufficient condition for the existence of $x^* \in \mathbf{X}^*$ such that $x^* x_n = c_n$ for every $n \in \mathbf{N}$ and $\|x^*\|_{\mathbf{X}^*} \le N$ is the following*

$$\left|\sum \alpha_i c_i\right| \le N \left\|\sum \alpha_i x_i\right\|_{\mathbf{X}^*}.$$

for every finite set of real numbers $\{\alpha_i\}$. □

Making use of this result we prove

Theorem 3. For every $(t,\omega) \in [0,T] \times \Omega'$, where $\Omega' \subset \Omega$ and $\mathbf{P}(\Omega')=1$, there exists a continuous linear functional $\tilde{\Phi}_t[\cdot](\omega)$ on $\mathbf{C}_b^0(\mathbf{R}^d)$ with these properties.

(a) For every $\psi \in \mathbf{C}_b^0(\mathbf{R}^d)$, $\tilde{\Phi}_t[\psi]$ is a continuous (in t), predictable version of $\Phi_t[\psi]$.

(b) For every $\psi \in \mathbf{C}_b^0(\mathbf{R}^d)$, $t \in [0,T]$ and $\omega \in \Omega'$

$$(3.7) \qquad \left|\tilde{\Phi}_t[\psi](\omega)\right| \le \sup_{x \in \mathbf{R}^d} |\psi(t,x,\omega)|\ \tilde{\Phi}_t[1](\omega) < \infty, \forall (t,\omega) \in [0,T] \times \Omega. \quad \square$$

Proof. We take a countable set $\{\eta_i\}$, $i \in \mathbf{N}$, dense in $\mathbf{C}_b^0(\mathbf{R}^d)$ in the uniform norm. For all $t \in [0,T]$ and collections of rational numbers r_i, only a finite number of which are different from zero, we have, ($\mathbf{P}$-a.s.),

$$(3.8) \qquad \left|\sum_i r_i \Phi_t[\eta_i]\right| \le \max_{x \in \mathbf{R}^d} \left|\sum_i r_i \eta_i(t,x)\right| \Phi_t[1].$$

Due to the continuity in t of the expression appearing above, and also because of the countability of the set of finite collections of rational numbers, the relations (3.8) are satisfied on one and the same set Ω' of probability one for all $t \in [0,T]$ and sets of rational numbers r_i under consideration. In addition, both parts of (3.8) are continuous in r_i. Therefore, this inequality holds for arbitrary η_i and arbitrary real number r_i only a finite number of which are different from zero for all $t \in [0,T]$ and $\omega \in \Omega'$. From this, by Theorem 3, it follows that for all $(t,\omega) \in [0,T] \times \Omega'$ there exists on $\mathbf{C}_b^0(\mathbf{R}^d)$ a continuous functional $\tilde{\Phi}_t[\cdot](\omega)$ such that for all $i \in \mathbf{N}$,

$$(3.9) \qquad \tilde{\Phi}_t[\eta_i] = \Phi_t[\eta_i],$$

and

$$(3.10) \qquad \left|\tilde{\Phi}_t[\psi]\right| \le \max_{x \in \mathbf{R}^d} |\psi(x)|\ \Phi_t[1]$$

for every $\psi\in \mathbf{C}_b^0(\mathbf{R}^d)$.

From these relations and the continuity of $\Phi_t[\eta_i]$ in t, it follows at once that $\tilde{\Phi}_t[\psi]$ is continuous in t for every $\psi\in \mathbf{C}_b^0(\mathbf{R}^d)$. In addition, choosing $\eta_{i(n)}$ such that $\eta_{i(n)} \rightarrow \psi$ uniformly, we obtain that for every $t\in[0,T]$, ($\mathbf{P}$-*a.s.*),

$$\tilde{\Phi}_t[\psi] = \lim_{n\to\infty} \tilde{\Phi}_t[\eta_{i(n)}] = \lim_{n\to\infty} \Phi_t[\eta_{i(n)}] = \Phi_t[\psi]. \tag{3.11}$$

Since $\Phi_t[\psi]$ is $\mathcal{F}_t$-measurable for every t, we obtain from (3.11) that $\Phi_t[\psi]$ is a predictable stochastic process (see §1.3.6). This completes the proof. □

From this theorem and Lemma 2 we obtain at once

Corollary 3. *If* $\psi\in \mathbf{C}_b^{1,2}([0,T]\times\mathbf{R}^d)$ *then* $\tilde{\Phi}_t[\psi(t,\cdot)]$ *is continuous in* t, ($\mathbf{P}$-*a.s.*) *and moreover it possesses the same stochastic differential* (3.5) *as* $\Phi_t[\psi(t,\cdot)]$. □

Remark 3. *An analysis of the proofs of Theorem 2 and 3 shows that the assumptions concerning the smoothness of the coefficients can be weakened. It suffices to assume that the coefficients* Σ^{il} *and* B^i *are Lipschitz-continuous in* x. □

3.4. Denote by d^0 the least integer greater than $d/2$.

Lemma 4. *In addition to the assumptions of Theorem* 1 *let the following conditions be satisfied.*

(*i*) *For all* $(t,\omega)\in[0,T]\times\Omega$, $i=1,2,\dots,d$, $k=1,2,\dots,d+d_1$, *and* $l=1,2,\dots,d_1$, $b^i(t,\cdot,\omega)$, $c(t,\cdot,\omega)$, $\Sigma^{ik}(t,\cdot,\omega)$, *and* $h^l(t,\cdot,\omega)\in \mathbf{C}_b^{d^0+2}(\mathbf{R}^d)$.

(*ii*) *Equation* (3.1) *is superparabolic.*

Then for every $\psi\in \mathbf{C}_b^\infty(\mathbf{R}^d)$ *the AOC formula* (3.3) *is valid.* □

Proof. We begin with the following result.

Proposition 4. *Under the assumptions of the lemma there exists a function* v: $[0,T]\times\Omega \rightarrow \mathbf{H}^{d^0}$ *belonging to* $\mathbf{L}_2^\omega([0,T];\ \mathcal{P}(\mathfrak{F});\ \mathbf{H}^{d^0})$ *and such that for every* $\psi\in \mathbf{C}_0^\infty(\mathbf{R}^d)$ *and all* $(t,\omega)\in[0,T]\times\Omega'$, *where* $\Omega'\subset\Omega$ *and* $\mathbf{P}(\Omega')=1$,

$$\tilde{\Phi}_t[\psi] = (\psi,\ v(t))_{d^0}. \quad \square \tag{3.13}$$

Proof. From Theorems 3 and 3.4.4 we obtain that there exists $n\in\mathbf{R}_+$ such that for all $t\in[0,T]$ and ω belonging to a set Ω' of probability one

$$(3.14)\qquad \left|\tilde{\Phi}_t[\psi]\right| \le \sup_{x\in\mathbf{R}^d} |\psi(x)|\ \Phi_t[1] \le N\|\psi\|_{d^0}\ \Phi_t[1],$$

and $\sup_{t\in[0,T]} \Phi_t[1] < \infty$.

From this by the Fréchet-Riesz theorem on the representation of a continuous linear functional on a Hilbert space we have that for all $(t,\omega)\in[0,T]\times\Omega'$ there exists a function $v(t,\omega)$ taking values in $\mathbf{H}^{d^0}$ and satisfying equality (3.13).

Since $\tilde{\phi}_t[\psi]$ is a $\mathcal{P}(\mathfrak{F})$-measurable for every $\psi\in\mathbf{C}_0^\infty(\mathbf{R}^d)$ and $\mathbf{C}_0^\infty(\mathbf{R}^d)$ is dense in $\mathbf{H}^{d^0}$, we obtain from Corollary 1.2.6 that $(v(t),\psi)_{d^0}$ is a $\mathcal{P}(\mathfrak{F})$-measurable stochastic process, for every $\psi\in\mathbf{H}^{d^0}$. By the Pettis theorem, this yields the $\mathcal{P}(\mathfrak{F})$-measurability of $v(t)$.

To complete the proof, observe that

$$\int_{[0,T]} \|v(t)\|_{d^0}^2\, dt = \int_{[0,T]} \sup_{\psi\in\mathbf{C}_0^\infty(\mathbf{R}^d)} \frac{|\tilde{\Phi}_t[\psi]|^2}{\|\psi\|_{d^0}^2}\, dt \le$$

$$\le N \int_{[0,T]} \left|\Phi_t[1]\right|^2 dt \le \infty. \quad \square$$

By Proposition 4 and Theorem 3 we have that for all $t\in[0,T]$ and $\psi\in\mathbf{C}_0^\infty(\mathbf{R}^d)$ on one and the same ω-set Ω'' of probability one the following equality holds

$$(v(t),\, \psi)_{d^0} = (v(0),\, \psi)_{d^0} + \int_{[0,t]} (v(s),$$

$$(3.15)\qquad \mathcal{L}(s)\psi)_{d^0} ds + \int_{[0,t]} (v(s),\, \mathcal{M}^l(s)\psi)_{d^0} dw^l(s).$$

Let Δ be the Laplace operator on $\mathbf{R}^d$ (see §.3.4.2) and $\Lambda := (\mathbf{I}-\Delta)^{1/2}$. Denote $\tilde{v}(t) := \Lambda^{-2}v(t)$. From Propositions 3.4.2 (i) and 1.2.9 (ii) it follows that $\tilde{u}\in\mathbf{L}_2^\omega([0,T];\ \mathcal{P}(\mathfrak{F});\ \mathbf{H}^{d^0+2})$. Again making use of Proposition 3.4.2 (i), we obtain that equation (3.15) can be transformed into the form

$$(\hat v(t), \psi)_{d^0+1} + (\hat v(0), \psi)_{d^0+1} + \int_{[0,t]} (\hat v(s), \mathcal{L}(s)\psi)_{d^0+1} ds + \int_{[0,t]} (\hat v(s), \mathcal{M}^l(s)\psi)_{d^0+1} dw^l(s),\ t\in[0,T]. \tag{3.16}$$

Note that for every $\eta\in\mathbf{H}^{d^0+2}$ and all $(s,\omega)\in[0,T]\times\Omega$,

$$\begin{aligned}(\eta, \mathcal{L}(s,\omega)\psi)_{d^0+1} &= -(\eta_j, a^{ij}(s,\omega)\psi_i)_{d^0+1} + \\ &+ (\eta, b^i(s,\omega) - a^{ij}_j(s,\omega))\psi_i + c(s,\omega)\psi)_{d^0+1} := \\ &= (\tilde{\mathcal{L}}(s,\omega)\eta, \psi)_{d^0+1}.\end{aligned} \tag{3.17}$$

Clearly it suffices to verify this equality for $\eta\in\mathbf{C}_0^\infty(\mathbf{R}^d)$. Under this additional assumption

$$(\eta, \mathcal{L}\psi)_{d^0+1} - (\Lambda^{2d^0+2}\eta, \mathcal{L}\psi)_0.$$

Integrating by parts the right hand side of the last equality, then changing the order of application of the operators $\partial/\partial x^i$ and $\Lambda^{2d^0+2} = (\mathrm{I}-\Delta)^{d^0+1}$ and coming back, by Proposition 3.4.2(i) to the scalar product in $\mathbf{H}^{d^0+1}$, we obtain equality (3.17).

From (3.17) and the Schwarz inequality, making use of Proposition 3.4.6, we obtain that there exists a constant N such that for every $(s,\omega)\in[0,T]\times\Omega$ and η, $\psi\in\mathbf{H}^{d^0+2}$,

$$\left|(\eta, \mathcal{L}(s,\omega)\psi)_{d^0+1}\right| \le N\|\eta\|_{d^0+2}\,\|\psi\|_{d^0+2}. \tag{3.18}$$

Since the CBF of the normal triple $(\mathbf{H}^{d^0+2}, \mathbf{H}^{d^0+1}, \mathbf{H}^{d^0})$* defines an isometric

---

*We denote it by $[\cdot,\cdot]_{d^0+1}$.

isomorphism between $\mathbf{H}^{d^0}$ and $(\mathbf{H}^{d^0+2})^*$ (see Corollary 3.4.2), it follows from inequality (3.18) that for every $(s,\omega)\in[0,T]\times\Omega$, there exists a linear operator

$A(s,\omega)$: $\mathbf{H}^{d^0+2}\rightarrow\mathbf{H}^{d^0}$ such that

$$[A(s,\omega)\ \eta,\ \psi]_{d^0+1} = (\widetilde{\mathcal{L}}(s,\omega)\ \eta,\ \psi)_{d^0+1} \tag{3.19}$$

for every η, $\psi\in\mathbf{H}^{d^0+2}$.

Next, from inequality (3.18) it follows that for all $(s,\omega)\in[0,T]\times\Omega$

$$\|A(x,\omega)\eta\|_{d^0} \leq N\|\eta\|_{d^0+2},\quad \forall\eta\in\mathbf{H}^{d^0+2}. \tag{3.20}$$

Arguing in the same way as in the proof of (3.17), it is easy to see that

$$(\eta,\ \mathcal{M}^l(s,\omega)\psi)_{d^0+1} = -(\eta_i,\ \sigma^{il}(s,\omega)\psi)_{d^0+1} +$$

$$+ (\eta,\ (h^l-\sigma^{il}_i)\ (s,\omega)\psi)_{d^0+1},\quad \forall(s,\omega).$$

This together with Proposition 3.4.6 gives us that for every $(s,\omega)\in[0,T]\times\Omega$

there exists an operator $B(s,\omega)$: $\mathbf{H}^{d^0+2}\rightarrow\mathcal{L}_2(\mathbf{R}^{d_1},\ \mathbf{H}^{d^0+1})$ defined by the following equality

$$(B^l(s,\omega)\ \eta,\ \psi)_{d^0+1} := (-(\eta_i,\ \sigma^{il}(s,\omega)\psi)_{d^0+1} +$$

$$+ (\eta,\ h^l-\sigma^{il}_i)\ (s,\omega)\psi)_{d^0+1} \tag{3.21}$$

$$\forall\eta\in\mathbf{H}^{d^0+2},\ \psi\in\mathbf{H}^{d^0+2},\ l=1,2,\ldots,d,$$

where B^i denotes the i-th coordinate of B.

As usual, applying the Pettis theorem, it is easy to show that for every

$\eta\in\mathbf{H}^{d^0+2}$, $A(s,\omega)\eta$ is a $\mathcal{P}(\mathfrak{F})$-measurable $\mathbf{H}^{d^0}$-process and $B(s,\omega)\eta$ is a $\mathcal{P}(\mathfrak{F})$-measurable $\mathfrak{L}_2(\mathbf{R}^{d_1},\ \mathbf{H}^{d^0+1})$-process.

Thus we have proved that $\hat{v}$ is a solution (in the normal triple $(\mathbf{H}^{d^0+2},\ \mathbf{H}^{d^0+1},$

$\mathbf{H}^{d^0}$)) of the LSES

$$(3.22)\qquad u(t) = v(0) + \int_{[0,t]} Au(s)ds + \int_{[0,t]} Bu(s)dw(s),\ t\in[0,T],$$

where the operators A and B are defined (3.19), (3.21).

Due to Proposition 3.1.3 and Warning 3.1.3 we can and shall suppose that

$\hat{v}\in \mathbf{L}_2^{\omega}([0,T];\ \mathcal{P}(\mathfrak{F}),\ \mathbf{H}^{d^0+2}) \cap \mathbf{C}([0,T];\ \mathcal{P}(\mathfrak{F});\ \mathbf{H}^{d^0+1})$.

Now we prove that $\hat{v}$ is the only solution of LSES (3.22). For this purpose we are going to apply Theorem 3.3.1. Condition (B_ω) of this theorem is already verified (see (3.19)). Let us now verify condition (A_ω).

First observe (see Theorem 2.1.11) that for all $(s,\omega)\in[0,T]\times\Omega$ and every $\eta\in\mathbf{H}^{d^0+2}$ we have

$$(3.23)\qquad |||B(s,\omega)\eta|||^2_{d^0+1} = \sum_{l=1}^{d_1} \|B^l(s,\omega)\eta\|^2_{d^0+1}.^*$$

Applying equality (4.1.10), it is easy to show that for every $l=1,2,\ldots,d_1$,

$$(B^l\eta,\ \psi)_{d+1} = \sum_{|\alpha|\le d^0+1} C^{\alpha}(-\sigma^{il}(\eta_i)_\alpha + \tilde{h}^l\eta_\alpha,\ \psi_\alpha)_0 + V(\eta,\psi),$$

where $\tilde{h}^l := h^l - \sigma^{il}_i$, $V(\eta,\psi) := \sum^{\alpha}_{d^0+1} C^{\alpha}(\eta_\alpha - \sigma^{il}_\beta\psi_\gamma)_i + \tilde{h}^l_\beta\psi_\gamma)_0$.

Here and in the sequel $\sum^{\alpha}_{d^0+1}$-denotes summation over all multiindices $\alpha = \beta+\gamma$ such that $|\alpha|\le d^0+1$ and $|\beta|\ge 1$.

From the last equality with the aid of the Schwarz inequality and the elementary inequality

---

* In this section the norm in $\mathfrak{L}_2(\mathbf{R}^{d_1},\ \mathbf{H}^{d^0+1})$ is denoted by $|||\cdot|||_{d^0+1}$.

$$(a+b)^2 \leq (1+\varepsilon)a^2 + (1+\varepsilon^{-1})b^2, \ \forall \varepsilon>0,$$

we find that

$$(3.24) \qquad \sum_{l=1}^{d_1} (B^l\eta,\ \psi)^2 \leq (1+\varepsilon) V_0^2(\eta)\ \|\psi\|^2_{d^0+1} + (1+\varepsilon^{-1})\ |V(\eta,\psi)|^2$$

where

$$V_0^2(\eta) := \sum_{|\alpha|\leq d^0+1} C^\alpha \|-\sigma^{il}(\eta_i)_\alpha + \tilde{h}^l\eta_\alpha\|_0^2.$$

From the Schwarz inequality and Proposition 3.4.6 it follows that for all $(s,\omega)\in[0,T]\times\Omega$ and $\eta,\ \psi\in\mathbb{H}^{d^0+1}$,

$$(3.25) \qquad |V(\eta,\psi)|^2 \leq N\|\eta\|^2_{d^0+1}\ \|\psi\|^2_{d^0+1},$$

where $N\in\mathbb{R}_+$.

So from relations (3.23) to (3.25) we obtain that for all $(s,\omega)\in[0,T]\times\Omega$ and every $\varepsilon\geq 0$

$$(3.26) \qquad |||B(s,\omega)\eta|||^2_{d^0+1} \leq (1+\varepsilon)\ V_0^2(\eta) + N(1+\varepsilon^{-1})\|\eta\|^2_{d^0+1}.$$

On the other hand, from (3.19) and equality (4.1.10) it follows that for every $\eta\in\mathbb{H}^{d^0+2}$ and $(s,\omega)\in[0,T]\times\Omega$

$$[A\eta,\ \eta]_{d^0+1} = \sum_{i=1}^{3} V_i(\eta),$$

where

$$V_1(\eta) := \sum_{|\alpha|\leq d^0+1} C^\alpha[-((\eta_j)_\alpha,\ a^{ij}(\eta_i)_\alpha)_0 +$$

$$+ (\eta_\alpha, (b^i - a_j^{ij})(\eta_i)_\alpha + c\eta_\alpha)_0],$$

$$V_2(\eta) := -\sum_{d^0+1}^{\alpha} C^\alpha((\eta_j)_\alpha, a_\beta^{ij}(\eta_i)_\gamma)_0,$$

and

$$|V_3(\eta)| \leq N\|\eta\|^2_{d^0+1}, \quad \forall \eta \in \mathbf{H}^{d^0+1}. \tag{3.7}$$

Applying the Schwarz inequality, Theorem 3.4.2, proposition 3.4.6 and inequality (3.1.10), we obtain that for every $\varepsilon > 0$, there exists a constant $N(\varepsilon)$ (depending on ε) such that for all $\eta \in \mathbf{H}^{d^0+2}$ and $(s,\omega) \in [0,T] \times \Omega$,

$$|V_2(\eta)| \leq \varepsilon\|\eta\|^2_{d^0+2} + N(\varepsilon)\|\eta\|^2_{d^0+1}. \tag{3.28}$$

From Lemma 4.1.4 and equality (4.1.10) it follows that there exist $\delta \geq 0$ and $N \in \mathbb{R}_+$ such that for all $\eta \in \mathbf{H}^{d^0+2}$ and $(s,\omega) \in [0,T] \times \Omega$,

$$2V_1(\eta) + V_0^2(\eta) \leq -\delta\|\eta\|^2_{d^0+2} + N\|\eta\|^2_{d^0+1}. \tag{3.29}$$

Lastly, by Proposition 3.4.6 and Theorem 3.4.2 we have that there exists $N \in \mathbf{R}_+$ such that for all $(s,w) \in [0,T] \times \Omega$,

$$V_0^2(\eta) \leq N\|\eta\|^2_{d^0+2}, \quad \forall \eta \in \mathbf{H}^{d^0+2}. \tag{3.30}$$

Collecting (3.26) - (3.30) and making ε sufficiently small, we obtain that there exist numbers $\gamma > 0$ and $N \in \mathbf{R}_+$ such that for all $(s,\omega) \in [0,T] \times \Omega$ and $\eta \in \mathbf{H}^{d^0+2}$, the following inequality holds,

$$2[A(s,\omega)\eta,\ \eta]_{d^0+1} + |||B(s,\omega)\eta|||^2_{d^0+1} \le -\gamma\|\eta\|^2_{d^0+2} + N\|\eta\|^2_{d^0+1}.$$

Thus we have proved that LSFS (3.22) satisfies the assumptions of Theorem 3.3.1, which implies the uniqueness of its solution. Hence $\hat{v}$ is the only function from $\mathbf{L}_2^\omega([0,T];\ \mathcal{P}(\mathfrak{F});\ \mathbf{H}^{d^0+2}) \cap \mathbf{C}([0,T];\ \mathcal{P}(\mathfrak{F});\ \mathbf{H}^{d^0+1})$ satisfying equality of (3.16) for all $t\in[0,T]$, $w\in\Omega'$, where $\Omega\subset\Omega'$ and $\mathbf{P}(\Omega') = 1$, and $\psi\in\mathbf{C}_0^\infty(\mathbf{R}^d)$.

On the other hand, in view of Proposition 4 and Proposition 3.4.2 for $\mathbf{P}$-a.a. ω,

$$(\varphi,\psi)_0 = (v(0)),\ \psi)_{d^0} = (\hat{v}(0),\ \psi)_{d^0+1} = (\Lambda^{-2(d^0+1)}\varphi,\psi)_{d^0+1}$$

$$\forall\psi\in\mathbf{C}_0^\infty(\mathbf{R}^d),$$

and consequently

$$\text{(3.31)}\qquad \hat{v}(0) = \Lambda^{-2d^0+1}\varphi$$

Problem (3.1), (3.2) as it was mentioned in §.1 has a unique solution $u\in\mathbf{L}_2([0,T];\ \mathcal{P}(\mathfrak{F});\ \mathbf{H}^1) \cap \mathbf{L}_2(\Omega;\ \mathbf{C}([0,T];\ \mathbf{L}_2))$.

Hence Proposition 3.4.2 (i) and (3.31) yield that $\hat{u}:= \Lambda^{-2(d^0+1)}u\in\mathbf{L}_2([0,T];$ $\mathcal{P}(\mathfrak{F});\ \mathbf{H}^{2d^0+3}) \cap \mathbf{L}_2(\Omega;\ \mathbf{C}([0,T];\ \mathbf{H}^{2d^0+2})$ and satisfies equality (3.16) for all $(t,\omega)\in[0,T]\times\Omega''$, where $\Omega''\subset\Omega$ and $P(\Omega'') = 1$, and $\psi\in\mathbf{C}_0^\infty(\mathbf{R}^d)$, which implies that $\hat{u} = \hat{v}$.

Thus the following equality is proved

$$\text{(3.32)}\qquad \Lambda^{2(d^0+1)}\hat{v} = u$$

which we understand as equality of elements in $\mathbf{L}_2([0,T];\ \mathcal{P}(\mathfrak{F});\ \mathbf{H}^1) \cap \mathbf{L}_2(\Omega;$ $\mathbf{C}([0,T];\ \mathbf{L}_2))$.

From the proposition of this paragraph, Proposition 3.4.2 (i), and (3.32) it follows that for every $t\in[0,T]$ and $\psi\in\mathbf{C}_0^\infty(\mathbf{R}^d)$, $\mathbf{E}[\psi(\mathfrak{X}(t))\rho(t)|\mathfrak{F}_t] = (\psi,\hat{v}(t))_{d^0} =$ $(\psi,\ u(t))_0$, ($\mathbf{P}$-$a.s.$). □

3.5. *Proof of Theorem* 1. We have already mentioned that it suffices to prove the theorem in the case of a smooth ψ. Thus in the sequel we shall assume that $\psi \in \mathbf{C}_b^0(\mathbf{R}^d)$. Let $\tilde{w}(t)$ be a d-dimensional Wiener process on $\mathbf{F}(\mathfrak{F})$ independent of $\nu(t)$ and $\tilde{\nu}(t)$ be a $(d+d_0+d_1)$-dimensional Wiener process whose first d components coincide with $\tilde{w}(t)$ and the latter d_0+d_1 with $\nu(t)$.

In this paragraph we use the notation (introduced in §4.1.5) $\xi_{(\varepsilon)}(s,x,\omega):=$

$(T_\varepsilon \xi(s,\cdot,\omega))(x)$, where T_ε is the Sobolev averaging operator (see §1.4.10).

Let δ^{ij} be the Kronecker symbol. Given $\varepsilon > 0$ denote by ${}_\varepsilon\Sigma$ the matrix $({}_\varepsilon\Sigma^{il})$, $i=1,2,\dots,d$, $l=1,2,\dots,d+d_0,d_1)$, where

$$
{}_\varepsilon\Sigma^{il} := \begin{cases} \sqrt{2\varepsilon}\,\delta^{il} & \text{for } l=1,2,\dots,d \\ \hat{\sigma}^{il}_{(\varepsilon)} & \text{for } l=d+1,d+2,\dots,d+d_0 \\ \sigma^{il}_{(\varepsilon)} & \text{for } l=d+d_0+1,d+d_0+2,\dots d+d_0+d_1 \end{cases}
$$

Consider the diffusion process ${}_\varepsilon\mathfrak{X}(t)$, which is a solution of the following system of Ito's equation.

$$
{}_\varepsilon\mathfrak{X}^i(t) = x_0^i + \int_{[0,t]} B^i_{(\varepsilon)}(s, {}_\varepsilon\mathfrak{X}(s))\,ds + \int_{[0,t]} {}_\varepsilon\Sigma^{il}(s, {}_\varepsilon\mathfrak{X}(s))\,d\tilde{\nu}^l(s),
$$

(3.33)

$$
i=1,2,\dots,d,\ t\in[0,T].
$$

By Lemma 1.4.10, $B^i_{(\varepsilon)}(s,x,\omega)$ and $\Sigma_{(\varepsilon)}(s,x,\omega)$ belongs to $C_b^\infty(\mathbf{R}^d)$, for all i, l, ε, s, ω and are uniformly bounded with respect to ε, s, ω together with their derivatives. From Theorem 1.4.3 (see also the footnote to this theorem) this yields that system (3.33) has a unique solution.

Write ${}_\varepsilon a^{ij} := 1/2\ {}_\varepsilon\Sigma^{il}\ {}_\varepsilon\Sigma^{jl}$ and consider the problem

$$
(3.34) \qquad du^\varepsilon(t,x) = [({}_\varepsilon a^{ij} u^\varepsilon(t,x))_{ij} - (b^i_{(\varepsilon)} u^\varepsilon(t,x))_i + c_{(\varepsilon)} u^\varepsilon(t,x)]\,dt +
$$

$$
+ [-(\sigma^{il}_{(\varepsilon)} u^\varepsilon(t,x))_i + h^l_{(\varepsilon)} u^\varepsilon(t,x)]\,dw^l(t),
$$

$$(t,x,\omega)\in]0,T]\times \mathbf{R}^d\times \Omega;$$

(3.5) $$u(0,x) = \varphi(x),\ (x,\omega)\in \mathbf{R}^d\times\Omega$$

Clearly, thc coefficients of equation (3.34) are infinitely differentiable in x and uniformly (in ε, t, ω) bounded together with their derivatives. Further, by making simple calculations and applying Jensen inequality (in the form $(T_\varepsilon f)^2 \le T_\varepsilon(f)^2)$ it is easy to show that for all t, x, ω

$$2\,{}_\varepsilon a^{ij}(t,x,\omega)\xi^i\xi^j - \sum_{l=1}^{d_1}\left|\sigma^{il}_{(\varepsilon)}(t,x,\omega)\xi^i\right|^2 \geq \varepsilon|\xi|_d^2 \geq 0,$$

$$\forall \xi\in\mathbf{R}^d,\ \varepsilon\in\mathbf{R}_+,$$

that is, equation (3.3.4) is superparabolic if $\varepsilon>0$ and parabolic if $\varepsilon=0$.

Thus problem (3.34), (3.35) satisfies the condition of Lemma 4 and, consequently, the following representation holds

(3.36) $$\mathsf{E}[\psi({}_\varepsilon\mathfrak{X}(t))\,{}_\varepsilon\rho(t)|\,\mathfrak{F}_t] = (\psi,\ u^\varepsilon(t))_0,\ \forall t\in[0,T],\quad (\mathbf{P}\text{-}a.s.)$$

where

$${}_\varepsilon\rho(t) := exp\Big\{\int_{[0,t]}\Big(c_{(\varepsilon)} - \frac{1}{2}\,h^l_{(\varepsilon)}h^l_{(\varepsilon)}\Big)(s,\ {}_\varepsilon\mathfrak{X}(s))ds +$$

$$\int_{[0,t]} h^l_{(\varepsilon)}(s,\ {}_\varepsilon\mathfrak{X}(s))dw^l(s)\Big\}.$$

From Lemma 1.4.10 it follows that for all t,x,ω,i,j, and l, ${}_\varepsilon a^{ij} \to a^{ij}$, ${}_\varepsilon\Sigma^{ij}\to\Sigma^{ij}$, $b^i_{(\varepsilon)}\to b^i$, $c_{(\varepsilon)}\to c$, $h^l_{(\varepsilon)}\to h^l$, and $B^i_{(\varepsilon)}\to B^i$ as $\varepsilon\to 0$. Next, by Corollary 4.2.4, we have

(3.37) $$\lim_{\varepsilon\to 0}\mathsf{E}\sup_{t\in[0,T]}\|u^\varepsilon(t) - u(t)\|_d^2 = 0,$$

where u is the generalized solution of problem (3.1), (3.2). Making use of these facts, we can pass to the limit in equality (3.36). To justify this passage to the limit we multiply both parts of equality (3.36) by a $\mathfrak{F}_t$-measurable bounded function ξ. Then taking expectation of both sides of the obtained equality, we find that

$$\mathsf{E}[\psi({}_{\varepsilon}\mathfrak{X}(t))_{\varepsilon}\rho(t)\xi] = \mathsf{E}[(\psi,\ u^{\varepsilon}(t))_0\xi]. \tag{3.38}$$

From (3.37) it follows that as $\varepsilon \to 0$

$$\mathsf{E}[(\psi,\ u^{\varepsilon}(t))_0\xi] \to \mathsf{E}[(\psi,\ u(t))_0\xi]. \tag{3.39}$$

It is well known (see e.g. [34], Part 1, Ch. 2, §7) that there exists a subsequence $\{\varepsilon_n\}$ tending to zero, (P-a.s.), as $n \to \infty$ such that $\lim\limits_{n\to\infty} \sup\limits_{t\in[0,T]} |{}_{\varepsilon_n}\mathfrak{X}(t) - \mathfrak{X}(t)|_d = 0$ and $\lim\limits_{n\to\infty} {}_{\varepsilon_n}\rho(t) = \rho(t)$.

The set $\{\psi({}_{\varepsilon_n}\mathfrak{X}(t)_{\varepsilon_m}\rho(t)\xi\}$ is uniformly integrable, because in view of the uniform boundness of $h_{(\varepsilon)}$, $c_{(\varepsilon)}$, ψ, and ξ

$$\sup_{\varepsilon} \mathsf{E}|\psi({}_{\varepsilon}\mathfrak{X}(t))_{\varepsilon}\rho(t)\xi|^{p} < \infty \quad \forall p\in[1,\infty[$$

(see [100], Ch. II, Theorem 22).

Thus by passing to the limit in both sides of (3.8) and taking into account (3.39) we obtain

$$\mathsf{E}[\psi(\mathfrak{X}(t))\rho(t)\xi] = \mathsf{E}[(\psi,\ u(t))_0\xi].$$

3.6. *Proof of Corollary* 1. *Let* v be a generalized solution of problem (3.1), (3.2). From the AOC formula (3.5) it follows in particular that for every $\psi\in C_0^{\infty}(\mathbf{R}^d)$ and $t\in[0,T]$ there exists an ω-set of probability one (depending on t) such that

$$\tilde{\Phi}_t[\psi] = (v(t),\ \psi)_0. \tag{3.40}$$

However, since $\tilde{\Phi}[\psi]$ is a continuous in t process and $v\in \mathsf{L}_2(\Omega;\ \mathsf{C}[0,T];\ \mathsf{L}_2))$ (see Theorem 4.2.1), for every ψ we can find an ω-set of probability one where

(3.40) is valid for all $t\in[0,T]$ simultaneously. On the other hand, there exists a countable subset of $\mathbf{C}_0^\infty(\mathbf{R}^d)$ which is dense in $\mathbf{C}_b^0(\mathbf{R}^d)$ for all t,ω (Theorem 5.3.3). Thus there exists an ω-set of probability one, call it Ω_1, where the equality (3.40) holds for all $t\in[0,T]$ and $\psi\in\mathbf{C}_0^\infty(\mathbf{R}^d)$ simultaneously.

Similarly it is easy to demonstrate that for every $(t,\omega)\in[0,T]\times\Omega_2$, where $\Omega_2\subset\Omega$ and $\mathbf{P}(\Omega_2)=1$, and $\psi\in\mathbf{C}_0^\infty(\mathbf{R}^d)$, $(v(t,\omega),\psi)_0\geq 0$.

Let $\psi_n\uparrow 1$ in $\mathbf{C}_b^0(\mathbf{R}^d)$ as $n\to\infty$. By the monotone convergence theorem, we can pass to the limit in the equality

$$\tilde{\Phi}_t[\psi_n]=(\hat{v}(t),\psi_n)_0,$$

which yields that for all $(t,\omega)\in[0,T]\times(\Omega_1\cap\Omega_2)$

$$\tilde{\Phi}_t[1]=(\hat{v}(t),1)_0. \tag{3.41}$$

Obviously, $\tilde{\Phi}_t[1]<\infty$ for all $(t,\omega)\in[0,T]\times\Omega_3$, where $\Omega_3\subset\Omega$ and $\mathbf{P}(\Omega_3)=1$.

Thus for all $(t,\omega)\in[0,T]\times\Omega'$, where $\Omega'=\bigcap_{i=1}^{3}$, $(v(t),1)_0<\infty$.

Let u^0 be a non-negative function from $\mathbf{C}_0^\infty(\mathbf{R}^d)$. Put

$$u(t,\omega):=\begin{cases} v(t,\omega & \textit{for } \omega\in\Omega',\\ u^0 & \textit{for } \omega\notin\Omega'.\end{cases}$$

Evidently, this fuction is a generalized solution of problem (3.1), (3.2) and for all $(t,\omega)\in[0,T]\times\Omega$ it belongs to the cone of non-negative functions from $\mathbf{L}_1$. Besides, u is a $\mathcal{P}(\mathfrak{F})$-measurable mapping of $[0,T]\times\Omega$ in $\mathbf{L}_1$. Actually, from the equality (3.40) if follows that $(u(t),\psi)_0$ is a $\mathcal{P}(\mathfrak{F})$-measurable function for every $\psi\in\mathbf{L}_\infty$. Since $\mathbf{L}_\infty$ and $\mathbf{L}_1^*$ are isometrically isomorphic, we obtain by the Pettis theorem that u is a $\mathcal{P}(\mathfrak{F})$-measurable mapping $[0,T]\times\Omega$ in $\mathbf{L}_1$.

We now show that u is weakly continuous, ($\mathbf{P}$-a.s.) as a function of t in $\mathbf{L}_1$. Let $N\in\mathbf{R}_+$ and ψ be a function from $\mathbf{L}_\infty$ such that $0\leq\psi\leq N$. Then there exists a sequence $\{\psi_n\}$, $n\in\mathbf{N}$, of finitary functions such that $\psi_n\uparrow\psi$ as $n\to\infty$. (We can take as ψ_n the restriction of ψ to the ball with the radius n and the center at origin). Since ψ_n satisfies the assumptions of Theorem 4.2.1, $(\psi_n, u(t))_0$ is continuous in t, ($\mathbf{P}$-a.s.). Next, due to the density-valued character of $u(t)$, $(u(t),\psi_n)\uparrow(u(t),\psi)$, ($\mathbf{P}$-a.s.), as $n\to\infty$ for every $t\in[0,T]$. Thus we find that $(u(t),\psi)_0$ is a lower semi-continuous function of t, that is, $(u(t_0),\psi)\leq \underline{lim}_{t\to t_0}(u(t),\psi)_0$, ($\mathbf{P}$-a.s.), for every $t_0\in[0,T]$.

On the other hand, it is clear that $0 \leq N-\psi \leq N$ and by the same arguments we obtain that $(u(t), N-\psi)_0$ also is a lower semi-continuous (in t) function. Since $(u(t),1)_0 = \tilde{\Phi}_t[1]$ is continuous in t, we get that $(u(t),\psi)_0 = N(u(t),1)_0 - (u(t), N-\psi)_0$ is upper semi-continuous in t.

Thus we have proved that for every $\psi \in \mathbf{L}_\infty$ such that $0 \leq \psi \leq N$ $(u(t),\psi)$ is continuous in t ($\mathbf{P}$-a.s.).

The general case can be reduced to that considered above by the representation $\psi = \max(0,\psi) + \min(0,\psi)$.

Since by (3.3) $\mathbf{E}(u(t),1)_0 = \mathbf{E}\rho(t)=1$, we find that $u \in \mathbf{L}_1([0,T]; \mathcal{P}(\mathfrak{F}); \mathbf{L}_1)$, which completes the proof. □

Remark 6. *We can weaken the assumptions on the smoothness of the coefficients and the free data of problems* (4.0.1), (4.1.2), (4.3.3), (4.3.4), *and* (3.1), (3.2) *that we have made to prove the AOC formulas* (1.2), (1.3), *and* (3.3), *respectively, if instead, we assume that the matrix* $(2a^{ij} - \sigma^{il}\sigma^{jl})$ *is uniformly positive definite. In this case, the above problems are superparabolic.* □

Chapter 6

FILTERING, INTERPOLATION AND EXTRAPOLATION OF DIFFUSION PROCESSES.

6.0. Introduction

0.1. The physical prerequisites for the filtering problem of diffusion processes were described in §1.1.1, therefore, we shall begin here directly with a formal statement of the problem.

Warning 1. *Warning* 4.0.1 *is to be in force throughout this chapter.* □

Suppose that $T\in\mathbb{R}_+$ is given. Let us fix a standard probability space $\mathbb{F}:=(\Omega, \mathcal{F}, \{\mathcal{F}_t\}_{t\in[0,t]}, \mathbb{P})$ and a standard Wiener $\mathbf{R}^{d+d_1}$-valued process $\nu(t)$ on this space.

Consider a diffusion process $Z(t)$ which is the solution of the following system of Ito's equations:

$$Z^i(t) = z_0^i + \int_{[0,t]} b^i(s, Z(s))ds + \int_{[0,t]} \Sigma^{il}(s, Z(s))d\nu^l(s),$$

$$i,\ l=1,2,\dots,d+d_1,\ t\in[0,T]. \tag{0.1}$$

Throughout what follows it is assumed that $z_0:=(z_0^1,\dots,z_0^{d+d_1})$ is an $\mathcal{F}_0$-measurable random variable taking values in $\mathbf{R}^{d+d_1}$, and coefficients B^i, Σ^{il} satisfy the conditions of Theorem 1.4.3, which guarantee the existence of a unique solution of system (0.1).

Suppose that one part of the components of the diffusion process $Z(t)$ is observable and the other is not. Denote the observable part by $Y(t)$ and the unobservable one by $X(t)$.

Without loss of generality it can be assumed that $X(t)$ consists of the first d

coordinates of $Z(t)$ and $Y(t)$ of the remaining d_1 coordinates.

Let $\mathcal{Y}_r^s$ (resp. $\mathcal{X}_r^s$, $\mathcal{Z}_r^s$) be the **P**-complete σ-algebra generated by $Y(t)$ (resp. $X(t)$, $Z(t)$) for all $t\in[s,r]$.

Suppose that the function f: $\mathbf{R}^d \to \mathbf{R}^1$ such that $\mathbf{E}|f(X(t))|^2 < \infty$, and also the numbers T_0, T_1, T_2 belonging to $[0,T]$, where $T_0 \le T_1 \wedge T_2$ are given.

It is well known that $\mathbf{E}[f(X(T_2))|\mathcal{Y}_{T_1}^0]$ is the best, in the mean square, $\mathcal{Y}_{T_1}^0$-measurable estimate for $f(X(T_2))$.

The problem of calculation $\mathbf{E}[f(X(T_2))|\mathcal{Y}_{T_1}^0]$ is called the problem of filtering if $T_2 = T_1$, the problem of interpolation if $T_2 < T_1$ and that of extrapolation if $T_2 > T_1$.

The conditional distribution $\mathbf{P}(X(T_2)\in\Gamma|\mathcal{Y}_{T_1}^0]$ is respectively called the filtering, interpolation or extrapolation measure.

In this chapter it will be proved that under mild smoothness conditions (in x) on the coefficients of equation (0.1), the measure $\mathbf{P}(X(T_2)\in dx|\mathcal{Y}_{T_1}^0)$ is absolutely continuous with respect to the Lebesgue measure.

Clearly, the calculation of the density $\pi^{T_2,T_1}(x) := dP(X(T_2)\in ds|\mathcal{Y}_{T_1}^0)/dx$ solves the corresponding problem of estimating $X(t)$ based on observations of $Y(t)$. Therefore in the present chapter considerable attention is payed to the evaluation and treatment of equations for $\pi^{T_2,T_1}(x)$. Apart from these equations, we also derive some equations that describe the dynamics of $\mathbf{E}[f(X(T_1))|\mathcal{Y}_{T_1}^{T_0}, Z(T_0)=z]$ with respect to T_0 when T_1 is fixed (backward filtering equations).

0.2. Let us introduce some additional notation and stipulations which, together with those introduced in §.1, we shall use throughout the chapter.

The process consisting of the first d coordinates of the Wiener process $\nu(t)$, is denoted by $\hat{w}(t)$. The remaining d_1 coordinates form the Wiener process $w(t)$, i.e. $w^l(t) := \nu^{d+l}(t)$ for $l=1,2,\ldots,d_1$.

Let ${}_Xb^i := b^i$, ${}_X\Sigma^{il} := \Sigma^{il}$ for $i=1,2,\ldots,d$, $l=1,2,\ldots,d+d_1$ and ${}_Yb^i := b^{i+d}$, ${}_Y\Sigma^{il} := \Sigma^{(i+d)l}$ for $i=1,2,\ldots,d_1$, $l=1,2,\ldots,d+d_1$. As usual, the matrices $({}_X\Sigma^{il})$, $({}_Y\Sigma^{il})$ will be denoted by ${}_X\Sigma$, ${}_Y\Sigma$. Set ${}_0\tilde{\sigma} := ({}_0\tilde{\sigma}^{il}) := ({}_Y\Sigma\, {}_Y\Sigma^*)^{1/2}$.

Assume that this matrix is uniformly non-singular (with respect to all variables) and depends only on $(t,\omega)\in[0,T]\times\Omega$ and the last d_1 coordinates of $z\in\mathbf{R}^{d+d_1}$ i.e. ${}_0\tilde\sigma(t,z,\omega) = {}_0\tilde\sigma(t,y,\omega)$ *for* $z = (x,y)$, $x\in\mathbf{R}^d$, $y\in\mathbf{R}^{d_1}$. Denote

$$\tilde\Sigma := \begin{pmatrix} {}_1\tilde\sigma & \tilde\sigma \\ 0 & {}_0\tilde\sigma \end{pmatrix},$$ where 0 is the zero $d_1\times d$-matrix,

$${}_1\tilde\sigma := [{}_X\Sigma_X\Sigma^* - {}_X\Sigma_Y\Sigma^*\,{}_0\tilde\sigma^{-2}\,{}_Y\Sigma_X\Sigma^*]^{1/2},$$

$$\tilde\sigma := ({}_X\Sigma_Y\Sigma^*)^{1/2}\,{}_0\tilde\sigma^{-1}.$$

Simple algebraic transformations show that matrix ${}_1\tilde\sigma$ is non-negatively definite (a probabilistic proof of this fact is presented in [90], Lemma 13.2). it is also clear that ${}_0\tilde\sigma$ and ${}_1\tilde\sigma$ are self-adjoint matrices.

There exists (e.g. [90], Lemma 10.4) a standard $\mathbf{R}^{d+d_1}$-valued Wiener process $\nu(t)$ such that for every t, ($\mathbf{P}$-a.s.),

$$\int_{[0,t]} \Sigma^{il}(s, Y(s))\,d\nu^l(s) = \int_{[0,t]} \hat\Sigma^{il}(s, Y(s))\,d\hat\nu^l(s).$$

Thus without loss of generality we can and shall assume that

$$\Sigma := \begin{pmatrix} {}_1\sigma & \sigma \\ 0 & {}_0\sigma \end{pmatrix},$$ where ${}_1\sigma := ({}_1\sigma^{il})$, $(i,l=1,2,\dots,d)$, $\sigma := (\sigma^{il})$, $(i=1,2,\dots,d$, $l=1,2,\dots,d_1)$, and 0 is the zero $d_1\times d$-matrix, while ${}_1\sigma$ and ${}_0\sigma$ are self-adjoint matrices, ${}_1\sigma$ is non-negatively definite, and ${}_0\sigma$ is uniformly non-singular and depends on the same variable as ${}_0\tilde\sigma$.

We will denote the elements of ${}_0\sigma^{-1}$ by ${}_0\sigma^{-il}$. Futhermore, we will use the following notations:

$(A^{ij} := 2^{-1}\,(\Sigma^{il}\Sigma^{jl})$ *where* $i,j=1,2,\dots,d+d_1$,

$(a^{ij} := 2^{-1}\,({}_X\Sigma^{il}\,{}_X\Sigma^{jl})$, *where* $i,j=1,2,\dots,d$,

$h^l := {}_Y b^i \, {}_0\sigma^{-il}$, *where* $l=1,2,\dots,d_1$,

$$D^{il} := \begin{cases} \sigma^{il}, & \text{for } i=1,2,\dots,d,\ l=1,2,\dots,d_1, \\ {}_0\sigma^{(i-d)l}, & \text{for } i=d+1,\dots,d+d_1,\ l=1,2,\dots,d_1 \end{cases}$$

Let us denote by z a point of the space $\mathbf{R}^{d+d_1}$, its first d coordinates by x and the remaining d_1 by y. Then, as usual, we write $dz := dz^1 \dots dz^{d+d_1}$ and $f_i(z) := \frac{\partial}{\partial z^i} f(z)$.

We assume that the following condition is fulfilled:

(*H*) *functions* $h^l(t,z,\omega)$, *where* $l=1,2,\dots,d_1$, $(t,z,\omega) \in [0,T] \times \mathbf{R}^{d+d_1} \times \Omega$ *are uniformly bounded.* □

6.1. Bayes' Formula and the Conditional Markov Property.

1.0. This section investigates the structure of conditional expectations $\mathbf{E}[f(X(T_2)) | \mathfrak{Y}^0_{T_1} \vee \mathfrak{U}]$, where $\mathfrak{U}$ is a sub-σ-algebra of $\mathfrak{F}^0_{T_0}$.*

A change of measure formula (a version of Bayes' formula) which connects these conditional expectations with AOC formula from the previous chapter, is derived.

Particular attention is given to the case when $Z(t)$ is a Markov process and $\mathfrak{U} := \mathfrak{F}^{T_0}_{T_0}$.

In the subsequent sections these formulas are used to derive equations which describe the dynamics of the densities of filtering, interpolation and extrapolation measures as well as backward filtering equations.

1.1. Denote:

$$(\rho^s_t)^{-1} := exp\Big\{- \int_{[s,t]} h^l(r, Z(r))dw^l(r) - \frac{1}{2} \int_{[s,t]} h^l h^l(r, Z(r))dr\Big\}.$$

---

*Note that $\mathfrak{U}$ might also be a trivial σ-algebra.

According to the assumption (H), $\mathbb{E}(\rho^0_T)^{-1}=1$. Therefore it is possible to introduce on $(\Omega,\mathfrak{F})$ a new probability measure by the formula

$$\tilde{\mathbb{P}}(\Gamma) = \int_\Gamma (\rho^0_T)^{-1} \, d\mathbb{P}(\omega), \quad \forall \Gamma\in\mathfrak{F}.$$

Since evidently $\mathbb{P}((\rho^0_T)^{-1}>0)=1$, measures $\mathbb{P}$ and $\tilde{\mathbb{P}}$ are mutually absolutely continuous and $d\mathbb{P}/d\tilde{\mathbb{P}}$, the Radon-Nikodym derivative of measure $\mathbb{P}$ with respect to measure $\tilde{\mathbb{P}}$ coincides with ρ^0_T. We shall henceforth denote the expectation with respect to $\tilde{\mathbb{P}}$ by $\tilde{\mathbb{E}}$.

According to the Girsanov theorem, there exists a standard Wiener process $\tilde{\nu}(t)$ on $\tilde{\mathbb{F}}:= (\Omega, \mathfrak{F}, \{\mathfrak{F}_t\}_{t\in[0,T]}, \mathbb{P})$, taking values in $\mathbb{R}^{d+d_1}$, such that the process $Z(t) = (X(t), Y(t))$, considered on $\mathbb{F}$, satisfies the system of Ito's equations:

$$X^i(t) = x^i_0 + \int_{[0,t]} {}_XB^i(s, Z(s))ds + \tag{1.1}$$

$$+ \int_{[0,t]} {}_X\Sigma^{il}(s, Z(s))d\tilde{\nu}^l(s), \; i=1,2,\dots,d,$$

$$Y^i(t) = y^i_0 + \int_{[0,t]} {}_0\sigma^{il}(s, Y(s))d\tilde{w}^l(s), \; i=1,2,\dots,d_1, \; t\in[0,T], \tag{1.2}$$

where $x^i_0:= z^i_0$ for $i=1,2,\dots,d$, $y^i_0:= z^{d+i}_0$ for $i=1,2,\dots,d_1$, ${}_XB^i:= {}_Xb^i\text{-}\sigma^{il}h^l$ for $i=1,2,\dots,d$ and $\tilde{w}^l:= \tilde{\nu}^{d+l}$ for $l=1,2,\dots,d_1$.

Theorem 1. Let ψ be a $\mathfrak{Z}^{T_0}_{T_2}$-measurable, $\mathbb{P}$-integrable random variable, $\mathfrak{U}$ sub-σ-algebra of $\mathfrak{F}_{T_1\vee T_2}$ and $\rho^0_{T_0}$ be a $\mathfrak{U}$-measurable function. Then $\mathbb{P}$-a.s.

$$\mathbb{E}[\psi|\mathfrak{Y}^0_{T_1}\vee\mathfrak{U}] = \frac{\tilde{\mathbb{E}}[\psi\rho(T_1\vee T_2, T_0)|\mathfrak{Y}^0_{T_1}\vee\mathfrak{U}]}{\tilde{\mathbb{E}}[\rho(T_1\vee T_2, T_0)|\mathfrak{Y}^0_{T_1}\vee\mathfrak{U}]}, \tag{1.3}$$

where

$$\rho(t,s):= exp\left\{ \int_{[s,t]} h^l(r, Z(r))d\tilde{w}^l(r) - \frac{1}{2} \int_{[s,t]} h^l h^l(r, Z(r))dr\right\}. \quad \square$$

Before we give the proof itself let us state a standard fact (*see* [91], §24.4).

Lemma 1. *(Bayes' formula, general version). Suppose there are given two mutually absolutely continuous measures Q and $\tilde{Q}$ on $(\Omega,\mathcal{F})$. If ξ is a Q-integrable random variable and $\mathcal{G}$ is some sub-σ-algebra of $\mathcal{F}$, then almost surely, with respect to the measures Q and $\tilde{Q}$,*

$$\mathbf{E}_Q[\xi|\mathcal{G}] = \frac{\mathbf{E}_{\tilde{Q}}[\xi dQ/d\tilde{Q}|\mathcal{G}]}{\mathbf{E}_{\tilde{Q}}[dQ/d\tilde{Q}|\mathcal{G}]}, \tag{1.4}$$

where $\mathbf{E}_{\mathbf{Q}}[\cdot|\mathcal{G}]$ (resp. $\mathbf{E}_{\mathbf{Q}}[\cdot|\mathcal{G}]$) is a conditional (relative to $\mathcal{G}$) expectation with respect to the measure $\mathbf{Q}$ (respectively $\tilde{\mathbf{Q}}$). $\square$

1.2. *Proof of Theorem* 1. By the definition of h, we obtain the identity

$$\int_{[s,t]} h^l(r, Z(r))dw^l(r) + \frac{1}{2} \int_{[s,t]} h^l h^l(r, Z(r))dr =$$

$$= \int_{[s,t]} h^i(r, Z(r))_0\sigma_0^{-il}\sigma^{lj}(r, Y(r))dw^j(r) +$$

$$+ \int_{[s,t]} h^l(r, Z(r))_0\, \sigma^{-il}(r, Y(r))_Y b^i(r, Z(r))dr \tag{1.5}$$

$$- \frac{1}{2} \int_{[s,t]} h^l h^l(r, Z(r))dr.$$

It is easy to show (by passage to the limit over simple integrands) that the sum of the first two terms in the right-hand side of the equality (1.5) coincides ($\tilde{\mathbf{P}}$-a.s.) with the stochastic integral

$$\int_{[s,t]} h^i(r, Z(r))_0\sigma^{-il}(r, Y(r))dY^l(r) = \int_{[s,t]} h^i(r, Z(r))d\tilde{w}^i(r).$$

From this it obviously follows that

$$\rho_t^s = \rho(t,s), \ (\tilde{\mathbf{P}}\text{-}a.s.), \ \forall \ s,t. \tag{1.16}$$

Therefore from Lemma 1, where $\mathbf{Q}:= \mathbf{P}$, $\tilde{\mathbf{Q}}:= \tilde{\mathbf{P}}$, $\xi:= \psi$ and $\mathfrak{G}:= \mathfrak{Y}_T^0 \vee \mathfrak{U}$ we get, ($\mathbf{P}$-a.s. and $\tilde{\mathbf{P}}$-a.s.),

$$\mathbf{E}[\psi|\mathfrak{Y}_{T_1}^0 \vee \mathfrak{U}] = \frac{\tilde{\mathbf{E}}[\psi\rho(T,0)|\mathfrak{Y}_{T_1}^0 \vee \mathfrak{U}]}{\tilde{\mathbf{E}}[\rho(T,0)|\mathfrak{Y}_{T_1}^0 \vee \mathfrak{U}]}. \tag{1.17}$$

Note now that $\rho(T,0) = \rho(T_0,0)\rho(T_1 \vee T_2, T_0)\rho(T, T_1 \vee T_2)$ and $\tilde{\mathbf{E}}[\rho(T, T_1 \vee T_2)|\mathfrak{F}_{T_1 \vee T_2}]=1$. Making use of this fact and the $\mathfrak{U}$-measurability of $\rho(T_0,0)$ we get the statement of the theorem from (1.7). □

1.3. *Warning* 3. *Henceforth in this section we will assume that the coefficients of system* (0.1) *are not random.* □

Denote the σ-algebra, generated by the increments of $\tilde{w}(t)$ for $t \in [T_0, T]$, and completed with respect to the measure $\mathbf{P}$, by $\tilde{\mathfrak{F}}_{T_1}^{T_0}$.

Theorem 3. *Let ψ be a $\mathfrak{F}_{T_1 \vee T_2}$-measurable,* $\mathbf{P}$*-integrable random variable. Then* $\mathbf{P}$*-a.s.*

$$\mathbf{E}[\psi|\mathfrak{Y}_{T_1}^0 \vee \mathfrak{B}_{T_0}^0] = \mathbf{E}[\psi|\mathfrak{Y}_{T_1}^0 \vee \mathfrak{B}_{T_0}^{T_0}] =$$

$$= \frac{\tilde{\mathbf{E}}[\psi\rho(T_1 \vee T_2, T_0)| \tilde{\mathfrak{F}}_{T_1}^{T_0} \vee \mathfrak{Z}_{T_0}^{T_0}]}{\tilde{\mathbf{E}}[\rho(T_1 \vee T_2, T_0)|\tilde{\mathfrak{F}}_{T_1}^{T_0} \vee \mathfrak{Z}_{T_0}^{T_0}]}. \qquad □$$

Later we will need the two auxiliary assertions presented below.

Suppose the $\mathbf{Q}$ and $\mathbf{E}_{\mathbf{Q}}$ are the same as in Lemma 1, and a random process $z(t)$ on $(\Omega, \mathcal{F}, \{\mathcal{F}_t\}_{t\in[0,T]}, \mathbf{Q})$, taking values in $\mathbf{R}^{d+d_1}$ is given. Let $x(t)$ denote the first d coordinates of this process, and $y(t)$ will denote the remaining d_1 coordinates.

Let also $\mathbf{X}(t)$, $\mathbf{Y}(t)$, and $\mathbf{Z}(t)$ be the σ-algebras generated respectively by $x(t)$, $y(t)$ and $z(r)$ when $r\in[s,t]$.

Assume that $z(t)$ possesses the Markov property, i.e. for an arbitrary $\mathbf{Q}$-integrable, $\mathbf{Z}_{T_1}^{T^0}$-measurable real-valued random variable ψ, ($\mathbf{Q}$-a.s.),

$$(1.8)\qquad \mathbf{E}_{\mathbf{Q}}[\psi|\, \mathbf{Z}_{T_0}^0] = \mathbf{E}_{\mathbf{Q}}[\psi|\, \mathbf{Z}_{T_0}^{T_0}].$$

Lemma 3. Let ξ be a $\mathbf{Z}_{T_2}^{T_0}$-measurable, $\mathbf{Q}$-integrable, real-valued random variable, then, ($\mathbf{Q}$-a.s.),

$$(1.9)\quad \mathbf{E}_{\mathbf{Q}}[\xi|\, \mathbf{Y}_{T_0}^0 \vee \mathbf{X}_{T_0}^0] = \mathbf{E}_{\mathbf{Q}}[\xi|\, \mathbf{Y}_{T_1}^{T_0} \vee \mathbf{X}_{T_0}^{T_0}] = \mathbf{E}_{\mathbf{Q}}[\xi|\, \mathbf{Y}_{T_1}^0 \vee \mathbf{X}_{T_0}^{T_0}]. \ \square$$

Proof. Let us check the first equality in (1.9). We assumed that ζ and η are bounded random variables measurable with respect to the σ-algebras $\mathbf{Z}_{T_0}^0$ and $\mathbf{Y}_{T_1}^0$, respectively. Then, by (1.8) we get

$$\mathbf{E}_{\mathbf{Q}}[\zeta\eta\, \mathbf{E}_{\mathbf{Q}}[\xi|\, \mathbf{Y}_{T_1}^0 \vee \mathbf{X}_{T_0}^0] = \mathbf{E}_{\mathbf{Q}}(\zeta\eta\xi) = \mathbf{E}_{\mathbf{Q}}(\zeta\mathbf{E}_{\mathbf{Q}}[\eta\xi|\, \mathbf{Z}_{T_0}^0]) =$$

$$= \mathbf{E}_{\mathbf{Q}}(\zeta\mathbf{E}_{\mathbf{Q}}[\eta\xi|\, \mathbf{Z}_{T_0}^{T_0}]) = \mathbf{E}_{\mathbf{Q}}(\zeta\mathbf{E}_{\mathbf{Q}}[\eta\mathbf{E}_{\mathbf{Q}}[\xi|\, \mathbf{Y}_{T_1}^{T_0} \vee \mathbf{Z}_{T_0}^{T_0}]|\, \mathbf{Z}_{T_0}^{T_0}]) =$$

$$= \mathbf{E}_{\mathbf{Q}}(\zeta\eta\mathbf{E}_{\mathbf{Q}}[\xi|\mathcal{Y}_{T_1}^{T_0} \vee \mathbf{X}_{T_0}^{T_0}]).$$

This proves the first equality. Using it, the reader can easily prove the second equality in (1.9). $\square$

Remark 3. *It is natural to call property* (1.9) *the conditional Markov property. Clearly, every Markov process possesses an analogous property. Neither the specific structure of the range of values, nor the time interval of* $z(t)$ *were used in the proof of the lemma.* □

1.4. *Lemma* 4. *σ-algebras* $\mathcal{Y}_{T_0}^{T_0}$ *and* $\mathcal{F}_{T_1}^{T_0} \vee \mathcal{Y}_{T_0}^{T_0}$ *coincide.* □

Proof. For $t \leq T_0$ the process $Y(t)$ coincides (**P**-a.s.) with $Y(t, Y(T_0), T_0)$ which is the solution of the system:

$$(1.11) \qquad Y^i(t, Y(T_0), T_0) = Y(T_0) + \int_{[T_0, t]} {}_0\sigma^{il}(s, Y(s, Y(T_0), T_0)) d\tilde{w}^l(s),$$

$$i = 1, 2, \ldots, d_1, \; t \in [T_0, T].$$

This follows from (1.2) and the uniqueness of the solution of system (1.11) (see Theorem 1.4.3). Since this solution can be constructed by the method of successive approximations (see, e.g. [63], Theorem ii.5.7), we conclude that it is adapted to the family of σ-algebras $\{\mathcal{F}_t^{T_0} \vee \mathcal{Y}_{T_0}^{T_0}\}$, and consequently $Y_{T_1}^{T_0} \subset \tilde{\mathcal{F}}_{T_1}^{T_0} \vee Y_{T_1}^{T_0}$. On the other hand, for any $t \in [T_0, T]$, $l = 1, \ldots d$, (**P**-*a.s.*),

$$\tilde{w}^l(t) - \tilde{w}^l(T_0) = \int_{[T_0, t]} {}_0\sigma^{-lj}(s, Y(s)) dY^j(s).$$

The integrals in the right-hand side of the latter equality, are regarded as integrals with respect to martingales, which implies their measurability with respect to $\mathcal{Y}_t^{T_0}$. Hence, the equality $\mathcal{Y}_{T_1}^{T_0} = \mathcal{F}_{T_1}^{T} \vee \mathcal{Y}_{T_0}^{T}$ is proved. □

Proof of Theorem 3. By Theorem 1 and Lemma 4 we obtain that, (**P**-a.s.),

$$\mathbf{E}[\psi | \mathcal{Y}_{T_1}^0 \vee \mathfrak{X}_{T_0}^0] = \mathbf{E}[\psi | \mathcal{Y}_{T_1}^0 \vee \mathfrak{Z}_{T_0}^0] =$$

$$(1.10)\qquad = \frac{\tilde{\mathsf{E}}[\psi\rho(T_1\vee T_2,T_0)|\mathcal{Y}^0_{T_1}\vee\mathcal{Z}^0_{T_0}]}{\tilde{\mathsf{E}}[\rho(T_1\vee T_2,T_0)|\mathcal{Y}^0_{T_1}\vee\mathcal{Z}^0_{T_0}]} =$$

$$= \frac{\tilde{\mathsf{E}}[\psi\rho(T_1\vee T_2,T_0)|\mathcal{Y}^0_{T_1}\vee\mathcal{S}^0_{T_0}]}{\tilde{\mathsf{E}}[\rho(T_1\vee T_2,T_0)|\mathcal{Y}^0_{T_1}\vee\mathcal{S}^0_{T_0}]}.$$

The assertion of the theorem readily follows from this formula and lemmas 3 and 4.

6.2. The Forward Filtering Equation.

2.0. In this section we prove the existence of the filtering density $\pi(t,x):= P(X(t)\in dx|\mathcal{Y}^0_t)/dx$ and derive and investigate equations which it satisfies.

Notation introduced in section 1.0 is used here also. The following conditions are satisfied throughout:

(i) $P_0(X(0)\in\cdot)$ is a regular conditional distribution of $X(0)$ relative to $\mathcal{Y}^0_0$, it is absolutely continuous with respect to the Lebesgue measure on $\mathbf{R}^d$ and the density $\pi_0(:= dP_0/dx))\in\mathbb{L}_2(\Omega;\ \mathcal{Y}^0_0;\ \mathsf{P};\ \mathbb{H}^1)$;

(ii) For every $(t,y)\in[0,T]\times\mathbf{R}^{d_1}$, $(i,j=1,...d,\ l=1,...,d_1)$, coefficients $a^{ij}(t,x,y)$, $\sigma^{il}(t,x,y)$ possess all the derivatives in x up to the order of 3, and coefficients ${}_Xb^i(t,x,y)$, $h^l(t,x,y)$ up to order of 2; all the functions mentioned above and their derivatives are uniformly bounded in $(t,x,y)\in[0,T]\times\mathbf{R}^d\times\mathbf{R}^{d_1}$.

We introduce the operators $\mathcal{L}^*(t,x,Y(t))$ and $\mathcal{M}^{l*}(t,x,Y(t))$ by formulas

$$\mathcal{L}^*(t,x,Y(t))u:= (a^{ij}(t,x,Y(t))u)_{ij} - ({}_Xb^i(t,x,Y(t))u)_i,$$

$$\mathcal{M}^{l*}(t,x,Y(t))u:= -(\sigma^{il}(t,x,Y(t))u)_i + h^l(t,x,Y(t))u.$$

2.1. Let us consider the following problem on the standard probability space $\tilde{\mathsf{F}}$:

$$(2.1)\qquad du(t,x) = \mathcal{L}^*(t,x,Y(t))u(t,x)dt + \mathcal{M}^{l*}(t,x,Y(t))u(t,x)d\tilde{w}^l(t),$$

$$(t,x,\omega)\in[0,T]\times\mathbf{R}^d\times\Omega,$$

(2.2) $$u(0,x) = \pi_0(x),\ (x,\omega)\in\mathbf{R}^d\times\Omega.$$

Denote $(\Omega;\ \mathcal{F};\ \mathcal{Y}_t^0;\ \tilde{\mathbf{P}})$ by $\tilde{\mathbf{F}}(\mathcal{Y})$ and the σ-algebra of predictable sets on $\tilde{\mathbf{F}}(\mathcal{Y})$ by $\mathcal{P}(\mathcal{Y})$.

Observe that by virtue of Lemma 1.4 $\mathcal{Y}_t^0 = \tilde{\mathcal{F}}_t^0\vee\mathcal{Y}_0^0$.

It is clear that the coefficients of the equation 2.1 are $\mathcal{P}(\mathcal{Y})$-measurable, for any $x\in\mathbf{R}^d$, and on $\tilde{\mathbf{F}}(\mathcal{Y})$, w is also a standard Wiener process. Therefore, according to Theorem 4.2.1 (see also Remark 5.3.0) problem (2.1), (2.2) has a unique generalized solution $u\in\mathbf{L}_2([0,T];\ \mathcal{P}(\mathcal{Y});\ \mathbf{H}^1)\cap\mathbf{L}_2(\Omega;\ \mathbf{C}[0,T];\ \mathbf{L}_2))$. Furthermore, problem (2.1), (2.2) satisfies the conditions of Theorem 5.3.1 on $\tilde{\mathbf{F}}(\mathcal{Y})$, for $\mathcal{G}_t = \mathcal{Y}_t^0$. Thus the following theorem holds.

Theorem 1. *The equality*

(2.3) $$\tilde{\mathbf{E}}[f(X(t))\rho(t,0)|\ \mathcal{Y}_t^0] = \int_{\mathbf{R}^d} f(x)u(t,x)\,dx,$$

where $u(t)$ is a density-valued generalized solution of problem (2.1), (2.2), *holds true, for any $f\in\mathbf{L}_\infty$ and every $t\in[0,T]$,* ($\mathbf{P}$-*a.s.*). □

From this and Theorem 1.1 we obtain at once

Corollary 1. *The equality*

(2.4) $$\mathbf{E}[f(X(t))|\ \mathcal{Y}_t^0] =$$

$$= \int_{\mathbf{R}^d} f(x)u(t,x)\,dx\left(\int_{\mathbf{R}^d} u(t,x)\,ds\right)^{-1},$$

where u is a density-valued generalized solution of problem (2.1), (2.2), *holds true for any $f\in\mathbf{L}_\infty$ and every $t\in[0,T]$* ($\mathbf{P}$-*a.s.*).

2.2. It follows from Theorem 5.3.2 that $\tilde{\mathbf{E}}[\rho(t,0)|\ \mathcal{Y}_t^0]$ and $\tilde{\mathbf{E}}[h^l(t,Z(t))\rho(t,0)|\ \mathcal{Y}_t^0]$ have $\mathcal{P}(\mathcal{Y})$-measurable versions (the same notations for them will be used henceforth) satisfying, ($\mathbf{P}$-a.s.), on $[0,T]$ the equality

$$\tilde{\mathbf{E}}[\rho(t,0)|\ \mathcal{Y}_t^0] = 1 + \int_{[0,t]} \tilde{\mathbf{E}}[h^l(s,Z(s))\rho(s,0)|\ \mathcal{Y}_s^0]\ d\tilde{w}^l(s).$$

From this and Theorem 1.1, it follows that **P**-a.s.

$$\tilde{\mathsf{E}}[\rho(t,0)|\, \mathcal{Y}_t^0] = 1 + \int_{[0,t]} \tilde{\mathsf{E}}[h^l(s,Z(s))\rho(s,0)|\, \mathcal{Y}_s^0] \times$$

$$\times\, (\tilde{\mathsf{E}}[\rho(s,0)|\, \mathcal{Y}_s^0])^{-1}\, \tilde{\mathsf{E}}[\rho(s,0)|\, \mathcal{Y}_s^0]\; d\tilde{w}^l(s) =$$

$$= 1 + \int_{[0,t]} \pi_s[h^l(s,Y(s))]\, \tilde{\mathsf{E}}[\rho(s,0)|\, \mathcal{Y}_s^0]\; d\tilde{w}^l(s),$$

where $\pi_s[h^l(s,Y(s))]:= \mathsf{E}[h^l(s,Z(s))|\, \mathcal{Y}_t^0]$. The equation obtained for $\tilde{\mathsf{E}}[\rho(t,0)|\, \mathcal{Y}_t^0]$ has the following unique solutions:

$$V(t):= exp\Big\{ \int_{[0,t]} \pi_s[h^l(s,Y(s))]\; d\tilde{w}^l(s)$$

$$- \frac{1}{2} \int_{[0,t]} \sum_{l=1}^{d_1} |\, \pi_s[h^l(s,Y(s))]|^2 ds\Big\}.$$

Thus the following proposition is proved.

Proposition 2. The conditional expectation $\tilde{\mathsf{E}}[\rho(t,0)|\, \mathcal{Y}_t^0]$ *is stochastically equivalent to* $V(t)$. □

Let us denote the space of probability measures on $\mathbf{R}^d$ by $\mathcal{M}(\mathbf{R}^d)$.

Theorem 2. There exists a function $P_{\mathcal{Y}}^t\colon \Omega \to \mathcal{M}(\mathbf{R}^d)$ *possessing the following properties.*

(*i*) *For every* $t\in[0,T]$, $P_{\mathcal{Y}}^t(\cdot,\omega)$ *is regular conditional probability of* $X(t)$ *relative to* $\mathcal{Y}_t^0$.

(*ii*) *For all* $(t,\omega)\in[0,t]\times\Omega'$, $(P(\Omega')=1)$, $P_{\mathcal{Y}}^t(\cdot,\omega)$ *is absolutely continuous with respect to the Lebesgue measure on* $\mathbf{R}^d$ *and the Radon-Nikodym derivative*

$$\pi(t,x,\omega) = u(t,x,\omega)\Big(\int_{\mathbf{R}^d} u(t,x,\omega)\,dx\Big)^{-1},$$

(in $\mathbb{L}_1$, $\forall t,\omega$), *where u is the generalized solution of problem* (2.1), (2.2), (*the existence of this solution follows from Corollary* 5.3.1).

(*iii*) *For every* $\Gamma \in \mathcal{B}(\mathbb{R}^d)$, $P^t_{\mathcal{Y}}(\Gamma,\omega)$ *is* $\mathcal{P}(\mathcal{Y})$-*measurable, continuous* ($\mathbf{P}$-*a.s.*) *stochastic process.* □

Proof. Let u be a generalized (density-valued) solution of problem (2.1), (2.2), described in Corollary 5.3.1. It follows from Theorem 1, the proposition of this paragraph and the weak continuity of $u(t)$ (Corollary 5.3.1), that there exists a set $\Omega' \subset \Omega$, ($\tilde{P}(\Omega')=1$), where $(u(t),1)_0 > 0$, for $t \in [0,T]$. Therefore,

$$P^t_{\mathcal{Y}}(\Gamma,\omega) := \begin{cases} \int_\Gamma u(t,x,\omega)(u(t,\omega),1)_0^{-1}dx & \text{for } \omega \in \Omega', \\ 1 & \text{for } \omega \in \Omega \setminus \Omega' \end{cases}$$

is a probability measure ($\forall t,\omega$) on $\mathcal{B}(\mathbb{R}^d)$. Because of the weak continuity and predictability of u in $\mathbb{L}_1$ (Corollary 5.3.1), $P^t_{\mathcal{Y}}(\cdot,\omega)$ possesses the properties listed in (iii). Finally, it follows from Theorems 1.2 and 2.1 that for every $\Gamma \in \mathcal{B}(\mathbb{R}^d)$, $t \in [0,T]$,

$$\mathbb{P}(X(t) \in \Gamma \mid \mathcal{Y}^0_t) = \int_\Gamma u(t,x)\,(u(t),\,1)_0^{-1}dx, \quad (\mathbf{P}\text{-a.s.})$$

which completes the proof. □

Problem (2.1), (2.2) is called the (direct) linear filtering equation. The $\mathcal{P}(\mathcal{Y})$-measurable solution $u \in \mathbb{L}_1([0,T]\times\Omega;\, \mathbf{C}_w\mathbb{L}_1)$ of this problem, which belongs to the cone of non-negative functions in $\mathbb{L}_1$ (see Corollary 5.3.1) is said to be the non-normalized filtering density. The function π of the theorem is called the filtering density.

The non-normalized filtering density as a generalized solution of problem (2.1), (2.2) has a variety of analytical properties similar to those described in Theorems 4.2.1, 4.3.2 and Corollaries 4.2.1, 4.2.2. 4.3.2, and 5.3.1. In particular, from Corollary 4.3.2 it follows that under some additional assumptions on the smoothness of the coefficients and the initial data of problems (2.1), (2.2) the function u is its solution in the classical sense as well.

Remark 2. *In applications it is more natural to consider the linear filtering equation in the following somewhat different (but equivalent) form*

$$(2.1') \qquad du(t,x) = \mathcal{L}^*(t,x,Y(t))u(t,x)dt + \mathcal{M}^{l}(t,x,Y(t))\,dY^l(t),$$

$$(t,x,\omega)\in]0,T]\times\mathbf{R}^d\times\Omega,$$

(2.2′) $$u(0,x) = \pi_0(x),\ (x,\omega)\in\mathbf{R}^d\times\Omega,$$

where

$$\hat{\mathcal{M}}^{l*}(t,x,Y(t))v := {}_0\sigma^{-jl}(t,Y(t))\ [-\sigma^{ij}(t,Y(t))v)_i +$$

$$+ h^j(t,x,Y(t))v].$$

The stochastic integral with respect to $Y(t)$ can be defined, for example, as follows

$$\int_{[0,t]} \hat{\mathcal{M}}^{l*}(t,x,y(t))v(t,x)\ dY^l(t) :=$$

$$= \int_{[0,t]} \mathcal{M}^{l*}(t,x,Y(t))\ v\ (t,x)\ h^l(t,Z(t))dt$$

$$+ \int_{[0,t]} \mathcal{M}^{l*}(t,x,Y(t))\ v\ (t,x)dw^l(t).$$

It can be shown that if the integrand is continuous in t (with probability 1), then the following equality holds

$$\int_{[0,t]} \hat{\mathcal{M}}^l(t,x,Y(t))\ v\ (t,x)\ dY^l(t) =$$

$$= \lim_{n\to\infty} \sum_{i(n)=0}^{k(n)-1} \hat{\mathcal{M}}^{l*}(t_{i(n)},x,Y(t_{i(n)}))v(t_{i(n)},x)[Y^l(t_{i(n)+1})-Y^l(t_{i(n)})],$$

where $\{0=t_0<t_{1(n)}<\ldots<t_{k(n)}=t\}$ is a sequence of imbedded partitions of the

interval $[0,t]$, *whose diameter tends to zero, and the limit above is a limit in probability.*

The last formula is of particular importance for computing, because it is the trajectory of $Y(t)$ *that one observes, not of* $w(t)$. *From the analytical point of view the linear filtering equation in the form* (2.1), (2.2) *is more convenient than in the form* (2.1′), (2.2′). *Therefore, since the two forms are equivalent, in the sequel we shall consider problem* (2.1), (2.2). □

2.3. In this item we derive an equation for the filtering density $\pi(t,x)$ and investigate its main analytical properties.

Theorem 3. (*i*) *The filtering density* π *belongs to* $\mathbb{L}_2^{\omega}[0,T]; \mathcal{P}(\mathcal{Y}); \mathbf{H}^1) \cap \mathbf{C}([0,T]; \mathcal{P}(\mathcal{Y}); \mathbb{L}_2)$, *it is* $\mathcal{P}(\mathcal{Y})$-*measurable as a function in* $\mathbb{L}_1$, *weakly continuous (in t) in* $\mathbb{L}_1$, *and for all* $t\in[0,T]\times\Omega$ *and* $\eta\in\mathbf{C}_0^{\infty}(\mathbf{R}^d)$ *on one and the same* ω-*set of probability* 1 *satisfies the equality,*

$$(2.5)\qquad (\pi(t),\eta)_0 = (\pi_0,\eta_0) + \int_{[0,t]} (\pi(s), \mathcal{L}(s, Y(s))\eta)_0 ds +$$

$$+ \int_{[0,t]} (\pi(s),\mathcal{M}^l(s, Y(s))\eta-(\pi(s), h^l(s, Y(s)))_0\eta)_0 d\bar{w}(s),$$

where $d\bar{w}^l(s) := d\tilde{w}^l(s) - (\pi(s), h^l(s, Y(s))_0 ds$.

(*ii*) *If* $\pi^1(s)$ *and* $\pi^2(s)$ *are elements of* $\mathbb{L}_2^{\omega}([0,T]; \mathcal{P}(\mathcal{Y}); \mathbf{H}^1) \cap \mathbb{L}_2^{\omega}([0,T]; \mathcal{P}(\mathcal{Y}); \mathbb{L}_1) \cap \mathbf{C}([0,T]; \mathcal{P}(\mathcal{Y}); \mathbb{L}_2)$ *satisfying equation* (2.5) *for every* $t\in[0,T]$, $\eta\in\mathbf{C}_0^{\infty}(\mathbf{R}^d)$ ($\mathbf{P}$-*a.s.*), *then*

$$\mathbf{P}\left(\sup_{t\in[0,T]} \|\pi^1(t) - \pi^2(t)\|_0^2 > 0\right) = 0. \quad \square$$

Proof. By Theorems 1 and 2 the analytical properties of the filtering density mentioned in the statement of the theorem follow at once from the corresponding properties of the non-normalized filtering density u (see Theorem 4.2.1 and Corollary 5.3.1). We now prove that π satisfies equation (2.5).

Let u be the non-normalized filtering density and $V(t)$ be the stochastic process defined in §.2. By the definition, for all $(t,\omega)\in[0,T]\times\Omega'$, where $\Omega'\subset\Omega$ and $\mathbf{P}(\Omega')=1$, $u(t)V^{-1}(t) = \pi(t)$ (as elements of $\mathbb{L}_1$).

Making use of the fact that u is the generalized solution of problem (2.1), (2.2), we differentiate by the Ito formula the product $(u(t),\eta)V^{-1}(t)$, where η is an arbitrary function from $C_0^\infty(\mathbf{R}^d)$, and obtain at once equality (2.5).

Let us now prove the uniqueness of solution of equation (2.5). Suppose that π^1 and π^2 are solutions of equation (2.5) having the properties listed in (ii) of the assertion of the theorem. Then

$$\rho^i(t) := exp\left\{ \int_{[0,t]} (\pi^i(s), h^l(s, Y(s)))_0 d\tilde{w}^l(s) - \right.$$

$$\left. - \frac{1}{2} \int_{[0,t]} \sum_{l=1}^{d_1} (\pi^i(s), h^l(s, Y(s)))_0^2 ds \right\}.$$

and

$$u^i(t) := \pi^i(t)\rho^i(t), \; i=1,2.$$

Applying Ito's formula to $u^i(t)$, we easily find that for both values of index i, this function is a generalized solution of the linear filtering equation (2.1), (2.2), whose solution is unique (see Theorem 4.2.1). Hence, for every $t\in[0,T]$, ($\mathbf{P}$-a.s.),

$$\rho^i(t) = 1 + \int_{[0,t]} (u^1(s), h^l(s, Y(s))_0 d\tilde{w}^l(s) =$$

$$= 1 + \int_{[0,t]} (u^2(s), h^l(s, Y(s)))_0 d\tilde{w}^l(s) = \rho^2(t),$$

which implies

$$\tilde{\mathbf{P}}(\sup_{t\in[0,T]} \|\pi^1(t) - \pi^2(t)\|_0^2 > 0) =$$

$$= \tilde{\mathbf{P}}(\sup_{t\in[0,T]} \|u^1(t)(\rho^1(t))^{-1} - u^2(t)(\rho^2(t))^{-1}\|_0^2 > 0) = 0. \quad \square$$

The filtering density π can be considered as a generalized solution of the problem

$$(2.6)\qquad dv(t,x) = [\mathcal{L}^*(t,x,Y(t))v(t,x) + \hat{h}^l(t)(\hat{h}^l(t) + \mathcal{M}^{l*}(t,x,Y(t)))v(t,x)dt+$$

$$+ [\mathcal{M}^{l*}(t,x,Y(t))v(t,x) + \hat{h}^l(t)v(t,x,)]d\tilde{w}^l(t),$$

$$(t,x,\omega)\in]0,T]\times\mathbb{R}^d\times\Omega$$

$$(2.7)\qquad v(0,x) = \pi_0(x),\ (x,\omega)\in\mathbb{R}^d\times\Omega,$$

where $\hat{h}^l(t) := -(\pi(t),\ h^l(t,\ Y(t)))_0$.

This problem is the one of the type (4.0.1), (4.1.2) (*see Remark* 5.3.0) *and satisfies the conditions of Theorem* 4.2.1 *and Collary* 4.2.1. *These results imply.*

Corollary 3. *In addition to conditions* (*i*) *and* (*ii*) *of* §.0, *assume that given* $m\in\mathbb{N}$ *the coefficients* $\sigma^{il}(t,x,y)$ *have* derivatives *in* x *up to the order of* $m+2$, $a^{ij}(t,x,y)$ *and* $h^l(t,x,y)$ *up to the order of* $m+1$ *and* ${}_Xb^i(t,x,y)$ *up to the order of* m, *where*

$i,j=1,2,\dots,d$, $l=1,2,\dots,d_1$, *and* $(t,y)\in[0,T]\times\mathbb{R}^{d_1}$. *Assume also that all the derivatives are uniformly bounded and* $\varphi\in\mathbb{L}_{p'}(\Omega;\mathbb{W}^m_{p'})$ *for* $p'=2$ and $p'=p\in]2,\infty[$, *then*

$\pi\in\mathbb{L}_2(\Omega;\ \mathbb{C}([0,T];\ \mathbb{H}^{m-1}))\ \cap\ \mathbb{L}_2([0,T]\times\Omega;\ \mathbb{C}_w\mathbb{H}^m)\ \cap\ \mathbb{L}_p([0,T]\times\Omega;\ \mathbb{C}_w\mathbb{W}^m_p)$.
Moreover, for $p'=p$ *and* 2 *the following inequality holds*

$$\tilde{\mathbb{E}}\sup_{t\in[0,T]}\|\pi(t)\|^{p'}_{m,p'}\le N\tilde{\mathbb{E}}\|\pi_0\|^{p'}_{m,p'},$$

where the constant N *depends only on* p', d, d_1, m, T, *and on the constant that majorizes the absolute values of the coefficients of problem* (2.6), (2.7) *and their derivatives.* □

From this corollary and the theorem of this item, arguing in the same way as in the proofs of Theorem 4.1.3 and Corollary 4.1.4, it is easy to obtain the following result.

2.4. *Corollary* 4. *Suppose that the assumptions of Corollary* 3 *are satisfied,* $n\in\mathbb{N}\cup\{0\}$, *and* $(m-n)p>d$. *Then the filtering density* $\pi(t,x)$ *has a version (the same notation for it will be used henceforth) with these properties:*

(*a*) *For every* $x\in\mathbb{R}^d$, $\pi(t,x,\omega)$ *is a predictable real stochastic process.*

(*b*) *For all* ω, $\pi(t,x,\omega)\in\mathbf{C}_b^{0,n}([0,T]\times\mathbf{R}^d)$.

(*c*) π *possesses the properties enumerated in the theorem and the corollary of* §.3.

(*d*) $\mathbf{E}\sup_{t\in[0,T]}\|\pi(t)\|^p_{\mathbf{C}_b^n(\mathbf{R}^d)}<\infty.$

(*e*) *If* $n\geq 2$, *then for every* $x\in\mathbf{R}^d$ *and for all* $(t,\omega)\in[0,T]\times\Omega_x$, *where* $\Omega_x\subset\Omega$ *and* $\mathbf{P}(\Omega_x)=1$, *the following equality holds*

$$\pi(t,x) = \pi_0(x) + \int_{[0,t]}\mathcal{L}_0^*(s,x,Y(s))\pi(s,x)ds$$

$$+\int_{[0,t]}[\mathcal{M}^{l*}(s,x,Y(s))\pi(s,x)-\pi(s,x)(\pi(s)),\ h^l(s,Y(s)))_0]d\bar{w}^l(s),$$

where $d\bar{w}^l(s):= d\tilde{w}^l(s) - (\pi(s),\ h^l(s,\ Y(s)))_0ds.$

(*f*) *If* π^1 *and* π^2 *are functions having properties* (*a*), (*b*), *and* (*e*) *and belonging to* $\mathbb{L}_2^{\omega}([0,T];\ \mathbb{H}^1)\cap\mathbb{L}_2^{\omega}([0,T];\ \mathbb{L}_1)$, *then*

$$\mathbf{P}\Big(\sup_{(t,x)\in[0,T]\times\mathbf{R}^d}\left|\pi^1(t,x,\omega) - \pi^2(t,x,\omega)\right| > 0\Big)=0.\quad\square$$

2.5. In this paragraph it is supposed that the assumptions of Corollary 4 are satisfied for $n\geq 2$.

Consider the system of Ito's equations,

$$\eta^i(t,x) = x^i + \int_{[0,t]}\sigma^{il}(s,\ \eta(s,x),\ Y(s))d\tilde{w}^l(s),\ t\in[0,T],\ i=1,2,\dots,d.$$

From Corollary 5.2.1 it follows that the process $\eta(t,x):= (\eta^1(t,x),\dots,\eta^d(t,x))$ is a flow of $\mathbf{C}^{0,2}$-diffeomorphisms of $\mathbf{R}^d$ onto $\mathbf{R}^d$.

Put

$$_1\psi(t,x):= exp\Big\{-\int_{[0,t]}h^l(s,\ \eta(s,x),\ Y(s))d\tilde{w}^l(s) +$$

$$+\frac{1}{2}\int_{[0,t]}h^lh^l(s,\ \eta(s,x),\ Y(s))ds\Big\},$$

$$ {}_1\mathcal{L}(t,x)u := a^{ij}(t,x,\ Y(t))u_{ij} + {}_1b^i(t,x,\ Y(t))u_i + {}_1c(t,x,\ Y(t)), $$

$$ {}_1\mathcal{M}^l(t,s)u := -\sigma^{ij}(t,x,\ Y(t))u_i + {}_1h^l(t,x,\ Y(t))u, $$

where ${}_1b^i := -b^i + 2a^{ij}_i$, ${}_1c := a^{ij}_{ij} - b^i_i$, ${}_1h^l := h^l - \sigma^{il}_i$.

Let ${}_1\tilde{\mathcal{L}}$ be a differential operator defined by the coefficients of ${}_1\mathcal{L}$, ${}_1\mathcal{M}^l$ and by ${}_1\psi$ similarly to the differential operator $\tilde{\mathcal{L}}$ as defined in §5.2.2 by the coefficients of $\mathcal{L}$, $\mathcal{M}^l$ and by ψ. From Theorem 5.2.2 and Corollary 4 we obtain the following

Corollary 5. *Given $f \in \mathbf{L}_\infty$ for all $t \in [0,T]$ and ω from some subset of Ω of probability one the following equality holds*

$$ \mathbf{E}[f(X(t))|\mathcal{Y}^0_t] = \int_{\mathbf{R}^d} f(\eta(t,x))\ {}_1\psi^{-1}(t,x)v(t,x)\left|\frac{D\eta(t,x)}{Dx}\right| dx\ \times $$

$$ \times \left(\int_{\mathbf{R}^d} {}_1\psi^{-1}(t,x)v(t,x)\left|\frac{D\eta(t,x)}{Dx}\right| dx \right)^{-1}, $$

where v is a classic solution of the problem

$$ dv(t,x) = {}_1\tilde{\mathcal{L}}v(t,x)\,dt, $$

$$ v(0,x) = \pi_0(x), $$

and $\left|\dfrac{D\eta(t,x)}{Dx}\right|$ *is the Jacobian of the transformation $x' = \eta(t,x)$.* □

Remark 5. *All the results of paragraph item carry over to the case of Ito's equation for the process $Z(t)$, whose coefficients depend (in a measurable way) on the whole trajectory of Y up to the moment. Of course it is necessary to assume the existence of a solution of this equation.* □

6.3. The Backward Filtering Equation. Interpolation and Extrapolation.

3.0. In this section we derive the backward filtering equation which describes the dynamics of $\mathsf{E}[f(X(T_1))|\mathcal{Y}_{T_0}^{T_0}\vee\mathfrak{B}_{T_0}^{T_0}]$ with respect to T_0 for a given T_1. We shall also prove the absolute continuity of the interpolation and the extrapolation measures with respect to the Lebesgue one and investigate the structure of the corresponding densities.

The notation introduced previously in this chapter is also in force for this section.

Warning 0. *It is supposed that the coefficients of equation* (0.1) *do not depend on* ω. □

3.1. In this section it is assumed that for every $t\in[0,T]$, i, $j=1,2,\ldots,d+d_1$, and $l=1,2,\ldots,d_1$, the coefficients $A^{ij}(t,z)$, $\mathfrak{D}^{il}(t,z)$, and $h^l(t,z)$ are two times differentiable in z, $b^i(t,z)$ is one time differentiable in z, and all the derivatives are uniformly (in t,z) bounded.

In addition to the process $Z(t)$ we consider on the space F the process $Z(t,z,s):=(X(t,x,s),\ Y(t,x,s))$ defined by the system of Ito's equation,

$$X^i(t,x,s) = x^i + \int_{[s,t]} {}_X B^{ij}(r,\ Z(r,x,s))\,dr +$$

$$(3.1)\qquad + \int_{[s,t]} {}_X\Sigma^{il}(r,\ Z(r,x,s))\,d\hat{v}^l(r),\ x^i\in\mathbf{R}^1,\ i=1,2,\ldots,d$$

$$(3.2)\qquad Y^i(t,y,s) = y^i + \int_{[s,t]} {}_0\sigma^{il}(r,\ Y(r,y,s))\,d\tilde{w}^l(r),$$

$$y^i\in\mathbf{R}^1,\ i=1,2,\ldots,d_1,$$

$t\in[s,T]$, and s is a given point of $[0,T[$.

Define

$$\rho_t^{z,s} := exp\Big\{ \int_{[s,t]} h^l(r,\ Z(r,z,s))\,d\tilde{w}^l(r)\ -$$

$$- \frac{1}{2} \int_{[s,t]} h^l h^l(r, Z(r,z,s))dr.$$

Theorem 1. Let $f \in C_b^1(\mathbb{R}^{d+d_1})$ *and* $r < -2^{-1}(d+d_1)$. *Then,* ($\tilde{\mathbf{P}}$-*a.s.*),

$$\mathbf{E}[f(Z(T_1)) | \mathcal{Y}_{T_1}^0 \vee \mathcal{X}_{T_0}^0] = v_f(T_0, Z(T_0))(v_1(T_0, Z(T_0))^{-1}$$

where $v_f(s,z)$ *and* $v_1(s,z)$ *are* $\tilde{\mathbf{P}}$-*a.s. continuous (in* s,z*) versions of r-generalized solutions of the equation,*

$$- dv(s,z) = (A^{ij}(s,z)\ v_{ij}(s,z) + b^i(s,z)\ v_i(s,z))ds +$$

$$+ (D^{il}(s,z)\ v_i(s,z) + h^l(s,z)\ v(s,z)) * d\tilde{w}^l(s), \tag{3.3}$$

$$(s,z,\omega) \in [0, T_1[\times \mathbb{R}^{d+d_1} \times \Omega,$$

corresponding to the terminal conditions

$$v_f(T_1,z) = f(z),\ (z,\omega) \in \mathbf{R}^{d+d_1} \times \Omega, \text{ and} \tag{3.4}$$

$$v_1(T_1,z) = 1,\ (z,\omega) \in \mathbf{R}^{d+d_1} \times \Omega \tag{3.5}$$

respectively.

Moreover, given ψ *equal to* f *or* 1, *for every* $(s,z) \in [0, T_1] \times \mathbf{R}^{d+d_1}$, ($\tilde{\mathbf{P}}$-*a.s*), *the following equality holds*

$$v_\psi(s,z) = \tilde{\mathbf{E}}[\psi(Z(T_1,z,s))\rho_{T_1}^{z,s} | \tilde{\mathfrak{F}}_{T_1}^s]. \quad \square$$

Proof. First observe that by Lemma 3.4.7, f and 1 belongs to $\mathbf{W}_p^1(r, \mathbf{R}^{d+d_1})$ for every $p \geq 2$ and in particular for $p > d+d_1$. Therefore by Theorem 4.2.1 both of the problems (3.3) (3.4) and (3.3), (3.5) have unique r-generalized solutions. By

Corollary 4.2.1 these solutions have continuous (**P**-a.s.), versions, which will be denoted by $v_f(s,z)$ and $v_1(s,z)$ respectively.

From theorem 5.1.1 it follows * that for every (s,z), (**P**-a.s.),

$$v_f(s,z) = \tilde{E}[f(Z(T_1,z,s))\rho_{T_1}^{z,s} | \tilde{\mathfrak{F}}_{T_1}^s], \tag{3.6}$$

$$v_1(s,z) = \tilde{E}[\rho_{T_1}^{z,s} | \tilde{\mathfrak{F}}_{T_1}^s]. \tag{3.7}$$

Consider the continuous diffusion process $Z(s,Z_{T_0})$ satisfying on **F** the following system of Ito's equations:

$$Z^i(s,Z_{T_0}) = Z_{T_0} + \int_{[T_0,s]} B^i(r, Z(r,Z_{T_0}))dr +$$

$$+ \int_{[T_0,s]} \Sigma^{il}(r, Z(r, Z_{T_0}))d\tilde{v}^l(r), \tag{3.8}$$

$$i=1,2,\dots,d+d_1, \; s\in[T_0,T],$$

where $B^i := {}_X B^i$ for $i=1,2,\dots,d$, $B^i:=0$ for $i=d+1,\dots,d+d_1$, and Z_{T_0} is $\mathfrak{F}_{T_0}$-measurable random variable on $\mathbf{R}^{d+d_1}$.

Note that due to the uniqueness of solution of system (3.8) (Theorem 1.4.3) if $Z_{T_0} = Z(T_0)$, then $Z(s, Z_{T_0})$ coincides with $Z(s)$ on $[T_0,T]$, (**P**-a.s.).

---

*Evidently $A^{ij} := 2^{-1}(\mathfrak{D}^{il}\mathfrak{D}^{jl} + \hat{\mathfrak{D}}^{il}\hat{\mathfrak{D}}^{jl})$, where

$$\hat{\mathfrak{D}}^{il} := \begin{cases} {}_1\sigma^{il} & for \; i=1,\dots,d \\ 0 & for \; i=d+1,\dots,d+d_1 \end{cases}$$

Thus condition 5.0.1 required in Theorem 5.1.1 is satisfied.

Let $x_n(a):= 2^{-n}[2^n a]$, where $[a]$ is the integral part of the number a, $n\in\mathbb{N}\cup\{0\}$ and Γ_n is the range of values of $x_n(Z(T_0)):= (x_n(Z^1(T_0),\dots,x_n(Z^{d+d_1}(T_0))$.

Define the process $Z^n(s):= Z(s,\ x_n(Z(T_0))$. A trajectory of this process we denote by Z_s^{n,T_0}.

Let $\rho^n(T_1,T_0)$ be the function obtained from $\rho^n(T_1,T_0)$ by the replacement of $Z(\cdot)$ by $Z^n(\cdot)$.

Clearly, there exists a sequence $\{n'\}$ such that

$$(3.9)\qquad \sup_{s\in[T_0,T]} \left|Z^{n'}(s)-Z(s)\right|_{d+d_1} \to 0 \qquad (\mathbf{P}\text{-}a.s.),$$

as $n\to 0$. Thus there exists a subsequence $\{k\}$ such that, ($\mathbf{P}$-a.s.),

$$(3.10)\qquad f(Z^k(T_1))\rho^k(T_1,T_0) \to f(Z(T_1))\rho(T_1,T_0) \quad as \quad k\to\infty.$$

Let η be a bounded $\tilde{\mathfrak{F}}_{T_1}^{T_0}\vee\mathfrak{Z}_{T_0}^{T_0}$-measurable random variable.

Arguing in the same way as in the end of the proof of Theorem 5.3.1, it is easy to show that the sequence $\{f(Z^k(T_1)\rho^k(T_1,T_0)\eta\}$ is uniformly $\tilde{\mathbf{P}}$-integrable and hence

$$(3.11)\qquad \lim_{k\to\infty} \tilde{\mathbf{E}}(f(Z^k(T_1))\rho^k(T_1,T_0)\eta) = \tilde{\mathbf{E}}(f(Z(T_1))\rho(T_1,T_0)\eta).$$

Further, due to the uniqueness of solution of system (3.8) it follows that for $t\in[T_0,T]$, ($\tilde{\mathbf{P}}$-$a.s.$),

$$Z^n(t) = \sum_{z\in\Gamma^n} 1_{\{x_n(Z(T_0))=z\}} Z(t,z,T_0).$$

This and (3.11) yields that, ($\mathbf{P}$-a.s.)

$$\tilde{\mathbf{E}}(f(Z(T_1))\rho(T_1,T_0)\eta) =$$

$$= \lim_{k\to\infty} \tilde{\mathbf{E}}\Big(\sum_{z\in\Gamma_k} 1_{\{x_k(Z(T_0))=z\}} f(Z(T_1,z,T_0)) \times$$

$$\times \rho_{T_1}^{z,T_0}\eta\Big) = \lim_{k\to\infty} \tilde{\mathbf{E}}\Big(\sum_{z\in\Gamma_k} 1_{\{x_k(Z(T_0))=z\}} \times \tag{3.12}$$

$$\times \tilde{\mathbf{E}}[f(Z(T_1,z,T_0))\rho_{T_1}^{z,T_0} | \mathfrak{F}_{T_1}^{T_0} \vee \mathfrak{Z}_{T_0}^{T_0}]\ \eta\Big) =$$

Since $v_f(s,z)$ is continuous, $\lim_{k\to\infty} v_f(T_0, Z^k(T_0))\eta = v_f(T_0, Z(T_0))\eta$, ($\mathbf{P}$-*a.s.*). Moreover, ($\mathbf{P}$-a.s),

$$v_f(T_0, Z^k(T_0))\eta = \sum_{z\in\Gamma_k} 1_{\{x_k(Z(T_0))=z\}} \times$$

$$\times \tilde{\mathbf{E}}[f(Z(T_1,z,T_0))\rho_{T_1}^{z,T_0} | \tilde{\mathfrak{F}}_{T_1}^{T_0} \vee \mathfrak{Z}_{T_0}^{T_0}]\ \eta.$$

Consequently

$$\tilde{\mathbf{E}}\Big|v_f(T_0, Z^k(T_0))\eta\Big|^2 = \tilde{\mathbf{E}} \sum_{z\in\Gamma_k} 1_{\{x_k(Z(T_0))=z\}} \times$$

$$\times \Big|\tilde{\mathbf{E}} f(Z(T_1,z,T_0))\rho_{T_1}^{z,T_0} | \tilde{\mathfrak{F}}_{T_1}^{T_0} \vee \mathfrak{Z}_{T_0}^{T_0}]\ \eta\Big|^2. \tag{3.13}$$

Making use of the generalized Jensen's inequality (see [24], Ch.1, §9) and taking into account the boundedness of h^l, we obtain from (3.13) that

$$\max_k \tilde{\mathbf{E}}|v_j(T_0,Z^k(T_0))\eta|^2 \le \max_k N\,\tilde{\mathbf{E}}\Big(\sum_{z\in\Gamma_k} 1_{\{x_k(Z(T_0))=x\}} \times$$

$$\times \tilde{\mathbf{E}}[|f(Z(T_1,z,T_0))|^2\ \rho_{T_1}^{z,T_0} | \tilde{\mathfrak{F}}_{T_1}^{T_0} \vee \mathfrak{Z}_{T_0}^{T_0}]\Big) =$$

$$= \max_k N\, \tilde{\mathsf{E}}\tilde{\mathsf{E}}[|f(Z^k(T_1))|^2\, \rho^k(T_1,T_0)|\ \tilde{\mathfrak{F}}^{T_0}_{T_1} \vee \mathfrak{Z}^{T_0}_{T_0}]] < \infty.$$

Therefore the set $\{v_f(T_0,\ Z^k(T_0))\eta\}$ is bounded in $\mathsf{L}_2(\Omega;\ \mathfrak{F};\ \tilde{\mathsf{P}};\ \mathbf{R}^1)$ and so is uniformly integrable (see [100], Ch. II, Theorem 22). Thus (3.12) yields,

$$\tilde{\mathsf{E}}(f(Z(T_1))\rho(T_1,T_0)\eta) = \tilde{\mathsf{E}}(v_f(T_0,\ Z(T_0))\eta).$$

Hence the following two equalities hold, (P-a.s.):

$$\tilde{\mathsf{E}}[f(Z(T_1))\rho(T_1,T_0)|\ \tilde{\mathfrak{F}}^{T_0}_{T_1} \vee \mathfrak{Z}^{T_0}_{T_0}] = v_f(T_0,\ Z(T_0)),$$

and

$$\tilde{\mathsf{E}}[\rho(T_1,T_0)|\ \tilde{\mathfrak{F}}^{T_0}_{T_1} \vee \mathfrak{Z}^{T_0}_{T_0}] = v_1(T_0,\ Z(T_0)).$$

Theorem (1.3) follows at once from (3.6), (3.7), (3.14) and (3.15). □

Remark 1. *From* (3.14) *and Lemma* 1.3 *and* 1.4 *we obtain that under the assumptions of the theorem of this paragraph,* (P-*a.s.*),

$$\tilde{\mathsf{E}}[f(Z(T_1))\rho(T_1,T_0)|\ \mathfrak{Y}^0_{T_1} \vee \mathfrak{X}^0_{T_0}] =$$

$$= \tilde{\mathsf{E}}[f(Z(T_1))\rho(T_1,T_0)|\ \mathfrak{Y}^0_{T_1} \vee \mathfrak{X}^{T_0}_{T_0}] = v_f(T_0,\ Z(T_0)). \quad \square$$

Corollary 1. *Suppose that the assumptions of the theorem are satisfied, and in addition the coefficients of the process* $Z(t)$ *and the function* f *do not depend on* y^*. *Then*

$$\mathsf{E}[f(X(T_1))|\ \mathfrak{Y}^0_{T_1} \vee \mathfrak{X}^0_{T_0}] = v^X_f(T_0,\ X(T_0))\ (v^X_1(T_0,\ X(T_0))^{-1},$$

---

*In this case X(t) is a Markov process. $\rho^{z,s}$ does not depend on y either and therefore is denoted by $\rho^{x,s}_t$.

($\mathbf{P}$-a.s.), where v_f^X and v_1^X are continuous in (s,x) versions of r-generalized solutions of the equation

$$-dv(s,x) = (a^{ij}(s,x)v_{ij}(s,x) + {}_Xb^i(s,x)v_i(s,x))ds +$$

(3.16)

$$+ (\sigma^{il}(s,x)v_i(s,x) + h^l(s,x)v(s,x))_* d\tilde{w}^l(s), (s,x,\omega)\in[0,T_1[\times\mathbf{R}^d\times\Omega$$

corresponding to the terminal conditions

(3.17) $$v_f(T_1,x) = f(x) \quad (x,\omega)\in\mathbf{R}^d\times\Omega,$$

and

(3.18) $$v_1(T_1,x) = 1 \qquad (x,\omega)\in\mathbf{R}^d\times\Omega,$$

respectively. Moreover, for every $(s,x)\in[0,T_1]\times\mathbf{R}^d$ letting ψ be equal to f or 1, we have, ($\mathbf{P}$-a.s.),

$$v_\psi^X(s,x) = \tilde{\mathbf{E}}[\psi(X(T_1,x,s))\rho_{T_1}^{x,s}|\,\tilde{\mathfrak{F}}_{T_1}^s]. \quad \square$$

Proof. Lemma 3.4.7 yields that f and 1 belong to $\mathbf{W}_{p'}^1(r,\mathbf{R}^d)$ for $p'=2$ and $p'=p>2$. Thus by Theorem 4.2.1, each of the problems (3.16), (3.17) and (3.16), (3.18) has a unique r-generalized solution. In view of Corollary 4.2.1 these solutions have continuous (in s,x) versions, which henceforth will be denoted by $v_f^X(s,x)$ and $v_1^X(s,x)$, respectively.

We can consider $v_f^X(s,x)$ and $v_1^X(s,x)$ as elements of $\mathbf{L}_2^\omega([0,T]; \mathcal{P}(\mathcal{Y}); \mathbf{H}^1(r,\mathbf{R}^{d_1+d}))$. Taking into account that $A^{ij}\equiv a^{ij}$, $b^i\equiv {}_Xb^i$ and $\mathfrak{D}^{il}\equiv\sigma^{il}$ for $i, j=1,2,\dots,d$ and $l=1,2,\dots,d_1$, we easily see that $v_f^X(s,x)$ and $v_1^X(s,x)$ are also r-generalized solutions of problems (3.3), (3.4) and (3.3), (3.5) respectively. Thus due to the uniqueness of generalized solutions of these problems we have completed the proof of the corollary. $\square$

3.2. Henceforth we suppose that the assumptions of §.1 and section 1.0 are in force. We also keep all the notation used in this chapter. In particular we denote the non-normalized filtering density by $u(t,x)$.

We can and shall suppose that the continuous version of the r-generalized solution of the backward filtering equation (3.3), (3.5) is continuous non-negative function of all $\omega\in\Omega$.

Theorem 2. There exists a function $P_{\mathcal{Y}}^{T_1,T_0}:\Omega\to\mathcal{M}(\mathbf{R}^d)$ *with these properties:*

(a) $P_{\mathcal{Y}}^{T_1,T_0}$ *is the regular conditional probability distribution of* $X(T_0)$ *with respect to* $\mathcal{Y}_{T_1}^0$.

(b) $P_{\mathcal{Y}}^{T_1,T_0}$ *is absolute continuous with respect to the Lebesgue measure,* (**P**-*a.s*) *with the Radon-Nikodym derivative*

$$\pi^{T_1,T_0}(x):= v_1(T_0,x,Y(T_0))\ u(T_0,x)\ \times$$

$$\times\Big(\int_{\mathbf{R}^d} v_1(T_0,x,Y(T_0))\ u(T_0,x)\,dx)^{-1}.\quad\square$$

Proof. Denote by $\varphi(\cdot,t)$ the measure defined on $\mathbf{R}^d$ by the equality,

$$\Phi(\Gamma,t)=\int_\Gamma u(t,x)\,dx,\quad \forall\Gamma\in\mathcal{B}(\mathbf{R}^d).$$

Lemma 2. Let ψ *be a* $\mathcal{B}(\mathbf{R}^d)\otimes\mathcal{Y}_{T_1}^0$*-measurable mapping of* $\mathbf{R}^d\times\Omega$ *in* $\overline{\mathbf{R}}^1$ *and* $\mathbf{E}|\psi(X(T_0))|<\infty$.

Then $\psi(x)$ *is a* $\Phi(\cdot,T_0)$*-integrable function,* (**P**-*a.s.*) *and*

$$\tilde{\mathbf{E}}[\psi(X(T_0))\rho(T_0,0)|\mathcal{Y}_{T_1}^0]=(\psi,\ u(T_0))_0 \tag{3.19}$$

Proof. First observe that the conditional expectation in the left side of (3.19) is well defined, because, as it is easy to verify,

$$\tilde{\mathbf{E}}|\psi(X(T_0))\rho(T_0,0)|=\tilde{\mathbf{E}}|\psi(X(T_0))\rho(T,0)|=$$

$$=\tilde{\mathbf{E}}|\psi(X(T_0)\rho_T^0|=\mathbf{E}|\psi(X(T_0))|<\infty.$$

We will prove formula (3.19) under the additional assumption that ψ is a non-negative, bounded and continuous function on $\mathbf{R}^d$, ($\mathbf{P}$-a.s.). Making use of this fact the reader will easily complete the proof of the lemma.

Due to the independence of $\mathfrak{Z}^0_{T_0}$ and $\tilde{\mathfrak{F}}^{T_0}_{T_1}$, we obtain by Theorem 2.1 that for $\mathbf{P}$-a.a. ω the following equalities are valid

$$\tilde{\mathbf{E}}[\psi(X(T_0))\rho(T_0,0)|\ \mathcal{Y}^0_{T_1}] =$$

$$= \lim_{n\to\infty} \tilde{\mathbf{E}}[\psi(x_n(X(T_0)))\rho(T_0,0)|\ \mathcal{Y}^0_{T_1}] =$$

$$= \lim_{n\to\infty} \tilde{\mathbf{E}}\Big[\sum_{\alpha\in\Gamma_n} \psi(\alpha)\, 1_{\{x_n(X(T_0))=\alpha\}}\rho(T_0,0)|\mathcal{Y}^0_T \vee \tilde{\mathfrak{F}}^{T_0}_{T_1}\Big] =$$

$$= \lim_{n\to\infty} \sum_{\alpha\in\Gamma_n} \psi(\alpha)\, \tilde{\mathbf{E}}\Big[1_{\{x_n(X(T_0))=\alpha\}}\rho(T_0,0)|\mathcal{Y}^0_T\Big]$$

$$= \lim_{n\to\infty} \sum_{\alpha\in\Gamma_n} \psi(\alpha) \int_{\mathbf{R}^d} 1_{\{x_n(x)=\alpha\}} u(T_0,x)\,dx =$$

$$= \lim_{n\to\infty} (\psi(x_n),\, u(T_0))_0 = (\psi,\, u(T_0))_0. \quad \square$$

By Theorem 1.1 for every function $f\in\mathbf{L}_\infty$ the following equality holds, ($\mathbf{P}$-a.s.),

$$\mathbf{E}[f(X(T_0))|\ \mathcal{Y}^0_{T_1}] = \tilde{\mathbf{E}}[f(X(T_0))\rho(T_1,0)|\ \mathcal{Y}^0_{T_1}]\ (\tilde{\mathbf{E}}[\rho(T_1,0)|\mathcal{Y}^0_{T_1}])^{-1}. \tag{3.20}$$

Further, it is easy to verify that

$$I := \tilde{E}[f(X(T_0))\rho(T_1,0)|\ \mathcal{Y}^0_{T_1}] =$$

$$= \tilde{\mathbf{E}}[f(X(T_0))\tilde{\mathbf{E}}[\rho(T_1,T_0)|\ \mathcal{Y}^0_{T_1} \vee \mathfrak{X}^0_{T_0}]\rho(T_0,0)\mathcal{Y}^0_{T_1}].$$

From this equality, Remark 1 and the lemma of this paragraph we obtain that

$$I = \tilde{\mathbf{E}}[f(X(T_0))v_1(T_0, Z(T_0))\rho(T_0,0)|\, \mathcal{Y}^0_{T_1}] =$$

(3.21) ($\mathbf{P}$-*a.s.*)

$$= \int_{\mathbf{R}^d} f(x)v_1(T_0, x, Y(T_0))\, u(T_0,x)\,dx.$$

Theorem 2.1 and (3.21) yield that $\tilde{\mathbf{P}}$-a.s.,

$$\tilde{\mathbf{E}}[\rho(T_1,0)|\, \mathcal{Y}^0_{T_1}] = (u(T_1),\, 1)_0 = \int_{\mathbf{R}^d} v_1(T_0,x,Y(T_0)u(T_0,x)\,dx > 0. \tag{3.22}$$

Collecting (3.20) to (3.22), we obtain that

$$\mathbf{P}(X(T_0)\in\Gamma|\, \mathcal{Y}^0_{T_1}) = \frac{\int_{\mathbf{R}^d} v_1(T_0,x,Y(T_0))u(T_0,x)\,dx}{\int_{\mathbf{R}^d} v_1(T_0,x,Y(T_0))u(T_0,x)\,dx}$$

(3.23)

$$\forall\Gamma\in\mathcal{B}(\mathbf{R}^d)\ (\tilde{\mathbf{P}}\text{-}a.s.).$$

Denote by Ω' the set where $0 < (v_1(T_0, Y(T_0)),\ u(T_0)) < \infty$. From (3.22) it follows that $\mathbf{P}(\Omega') = 1$.

Define the function $P_{\mathcal{Y}}^{T_1,T_0}(\Gamma,\omega)$ as the right hand side of (3.23) for $\omega\in\Omega'$ and as 1 *for* $\omega\notin\Omega'$. Evidently, this function meets the requirements of the theorem.

Remark 2. *From formula* (3.22) *it follows that the interpolation density can also be written in the form*

$$\pi^{T_1,T_0}(x) = v_1(T_0,x,Y(T_0))u(T_0,x)\ (u(T_1),\, 1)_0^{-1}. \quad \square$$

3.3. *Remark* 3. *The interpolation problem can be reduced to the filtering problem by artificially increasing the dimensionality of the non-observable (sigmal) process. Specifically, we can put*

$$\hat{X}(t) := 1_{\{[0,T_0]\}}(t)\, X(t) + 1_{\{[T_0,\infty[\}}(t)\, X(T_0),$$

$$\tilde{X}(t) := (X(t),\ \hat{X}(t)) \text{ and } Z(T) := (\tilde{X}(t),\ Y(t)).$$

Then for $f \in \mathbb{L}_\infty$,

$$\mathbf{E}[f(X(T_0)|\ \mathcal{Y}^0_{T_1}] = \mathbf{E}[f(\hat{X}(T_1))|\ \mathcal{Y}^0_{T_1}].$$

3.4. The extrapolation problem can also be reduced to the filtering problem. Actually, since $Z(t)$ is a Markov process, we have for every $f \in \mathbb{L}_\infty$,

$$\mathbf{E}[f(X(T_1))|\ \mathcal{Y}^0_{T_0}] = \mathbf{E}[\mathbf{E}[f(X(T_1))|\ \mathcal{Z}^0_{T_0}]|\mathcal{Y}^0_{T_0}] =$$

$$= \mathbf{E}[\mathbf{E}[F(X(T_1))|\ \mathcal{Z}^{T_0}_{T_0}]|\mathcal{Y}^0_{T_0}].$$

Suppose that f satisfies the assumptions of Theorem 1. Then it is well known (see for example, Theorem 5.1.1) that

$$\mathbf{E}[f(X(T_1))|\ \mathcal{Z}^{T_0}_{T_0}] = v(T_0,\ Z(T_0)), \qquad (\mathbf{P}\text{-}a.s.),$$

where $v(t,z)$ is a continuous in (t,z) version of r-generalized solution of the backward Kolmogorov equation

$$-\ dv(t,z) = (A^{ij}(t,z)v_{ij}(t,z) + b^i(t,z)v_i(t,z))dt,$$

$$(t,z) \in [0,\ T_1[\times\mathbf{R}^{d+d_1},$$

$$v(T_1,z) = f(z),\ z \in \mathbf{R}^{d+d_1}.$$

Making use of this it is easy to prove the following

Theorem 4. Under the assumptions of Theorem 1, for **P**-*a.a.* ω,

$$\mathsf{E}[f(X(T_1))|\,\mathcal{Y}^0_{T_0}] = \int_{\mathbf{R}^d} v(T_0,(x,Y(T_0)))\ \pi(T_0,x)dx. \quad \square$$

Remark 4. From the last theorem it is easy to deduce that the extrapolation measure is absolute continuous with respect to the Lebesgue measure and the Radon-Nikodym derivative is as follows

$$\pi^{T_0,T_1}(x) = \int_{\mathbf{R}^{d_1}} \int_{\mathbf{R}^d} p((x,y);\ T_1);\ (q,\ Y(T_0));\ T_0) \times \pi(T_0,q)dqdy,$$

where $p(z;\ T_1;\ r;\ T_0)$ *is the transition density of the process* $Z(t)$. $\square$

3.5. Note in conclusion that the assumptions on the smoothness of the coefficients, the initial and the terminal values of the direct and backward filtering equations could be made less restrictive, if instead we required uniform non-singularity of the matrix ${}_1\tilde{\sigma}$. In this case the equations mentioned above are of the superparabolic type (see, in particular [68]).

However, the assumption of uniform non-singularity of the matrix ${}_1\tilde{\sigma}$ is a quite restrictive one. It would be impossible to consider a degenerate $X(t)$ (for example, a purely deterministic signal process). In this case it is also impossible to reduce the interpolation problem to the filtering one as it was done in §3.

Chapter 7

HYPOELLIPTICITY OF ITO'S SECOND ORDER PARABOLIC EQUATIONS

7.0. Introduction

0.1. Smoothness of solutions of deterministic parabolic equations increases as the smoothness assumptions on their coefficients increase. This is a typical feature of parabolic equations. Moreover, under wide assumptions, the smoothness of solutions for $t>0$ depends only on the smoothness of coefficients and does not depend on the smoothness of the initial functions. This is important for example, in the study of the fundamental solution of a parabolic equation, since we can consider this solution as a solution of the corresponding Cauchy problem where the initial function is the Dirac-function. Hypoellipticity is a particular case of the growth of smoothness property mentioned above.

Say that a parabolic equation possesses the hypoellipticity property if its generalized solution has a C^{∞} version.

As is well known (e.g. [87]) a non-degenerate deterministic parabolic equation does possess this property. L. Hörmander [46] obtained a general condition for hypoellipticity of second-order degenerate deterministic parabolic equations.

In this chapter we consider the growth of smoothness and hypoellipticity of Ito's second-order parabolic and superparabolic equations. We give a generalization of Hörmander's condition for hypoellipticity of Ito's second-order parabolic equation. We also prove that Ito's second-order superparabolic equation has the growth of smoothness property. Opposed to previous chapters, it is not assumed here that the initial function belongs to some Sobolev space $\mathbf{H}^{\alpha}$ with non-negative α*, it may be an arbitrary finite measure on $\mathbf{R}^{d}$.

Because of the well-known connection between second-order parabolic equations and diffusion processes (see e.g. [34]) the hypoellipticity of second-order parabolic equation implies the existence of the infinitely smooth transition density of the corresponding diffusion process. In section 1 we prove a similar result in the case

*We remind the reader that in Chapter 3 under such assumption we have already investigated the growth of smoothness problem for Ito's second-order parabolic and superparabolic equations.

of Ito's second-order parabolic equation.

In section 2 we apply this result to prove hypoellipticity of the filteirng equation and prove that under Hörmander's type condition there exists a smooth conditional transition density for the corrooponding diffuoion procceo.

7.1. Measure-valued Solution and Hypoellipticity under Generalized Hörmander's Condition.

1.0. In this section as well as in section 5.1 we consider the following equation

$$du(t,x,\omega) = \mathcal{L}^* u(t,x,\omega)dt + \mathcal{M}^{l*}u(t,x,\omega)dw^l(t), \tag{1.1}$$

$$(t,x,\omega)\in[0,T]\times\mathbf{R}^d\times\Omega$$

where

$$\mathcal{L}^* u(t,x,0):= (a^{ij}(t,x,\omega)u)_{ij} - (b^i(t,x,\omega)u)_i + c(t,x,\omega),$$

$$\mathcal{M}^{l*}u(t,x,\omega):= (\sigma^{il}(t,x,\omega)u)_i + h^l(t,x,\omega)u$$

Warning 0. *Throughout this chapter, unless otherwise stated, the assumptions and notations in* §§5.0.2, 5.3.0 *and* 5.3.1 *will be in force. In particular, we suppose that some* σ*-algebra* $\mathfrak{F}_0\subset\mathfrak{F}$ *completed with respect to* $\mathbf{P}$ *is fixed.*

Contrary to §5.3.0, 5.3.1 *we do not assume here that the regular conditional probability* $P_{\mathfrak{F}_0}(\cdot)$ *has a density with respect to Lebesgue measure.* □

Denote by $\mathcal{M}_f(\mathbf{R}^d)$ *the space of countably additive finite measures on* $\mathbf{R}^d$.

Definition 0. *A family of measures* $\mu_t(\omega,\cdot)$, *where* $(t,\omega)\in[0,T]\times\Omega$, *is called a measure-valued solution of equation* (1.1), *corresponding to the initial condition*

$$\mu_0(\omega,\cdot) = P_{\mathfrak{F}_0}(\cdot) \qquad (\mathbf{P}\text{-}a.s.), \tag{1.2}$$

if it satisifes the following conditions:

(*i*) *For every* $(t,\omega)\in]0,T]\times\Omega$, $\mu_t(\omega,\cdot)\in\mathcal{M}_f(\mathbf{R}^d)$, *and* $\mu_t(\cdot,\mathbf{R}^d)\in\mathbf{L}_2([0,T]\times\Omega;\ \mathbf{R}^1)$.

(*ii*) *For every* $f\in\mathbf{L}_\infty$, $\mu_t[f(\cdot)]:= \int_{\mathbf{R}^d} f(x)\ \mu_t(dx)$ *has a predictable version which is continuous in* t, ($\mathbf{P}$-*a.s.*).

(*iii*) *For every function* $f \in \mathbf{C}_b^2(\mathbf{R}^d)$ *and for all* $t \in [0,t]$ *and* $\omega \in \tilde{\Omega}$, *where* $\mathbf{P}(\tilde{\Omega})=1$, *the following equality holds:*

$$\mu_t[f(\cdot)] = P_{\mathfrak{X}_0}[f(\cdot)] + \int_{[0,t]} \mu_s[\mathcal{L}(s,\cdot)f(\cdot)]ds +$$

(1.3)
$$+ \int_{[0,t]} \mu_s[\mathcal{M}^l(s,\cdot)f(\cdot)]dw^l(s). \quad \square$$

Here and below we denote by $\mathcal{L}$ and $\mathcal{M}^l$ the operators formally conjugated to $\mathcal{L}^*$ and $\mathcal{M}^{l*}$ respectively.

Note that the integrals on the right-hand side of (1.3) are well defined, since $\mathcal{L}(s,x,\omega)f(x)$ and $\mathcal{M}^l(s,x,\omega)f(x)$ are bounded, continuous in x and predictable for every x functions. Thus there exist predictable versions of $\mu_s[\mathcal{L}(s,\cdot)f(\cdot)]$ and $\mu_s[\mathcal{M}^l(s,\cdot)f(\cdot)]$.

The density $\pi(t,x,\omega)$ of the measure-valued solution $\mu_t(\omega,\cdot)$ with respect to the Lebesgue measure, provided it does exist, can be considered as a generalized solution of the equation

$$(\pi(t),\eta)_0 = P_{\mathfrak{X}_0}[\eta(\cdot)] + \int_{[0,t]} (\pi(s), \mathcal{L}(s)\eta)_0 ds +$$

$$+ \int_{[0,t]} (\pi(s), \mathcal{M}^l(s)\eta)_0 dw^l(s), \ \forall \eta \in \mathbf{C}_0^\infty(\mathbf{R}^d).$$

So, to prove the hypoellipticity of equation (1.1), it suffices to verify the infinite differentiability of this density.

In this paragraph we shall prove the existence of the measure-valued solution and in paragraphs 1 to 4 it will be shown that under condition of Hörmander type this solution possesses an infinitely differentiable density for $t>0$.

Let $\mathfrak{X}(t)$ be a diffusion process which is a solution of the following system of Ito's equations

(1.4) $$d\mathfrak{X}^i(t) = B^i(s, \mathfrak{X}(s))ds + \Sigma^{il}(s, \mathfrak{X}(s))d\nu^l(s), \ i=1,2,\dots,d, \ t \in]0,T],$$

(1.5) $$\mathfrak{X}^i(0) = x_0^i, \ i=1 \div d$$

where ν is a standard Wiener process in $\mathbf{R}^{d_0+d_1}$, x_0 is a $\mathcal{F}_0$-measurable random variable, and the coefficients B^i and Σ^{il} are defined by formulas

(1.6) $$B^i := b^i - \sigma^{il} h^l \quad for \quad i=1,2,...,d$$

(1.7) $$\Sigma^{il} := \begin{cases} \hat{\sigma}^{il} & for \quad l=1,2,...,d_0,\ i=1,2,...,d \\ \sigma^{i(l-d)} & for \quad l=d_0+1,...,d_0+d_1,\ i=1,2,...,d. \end{cases}$$

Recall that in accordance with the Warning the hypothesis (5.0.1) is assumed to be fulfilled. Just as in §5.0.2 we shall suppose that the latter d_f coordinates of the process $\nu(t)$ coincide with $w(t)$, and denote by $\tilde{w}(t)$ a Wiener process composed of the first d_0 coordinates of $\nu(t)$.

As in §5.3.0 we write $\mathfrak{F}_t := \mathcal{F}_0 \vee \mathcal{F}_t^0$, where $\mathcal{F}_t^0$ is a σ-algebra generated by the Wiener process $w(s)$ for $s\in[0,t]$ and completed with respect to $\mathbf{P}$. We shall also suppose that $P_{\mathfrak{F}_0}(\cdot)$ is a regular conditional distribution of x_0 relative to $\mathfrak{F}_0$.

Let

(1.8) $$\rho(t) = exp\Big\{ \int_{[0,t]} c(s, \mathfrak{X}(s))ds + \int_{[0,t]} h^l(s, \mathfrak{X}(s))dw^l(s) - \frac{1}{2} \int_{[0,t]} h^l h^l(s, \mathfrak{X}(s))ds\Big\}.$$

Theorem 0. *The problem* (1.1), (1.2) *has a measure-valued solution* μ_t, *and for any function* $\psi\in\mathbf{L}_\infty$ *and for every* $t\in[0,T]$ *the following representation holds* ($\mathbf{P}$-*a.s.*),

(1.9) $$\mu_t[\psi(\cdot)[= \mathbf{E}[\psi(\mathfrak{X}(t))\rho(t)|\,\mathfrak{F}_t]. \quad \square$$

Remark 0. *This theorem is closely connected with Theorem* 5.3.1. *In particular, if the initial condition* $P_{\mathfrak{F}_0}$ *has a density* $\varphi\in\mathbf{L}_2(\Omega;\ \mathfrak{F}_0;\ \mathbf{H}^1)$ *then Theorem* 5.3.1 *yields the existence of a measure-valued solution* $\mu_t(\omega,\cdot)$ *of the form* $\int_\bullet u(t,x)dx$, *where* $u(t,x)$ *is a generalized solution of* (5.3.1), (5.3.2). *Formula* (1.9) *is a generalization of AOC formula* (5.3.3) □

Proof. By Theorem 5.3.3, there exists a linear continuous functional $\tilde{\Phi}[\cdot](\omega)$ on $\mathbf{C}_b^0(\mathbf{R}^d)$ such that for every $\psi \in \mathbf{C}_b^0(\mathbf{R}^d)$, $\tilde{\Phi}_t[\psi]$ is continuous (in t) predictable version of $\mathbf{E}[\psi(\mathfrak{X}(t))\rho(t)|\ \mathfrak{F}_t]$.

Let K_i be a sequence of compact sets in $\mathbf{R}^d$ such that $K_i \subseteq K_{i+1}$ and $\bigcup_{i=1}^{\infty} K_i = \mathbf{R}^d$. Consider a restriction of $\tilde{\Phi}_t[\cdot](\omega)$ in $\mathbf{C}^0(K_i)$. By a theorem of Riesz on the representation of a linear functional on the space of continuous functions (see e.g. [25], Theorem IV. 6.3) for every $(t,\omega) \in]0,T] \times \Omega$, there exists a countable additive measure $\pi_t^i(\omega,dx)$ on K_i such that for every $\psi \in \mathbf{C}^0(K_i)$

$$\tilde{\Phi}_t[\psi](\omega) = \int_{K_i} \psi(x)\ \mu_t^i(\omega,dx). \tag{1.10}$$

Since $\tilde{\Phi}_t[\psi]$ is a version of $\mathbf{E}[\psi(\mathfrak{X}(t))\rho(t)|\ \mathfrak{F}_t]$, $\tilde{\Phi}_t[\psi](\omega) \geq 0$, for all t and all non-negative $\psi \in \mathbf{C}^0(K)$ on some set Ω' of probability 1. Therefore the measure $\mu_t^i(\omega,dx)$ is non-negative for for $\omega \in \Omega'$.

For $\omega \in \Omega'$ and $\Gamma \in \mathfrak{B}(\mathbf{R}^d)$ we set $\mu_t(\omega,\Gamma) := \lim_{i\to\infty} \mu_t^i(\omega,\ \Gamma \cap K_i)$.

Clearly, for all $\omega \in \Omega''$, where $\mathbf{P}(\Omega'')=1$, $\mu_t'(\omega,\mathbf{R}^d) = \mathbf{E}[\rho(t)|\ \mathfrak{F}_t] = \tilde{\Phi}_t[1] < \infty$. Denote $\Omega' \cap \Omega''$ by Ω''' and define $\mu_t(\omega,\cdot)$ by the following equality

$$\mu_t(\omega,\Gamma) = \begin{cases} \mu_t'(\omega,\Gamma) & \textit{for} \quad \omega \in \Omega''' \\ 0 & \textit{for} \quad \omega \notin \Omega''' \end{cases}$$

For every t, ω, $\mu_t(\omega,\cdot)$ is a countably additive, positive, finite measure.

Let us check for example that it is countably additive. Let $\{\Gamma_n\}$ be a sequence of disjoint sets from $\mathfrak{B}(\mathbf{R}^d)$. Then by a theorem on monotone passage to the limit we get for all t,ω

$$\mu_t(\omega,\ \bigcup_{n=1}^{\infty} \Gamma_n) = \lim_{i\to\infty} \mu_t^i(\omega,\ \bigcup_{n=1}^{\infty} \Gamma_n \cap K_i) =$$

$$= \lim_{i\to\infty} \sum_{n=1}^{\infty} \mu_t^i(\omega,\ \Gamma_n \cap K_i) = \sum_{n=1}^{\infty} \mu_t(\omega,\ \Gamma_n).$$

Note that the measure μ_t does not depend on the choice of the sequence of compacts K_i used in its definition. To see this, let $\{K'_i\}$ by another sequence of expanding compacts such that $\bigcup_{i=1}^{\infty} K'_i = \mathbf{R}^d$. By the Riesz theorem there exists a sequence of countably additive measures $\nu^i_t(\omega,\cdot)$ on K'_i such that for every $\Gamma\in\mathcal{B}(\mathbb{R}^d)$,

$$\nu^n_t(\Gamma\cap K_n\cap K'_n) = \mu^n_t(\Gamma\cap K_n\cap K'_n) \qquad \forall n\in\mathbb{N}. \tag{1.11}$$

Making use of ν^n_t we define a measure ν'_t in the same way as we did it for μ'_t. From (1.11) and countable additiveness of both measures it follows that for all $\Gamma\in\mathcal{B}(\mathbb{R}^d)$ and $\mathbb{P}$-a.a. ω

$$\mu'_t(\omega,\Gamma) = \nu'_t(\omega,\Gamma).$$

Note also that by definition for all $t\in[0,\Gamma]$ and $\Gamma\in\mathcal{B}(\mathbf{R}^d)$, $\mu_t(\omega,\Gamma)$ is $\mathfrak{F}_t$-measurable.

Suppose now that $\psi\in\mathbb{L}_\infty$ and $\{\psi_n\}$ is a sequence of functions from $\mathbf{C}^0_b(\mathbf{R}^d)$ with compact supports and such that $\psi_n(x)\uparrow\psi(x)$ for every $x\in\mathbf{R}^d$.

From(1.10) we get that

$$\mathbb{E}[\psi_n(\mathfrak{S}(t))\rho(t)|\,\mathfrak{F}_t] = \int_{\mathbf{R}^d} \psi_n(x)\,\mu_t(\omega,dx) \quad (\mathbf{P}\text{-}a.s.) \tag{1.12}$$

By definition, for a bounded smooth function ψ, $\mu_t[\psi(\cdot)]$ has a version which is continuous in t. The same property for an arbitrary bounded function ψ can be proved exactly as we proved the weak continuity of a solution of problem (5.3.1), (5.3.2) in §5.9.6.

This together with (1.12) yield that there exists a version of $\mu_t[\psi(\cdot)]$ which is continuous in t and $\mathfrak{F}_t$-adapted, and consequently $\mathcal{P}(\mathfrak{F})$-measurable (see [100], Ch. 4. §2 Theorem 22).

On the other hand, by the well known property of conditional expectation (see e.g. [35] Ch.1, §3), if φ: $[0,T]\times\mathbf{R}^d\times\Omega \to \mathbf{R}^1$ is a predictable (for every $x\in\mathbf{R}^d$) function which belongs to $\mathbf{C}^0_b(\mathbf{R}^d)$ for all t,ω then for every t, ($\mathbf{P}$-a.s.),

$$\begin{aligned}&\mathbb{E}[\varphi(t,\,\mathfrak{S}(t))\rho(t)|\,\mathfrak{F}_t] = \\ &= \mathbb{E}[\varphi(t,\,\mathfrak{S}(t),\,\omega_0)\rho(t)|\,\mathfrak{F}_t]|_{\omega_0=\omega}.\end{aligned} \tag{1.13}$$

From this, in view of (1.12), we obtain that for evey t, ($\mathbf{P}$-a.s.)

$$\mathbf{E}[\varphi(t, \mathfrak{X}(t))\rho(t)|\ \mathfrak{F}_t]\ (\omega) = \int_{\mathbf{R}^d} \varphi(t,x,\omega)\mu_t(\omega,dx). \tag{1.14}$$

It is clear that the right-hand side of (1.14) is a predictable stochastic process.

From (1.13), (1.14) it follows that μ_t satisfies equality (1.3). The equality $\mu_t[1] = \mathbf{E}[\rho(t)|\mathfrak{F}_t]$, ($\mathbf{P}$-a.s.), gives that $\mu_t(\omega,\mathbf{R}^d) \in \mathbf{L}_2([0,T]\times\Omega;\ \mathbf{R}^d)$. Thus the theorem is proved. □

We note that under wide assumptions a natural uniqueness theorem for a measure-valued solution of problem (1.1), (1.2) is also valid. This theorem is beyond the scope of the book. However, we shall prove later that under some additional hypotheses, this measure has a density satisfying equality (5.3.1). The uniqueness of a solution of equation (5.3.1) (Theorem 4.2.1) implies the uniqueness of the measure-valued solution of problem (1.1), (1.2) in the class of measures having density in $\mathbf{L}_2^\omega([0,T];\ \mathcal{P}(\mathfrak{F});\ \mathbf{H}^1)$.

1.1. In this paragraph we introduce some additional hypotheses on the structure of the coefficients of equation (1.1) and formulate the main result of the section.

We suppose that ${}_YB^i(y)$ and ${}_Y\Sigma^{il}(y)$, where $i,l=1,2,\dots,d_1$ are functions from $\mathbf{C}_b^\infty(\mathbf{R}^{d_1})$.

Denote by $\mathcal{Y}(t)$ the diffusion process which is a solution of the following system of Ito's equations

$$\mathcal{Y}^i(t) = y_0^i + \int_{[0,t]} {}_YB^i(\mathcal{Y}(s))ds + \int_{[0,t]} {}_Y\Sigma^{il}(\mathcal{Y}(s))dw^l(s),\ t\in[0,T],\ y_0\in\mathbf{R}^d. \tag{1.15}$$

The following hypothesis is assumed to be valid throughout the chapter.

(D) *Every coefficient* $F(t,x,\omega)$ *of equation* (1.1) *has the form* $\tilde F(x,\mathcal{Y}(t,\omega))$, *where* $\tilde F(z)$ is a function from $\mathbf{C}_b^\infty(\mathbf{R}^{d+d_1})$.

Write $Z(t) := (\mathfrak{X}(t), \mathfrak{Y}(t))$. We consider this process as a solution of the following Ito's system

$$(1.16) \qquad Z^i(t) = z_0^i + \int_{[0,t]} V^{i0}(Z(s))ds + \int_{[0,t]} \hat{V}^{il}(Z(s))d\nu^l(s),$$

where

$$V^{i0}(z) := \begin{cases} B^i(z) & \text{for } i=1,2,\dots,d \\ {}_YB^i(y) & \text{for } i=d+1,\dots,d+d_1 \end{cases}$$

$$\hat{V}^{il}(z) := \begin{cases} \Sigma^{il}(z) & \text{for } i=1,2,\dots,d,\ l=1,2,\dots,d_0+d_1 \\ 0 & \text{for } i=d+1,\dots,d+d_1,\ l=1,2,\dots,d_0 \\ {}_Y\Sigma^{(i-d)l}_{(y)} & \text{for } i=d+1,\dots,d+d_1,\ l=d_0+1,\dots,d_0+d_1. \end{cases}$$

We write also

$$\hat{V}^{i0}(z) = V^{i0}(z) - \frac{1}{2}\, \hat{V}^{il}_j(z)\, \hat{V}^{jl}(z), \quad i=1,2,\dots,d,$$

and

$$V^{il}(z) = \hat{V}^{il}(z), \quad i=1,2,\dots,d+d_1,\ l=1,2,\dots,d_0$$

Let $X^1,\dots,X^m$ be mappings from $\mathbf{R}^d$ into $\mathbf{R}^d$ and suppose that their coordinates $X^{ij}(x)$, $i=1,2,\dots,d$, $j=1,2,\dots,m$ belong to $C^\infty(\mathbf{R}^d)$. We denote by $l(X^1,\dots,X^m)$ the linear space generated by $X^1,\dots,X^m$. In accordance with geometric terminology we call the functions X^i vector fields. Denote by $[X^i,X^j]$ the vector field with coordinates $[X^i,X^j]^l = (X^{ki}X^{lj}_k - X^{kj}X^{li}_k)$, $l=1,2,\dots,d$. The vector field $[X^i,X^j]$ is usually called the Lie bracket of the vectors fields X^i and X^j. The linear field $l(X^1,\dots,X^m)$ equipped with the operation of multiplication defined by the formula $XY = [X,Y]$ became an algebra over the field of real numbers. Clearly, the multiplication operator defined in such a way has the following properties

(i) $[Y,X] = -[X,Y]$ (anti-symmetry)

(ii) $[[X,Y],Z] + [[Y,Z],X] + [[Z,X], Y] = 0$ (Jacoby's identity)

Algebras possessing these properties are usually referred to as Lie algebras.

Definition 1. *An algebra* $l(X^1,...,X^m)$ *with the multiplication operation* $X^iX^j := [X^i,X^j]$ *is called a Lie algebra generated by the vector fields* $X^1,...,X^m$. □

In the future we shall use the following hypothesis.
(H_1) (*generalized Hormander's condition*): *The linear space generated by the vector fields* $V^1,...,V^{d_0}$ *and by Lie brackets of the vector fields* $\hat{V}^0,...,\hat{V}^{d_0+d_1}$ *at least one of which coincides with some* V^i, for $i=1,2,...,d_0$* *has at the point* x_0 *dimension equal to d.* □

Clearly, if the rank of the matrix $\{V^{il}\}$, $i=1,2,...,d$, $l=1,2,...,d_0$, is equal to d, then the generalized Hormander's condition fulfilled. Getting ahead of the story, we note that if the matrix VV^* is uniformly non-degenerate, then equation (1.1) is superparabolic and possesses the growth of smoothness property (see §7.2).

If $\mathcal{M}^l \equiv 0$, ${}_Y\Sigma \equiv 0$, and ${}_YB \equiv 0$, then equation (1.1) is transformed into the forward Kolmogorov's equation. In this case, hypothesis (H_1) coincides with the restricted Hormander's condition for a deterministic second-order elliptic-parabolic equation, which is as follows:
(H_2) *The dimension of the linear space generated by the vector fields* $V^1,...,V^{d_0}$, $[\hat{V}^0,V^1],....,[\hat{V}^0,V^{d_0}]$ at the point x_0 is equal to d. □

If $d_0=d=1$ the hypothesis (H_2) is equivalent to this one:
At least one of the following two conditions has to be fulfilled:

(i) $V^{11}(x_0) \neq 0$, (*ii*) $\hat{V}^{10}(x_0) \dfrac{\partial^n}{\partial x^n} V^{11}(x_0) \neq 0$ for some n.

Warning 1. *All the assumptions and notation of this paragraph will be in force throughout the section, in particular the generalized Hörmander's condition* (H_1) *is assumed to be fulfulled.* □

**i.e. by Lie brackets of the form* $[\hat{V}^{i_1}[\hat{V}^{i_2},...,[\hat{V}^{i_m},\hat{V}^{i_{m+1}}],...]$, where at least one of the vector fields $\hat{V}^{i_k}$ is identical to V^l for some $l=1,2,...,d_0$.

Theorem 1. A measure-valued solution of problem (1.1), (1.2) *is absolutely continuous with respect to Lebesgue measure for* $t\in]0,T]$ *and has density* $u(t,x,\omega)\in \mathbf{C}^{0,\infty}(]0,T]\times\mathbf{R}^d)$ *for* $\mathbf{P}$-*a.a.* ω. □

The idea of the proof is based on the following simple statement.

Proposition 1. Let μ_t *be a measure-valued solution of problem* (1.1), (1.2). *If for some* $t\in]0,T]$, *every* $f\in \mathbf{C}_b^\infty(\mathbf{R}^d)$ *and every (finite) multi-index* α

$$\left|\mu_t[D^\alpha f(\cdot)]\right| := \left|\int_{\mathbf{R}^d} D^\alpha f(x)\,\mu_t(\omega,dx)\right| \le$$

$$\le N_\alpha(t,\omega)\,\|f\|_{\mathbf{C}_b(\mathbf{R}^d)}, \tag{1.17}$$

where $N_\alpha(t,\omega)<\infty$, $\mathbf{P}$-*a.s. Then for* $\mathbf{P}$-*a.s.* ω *the measure* $\mu_t(\omega,\cdot)$ *is absolutely continuous with respect to Lebesgue measure and has density* $u(x_0,t,x,\omega)\in\mathbf{L}_2\,(\mathbf{R}^d)$ $\cap\ \mathbf{C}_b^\infty(\mathbf{R}^d)$ *for* $\mathbf{P}$-*a.a.* ω. □

Proof. Denote by $\tilde\Omega$ a subset of Ω, where $\mu_t[1]<\infty$, $N_\alpha(t,\omega)<\infty$, and (1.17) holds. By our assumptions, $P(\tilde\Omega)=1$. We denote by $(\cdot,\cdot)$ the scalar product in $\mathbf{R}^d$.

Let us fix some $\omega\in\Omega$. From (1.17) it follows that for every $k\in\mathbf{N}$ and $\lambda\in\mathbf{R}^d$

$$\left|((\lambda^1)^{2k}+\ldots+(\lambda^d)^{2k})\mu_t[e^{i(\cdot,\lambda)}]\right| \le N<\infty,$$

and consequently

$$\left|\mu_t[e^{i(\cdot,\lambda)}]\right| \le N\Big/\Big(\sum_{i=1}^d(\lambda^i)^{2k}\Big)$$

From this, making use of the inequality $|\mu_t[e^{i(\cdot,\lambda)}]|\le\mu_t[1]$, we obtain

$$|[\mu_t[e^{i(\cdot,\lambda)}]| \le (N+\mu_t[1])\Big/(1+\sum_{i=1}^d(\lambda^i)^{2k}). \tag{1.18}$$

(To prove the last inequality it will be helpful to consider two cases, namely

$N > \sum_{i=1}^{d} (\lambda^i)^{2k}\ \mu_t[1]$ and $N \leq \sum_{i=1}^{d} (\lambda^i)^{2k}\ \mu_t[1]$.)

Since k was arbitrary, from (1.18) it follows that $\mu_t[e^{i(\cdot,\lambda)^{2k}}] \in \mathbf{L}_2(\mathbf{R}^d)$ and therefore the following inverse Fourier transform is well defined:

$$u(x_0,t,x) := (2\pi)^{-d} \int_{\mathbf{R}^d} e^{-i(\lambda,x)} \mu_t[e^{i(\cdot,\lambda)}]d\lambda.$$

Making use of the inversion formula for a Fourier transform we easily obtain that $u(x_0,t,\omega)$ is a density for the measure $\mu_t(\omega,\cdot)$ with respect to Lebesgue measure.

In view of (1.18), $u(x_0,t,x)$ is a bounded, infinitely differentiable function and all the derivatives are also bounded.

Hence, the problem of a smooth density of a measure-valued solution is reduced to the proof of estimate (1.17).

Remark 1. *By Theorem* 0, $\mu_t[f(\cdot)] = \mathbf{E}[f(\mathfrak{X}(t))\rho(t)|\ \mathfrak{F}_t]$ ($\mathbf{P}$-*a.s.*). *Therefore, to prove estimate* (1.17) *it suffices to verify that for every* $\mathfrak{F}_t$*-measurable random variable* θ *and every* $f \in \mathbf{C}_b^\infty(\mathbf{R}^d)$ *the equality*

$$\mathbf{E}[\mathfrak{D}^\alpha f(\mathfrak{X}(t))\rho(t)\theta] = \mathbf{E}[f(\mathfrak{X}(t))\rho(t)G_\alpha(t)\theta], \tag{1.19}$$

where $G_\alpha(t)$ *is a* $\mathfrak{F}_t$*-measurable random variable such that* $\mathbf{E}|G_\alpha(t)|^2 < \infty$, *holds.*

In fact, we shall prove more than that. In the remaining part of the section formula (1.19) will be proved with G satisfying the inequality

$$\mathbf{E} \sup_{t\in[\varepsilon,T]} |G_\alpha(t)|^p < \infty \qquad \forall \varepsilon \in]0,T],\ p \geq 0.$$

From these relations it follows that

$$\mu_t[f(\cdot)] \leq \|f\|_{\mathbf{C}_b(\mathbf{R}^d)} \mathbf{E}[|\rho(t)G_\alpha(t)|\ |\ \mathfrak{F}_t]$$

and

$$\mathbf{E} \sup_{t\in[\varepsilon,T]} (\mathbf{E}[|\rho(t)G_\alpha(t)|\ |\ \mathfrak{F}_t])^p < \infty$$

$\forall \varepsilon \in]0,T], \ p \geq 0.$ □

Since

$$\mathbf{E} \sup_{t\in[0,T]} |\mu_t[1]|^p < \infty \qquad \forall p \geq 0,$$

we can consider the numerator of the right-hand side of (1.18) as an element of $\bigcap_{p\geq 0} \mathbf{L}_p(\Omega;\ \mathbf{C}([\varepsilon,T];\ \mathbf{R}^1))$.

Thus in view of continuity of μ_t in t (**P**-a.s.) we get that $\mathfrak{D}^\alpha u(x_0,t,x)$ is a continuous (in t, **P**-a.s.) function for every α, and

$$\mathbf{E} \sup_{t\in[\varepsilon,T]} |u(x_0,t,x)|^p < \infty \qquad \forall \varepsilon \in]0,T],\ p \geq 0.$$

This yields in particular that $u(x_0,t,x) \in \mathbf{C}^{0,\infty}(]0,T] \times \mathbf{R}^d)$. □

Evidently, it suffices to prove the theorem in the case of non-random x_0 and y_0 which we now assume.

1.2. In this item we prove a stochastic integration by parts formula. It will be used later as a main technical device in the proof of Theorem 1.1.

We consider a diffusion process $\xi(t)$ in $(\mathbf{R}^r)$ satisfying the equation

$$\text{(1.20)} \qquad d\xi(t) = U^0(t,\xi(t),\omega)dt + U^l(t,\xi(t),\omega)d\nu^l(t),\ t \in]0,T],$$

$$\text{(1.21)} \qquad \xi(0) = \xi_0 \in \mathbf{R}^r.$$

on a standard probability space $(\Omega,\ \mathcal{F},\ \{\mathcal{F}_t\},\ \mathbf{P})$, where $\nu(t)$ is the same Wiener process in $\mathbf{R}^{d_0+d_1}$ as that in previous items.

Definition 2. A diffusion process of the form (1.20), (1.21) *will be called a standard diffusion process in* $\mathbf{R}^r$ *if its coefficients* $U^{ij}(t,x,\omega)$ *are* $\mathcal{B}([0,T]\times\mathbf{R}^r)\otimes\mathcal{F}$-*measurable, predictable (for every* x, *relative to* $\mathcal{F}_t$) *and belonging to* $\mathbf{C}^\infty(\mathbf{R}^p)$ *for every* t,ω, *and their derivatives* $U_k^{ij}(t,x,\omega) \in \mathbf{C}_b^\infty(\mathbf{R}^r)$ *for every* t,ω, *and for every* $p>0$ *satisfy the inequality*

$$\sum_{i=1}^{d} \sum_{j=0}^{d_0+d_1} \int_{[0,T]} \mathbf{E}|U^{ij}(t,0)|^p dt < \infty. \ \square$$

Remark 2. *In the future it will be used heavily that every standard diffusion process* ξ *is in* $\bigcap_{p\geq 0} \mathbf{L}_p(\Omega;\ \mathbf{C}([0,T];\ \mathbf{R}^r))$ (*see Theorem* 1.4.3). □

Warning 2. *Throughout this item it is assumed that* $\xi(t)$ *is a standard diffusion process.* □

Let $g(t)$ be a stochastic process belonging to $\mathbf{L}_p([0,T]\times\Omega;\ \mathcal{P}(\mathfrak{F});\ \mathbf{R}^r\times\mathbf{R}^{d_0})$ for every $p\geq 0$. Denote by ε a fixed vector from $\mathbf{R}^r$.

In addition to the process $\xi(t)$, we also consider the process $\xi^{\varepsilon}(t)$ given by the following system

$$d\xi^{\varepsilon}(t) = U^0(t,\xi^{\varepsilon}(t))dt + U^l(t,\xi^{\varepsilon}(t))\ (d\hat{w}^l(t) + \tag{1.22}$$

$$+\ g^{il}(t)\varepsilon^i dt) + U^{d_0+l}(t,\xi^{\varepsilon}(t))dw^l(t),\ t\in]0,T],$$

$$\xi^{\varepsilon}(0) = \xi_0. \tag{1.23}$$

Obviously the process $\xi^{\varepsilon}(t)$ is also a standard diffusion process in $\mathbf{R}^r$. By an arguments similar to that in [63] Ch.II, §8, it is easy to verify that the process $\xi^{\varepsilon}(t)$ has a version (we shall identify $\xi^{\varepsilon}(t)$ with this version in the future), which

is differentiable in ε ($\mathbf{P}$-a.s.) and the derivative $\xi^{\varepsilon}_{gk}(t) = \frac{\partial}{\partial\varepsilon^k}\ \xi^{\varepsilon}(t)$ satisfies the equation

$$(\xi_{gk}(t))^i = \int_{[0,t]} U^{i0}_j(s,\xi^{\varepsilon}(s))\ (\xi^{\varepsilon}_{gk}(s))^j ds +$$

$$+ \int_{[0,t]} U^{il}_j(s,\xi^{\varepsilon}(s))\ (\xi^{\varepsilon}_{gk}(s))^j\ (d\hat{w}^l(s) +$$

$$\varepsilon^i g^{il}(s)ds)) + \int_{[0,t]} U^{il}(s,\xi^{\varepsilon}(s))g^{kl}(s)ds +$$

$$+ \int_{[0,t]} U^{i(d_0+l)}_j(s,\xi^{\varepsilon}(s))(\xi^{\varepsilon}_{gk}(s))^j dw^l(s),\ t\in[0,T],\ i=1,2,...,r.$$

Assume for the moment that $g(t)$ is a bounded function. Then by Girsanov's theorem (Theorem 1.4.6) the measures generated by the process $\xi^{\varepsilon}(t)$ and $\xi(t)$ are mutually absolutely continuous. Define by $\frac{d\mu}{d\mu^{\varepsilon}}(\xi^{\varepsilon})$ the density of the measure generated by μ with respect to the one generated by μ^{ε}. Let φ be a Borel mapping of $(\mathbf{C}[0,T];\ \mathbf{R}^r)\times\Omega;\ \mathfrak{B}(\mathbf{C}([0,T];\ \mathbf{R}^r))\otimes\mathfrak{F}$ to $\mathbf{R}^1$ such that $\mathbf{E}|\varphi(\xi)|<\infty$.

Proposition 2. (*Stochastic Integration by Parts Formula*). *If* $g\in\mathbf{L}_\infty([0,T];\ \mathbf{R}^1)$, *and at the point* ε *there exist derivatives* ($\mathbf{P}$-*a.s.*) $\frac{\partial}{\partial\varepsilon^k}\varphi(\xi^{\varepsilon})$ and $\frac{\partial}{\partial\varepsilon^k}\left(\frac{d\mu}{d\mu^{\varepsilon}}(\xi^{\varepsilon})\right)$ *such that for some* $p>1$

$$\sup_{|\gamma|_r<1}\mathbf{E}\Big|\Big(\varphi(\xi^{\varepsilon+\gamma})\frac{d\mu}{d\mu^{\varepsilon}}(\xi^{\varepsilon+\gamma})-\varphi(\xi^{\varepsilon})\frac{d\mu}{d\mu^{\varepsilon}}(\xi^{\varepsilon})\Big)\Big/|\gamma|_r\Big|^p<\infty \tag{1.25}$$

then

$$\mathbf{E}\Big[\Big(\frac{\partial}{\partial\varepsilon^k}\varphi(\xi^{\varepsilon})\frac{d\mu}{d\mu^{\varepsilon}}(\xi^{\varepsilon})+\varphi(\xi^{\varepsilon})\frac{\partial}{\partial\varepsilon^k}\Big(\frac{d\mu}{d\mu^{\varepsilon}}(\xi^{\varepsilon})\Big)\Big]=0. \quad \square \tag{1.26}$$

Proof. Obviously

$$\mathbf{E}\varphi(\xi)=\mathbf{E}\Big[\varphi(\xi^{\varepsilon})\frac{d\mu}{d\mu^{\varepsilon}}(\xi^{\varepsilon})\Big].$$

Differentiating this equality in ε, we obtain formula (1.26). The validity of this procedure follows from the well known uniform integrability criterion (see e.g. [100] Ch.2, §2, Theorem 22). $\square$

Denote by $(\xi_g^{ik}(t)$ the i-th coordinate of the vector $\xi^{\varepsilon}_{gk}(t)$. The matrix $\{\xi_g^{ik}(t)\}$ as usual will be denoted $\xi_g(t)$.

Let $a(t,x,\omega)$ be a $\overline{\mathfrak{B}([0,T]\times\mathbf{R}^r)\otimes\mathfrak{F}}$-measurable predictable (relative to the family $\{\mathfrak{F}_t\}$) for every x mapping of $[0,T]\times\mathbf{R}^r\times\Omega$ to $\mathbf{R}^{d_1}$ belonging to $\mathbf{C}^1_b(\mathbf{R}^r)$ for every t,ω, and θ_t be a bounded, $\mathfrak{F}_t$-measurable random variable.

We write

$$e(t):=\theta_t\ exp\{\int_{[0,t]}a^l(s,\xi(s))dw^l(s)-1/2\int_{[0,t]}a^la^l(s,\xi(s))ds\},$$

$$e^k(t) := \int_{[0,t]} a_i^l(s, \xi(s))\, \xi_g^{ik}(s)\, dw^l(s)$$

$$- \int_{[0,t]} a^l a_i^l(s, \xi(s))\, \xi_g^{ik}(s)\, ds,$$

and

$$H^k(t) := \int_{[0,t]} g^{kl}(s)\, d\hat{w}^l(s).$$

Lemma 2. *Let* $f \in \mathbb{C}^1(\mathbb{R}^r)$ *and* $|f(x)| + \sum_{k=1}^{r} |f_k(x)| \le N(1+|x|_r^p)$ *for some* $p>0$ *and* $N<\infty$. *Then the following formula holds:*

$$\mathbb{E}[f_i(\xi(t))\, \xi_g^{ik}(t)\, e(t)] =$$

$$= \mathbb{E}[f(\xi(t))\, (e^k(t) - H^k(t))\, e(t)]. \quad \square \tag{1.27}$$

Proof. We shall prove it in two steps.

Step 1. First, we prove formula (1.27) under the additional assumptions that $g \in \mathbb{L}_\infty([0,T]; \mathbb{R}^{d_0})$ and $f \in \mathbb{C}_b^1(\mathbb{R}^d)$. For this purpose we use Proposition 2.

We define the functional φ by the equality

$$\varphi(\xi) := f(\xi(t))\, e(t). \tag{1.28}$$

By Girsanov's theorem (Theorem 1.4.6)

$$\frac{d\mu}{d\mu^\varepsilon}(\xi^\varepsilon) = exp\{ - \int_{[0,T]} \varepsilon^j g^{jl}(s)\, d\hat{w}^l(s) -$$

$$- \frac{1}{2} \int_{[0,T]} \sum_{l=1}^{d_0} (\varepsilon^j g^{jl})^2\, ds\}.$$

Informally applying (1.26) at the point $\varepsilon=0$ to functional (1.28) we obtain (1.27).

The differentiability of $\varphi(\xi^{\varepsilon})$ and $\frac{d\mu}{d\mu^{\varepsilon}}(\xi^{\varepsilon})$ in ε is clear.

Therefore to justify the application of the proposition, it suffices to verify that the functional (1.28) satisfies inequality (1.25). The proof can be reduced to establishing several inequalities of the type

$$\mathcal{T} := \sup_{|\alpha|_r \le 1} \mathbf{E}|f(\xi^{\gamma}(t)) \frac{d\mu}{d\mu^{\gamma}}(\xi^{\gamma})\, e^{\gamma}(t) - \tag{1.29}$$

$$- e(t))\, |\gamma|_r^{-1}|^{p} < \infty, \qquad \forall p<1,$$

where $e^{\gamma}(t)$ is obtained by substituting the process $\xi^{\gamma}(t)$ for $\xi(t)$ in $e(t)$.

To prove inequality (1.29) we use Hadamard formula

$$e^{\gamma}(t) - e(t) = \int_{[0,1]} \frac{\partial}{\partial \varepsilon^{i}} e^{\varepsilon}(t)\Big|_{\varepsilon=\tau\gamma} \times \gamma^{i} d\tau.$$

By this formula we get

$$\mathcal{T} \le \sup_{|\alpha|_r \le 1} \int_{[0,1]} \mathbf{E}|f(\xi^{\gamma}(t)) \frac{d\mu}{d\mu^{\gamma}}(\xi^{\gamma})\, e^{\tau\gamma}(t) \times$$

$$\times \Big(\int_{[0,t]} a_i^{l}(s, \xi^{\tau\gamma}(s))\, \xi_{gk}^{\varepsilon}(s)|_{\varepsilon=\tau\gamma} \times dw^{l}(s) -$$

$$- \int_{[0,t]} a^{l} a_i^{l}(s, \xi^{\tau\gamma}(s))\, \xi_{gk}^{\varepsilon}(s)|_{\varepsilon=\tau\gamma} ds)|^{p} d\tau.$$

From Theorem 1.4.3 it follows that for every $p>0$

$$\sup_{|\varepsilon|_r \le 1} \sup_{t\in[0,T]} \mathbf{E}|\xi_{gk}^{\varepsilon}(t)|_r^{p} < \infty,$$

and consequently $\mathcal{T} < \infty$.

Arguing in a similar way the reader will easily complete the proof of the formula (1.27) (under the additional assumptions made above).

Step 2. We now give up the assumptions $g \in \mathbf{L}_\infty([0,T]; \mathbf{R}^{d_0})$ and $f \in \mathbf{C}_b^1(\mathbf{R}^d)$.
Let $\zeta(x)$ be the function defined in §1.4.10, R and N be elements of $\mathbf{R}_+$. Write

$f^R(x) := f(x)\zeta(x/R)$ and $g_N^{il}(t) := (g^{il}(t) \wedge N) \vee (-N)$. These functions satisfy the

boundedness hypothesis made on the first step. Hence formula (1.27) is valid for this functions and looks in this case as follows

$$\mathcal{T} := \mathbf{E}[f_i(\xi(t))\ \zeta(\xi(t)/R)\ \xi_{g_N}^{ik}(t)\ e(t)] =$$

$$= - \mathbf{E}[f(\xi(t))R^{-1}\ \zeta_i(\xi(t)/R)\ \xi_{g_N}^{ik}(t)\ e(t)] +$$
$$+ \mathbf{E}[f^R(\xi(t)\ (e_N^k(t) - H_N^k(t))\ e(t)] := \mathcal{T}_1 + \mathcal{T}_2,$$

where e_N^k, $\xi_{g_N}^{ik}$, and H_N^k are obtained from e^k, ξ_g^{ik}, and H^k, respectively, by substituting g_N for g. As we have already mentioned (see Remark 2)

$$\mathbf{E} \sup_{t \in [0,T]} |\xi(t)|_r^p < \infty, \tag{1.30}$$

and

$$\mathbf{E} \sup_{t \in [0,t]} |\xi_g(t)|_{r^2}^p < \infty. \tag{1.31}$$

Thus all the terms in (1.27) are well defined.
It is also clear that

$$\lim_{R \to \infty} \zeta(\xi(t)/R) = 1, \qquad (\mathbf{P}\text{-}a.s.),$$

and

$$\lim_{R\to\infty} R^{-1}\, \zeta_i(\xi(t)/R) = 0 \qquad (\mathbb{P}\text{-}a.s.)$$

From this, in view of (1.30), (1.31) and by the dominated convergence theorem it follows that

$$\lim_{R\to\infty} \mathcal{T} = \mathbb{E}[f_i(\xi(t))\, \xi^{ik}_{g_N}(t)\, e(t)],\quad \lim_{R\to\infty} \mathcal{T}_1 = 0$$

and

$$\lim_{R\to\infty} \mathcal{T}_2 = \mathbb{E}[f(\xi(t))\, (e^k_N(t) - H^k_N(t))\, e(t)].$$

Thus the equality

$$\mathbb{E}[f_i(\xi(t))\, \xi^{ik}_{g_N}(t)\, e(t)] = \tag{1.32}$$

$$= \mathbb{E}[f(\xi(t))\, (e^k_N(t) - H^k_N(t))\, e(t)]$$

is proved.

Making use of equation (1.24) for $\varepsilon=0$, the Burkholder-Davis inequality (Theorem 2.17) and Fubini's theorem we obtain the inequality

$$\mathbb{E} \sup_{s\in[0,t]} |\xi_g(s) - \xi_{g_N}(s)|^2_{r^2} \le$$

$$\le N\Big(\int_{[0,t]} \mathbb{E} \sup_{s\le t} |\xi_g(s) - \xi_{g_N}(s)|^2_{r^2} ds +$$

$$\int_{[0,t]} |g(s) - g_N(s)|^2_{rd_0}\, ds\Big)$$

From this by Gronwall-Bellman's lemma (§1.4.15) taking into account that

(1.33) $$\lim_{N\to\infty} \int_{[0,T]} |g(s) - g_N(s)|^2_{rd_0}\, ds = 0$$

we get

(1.34) $$\lim_{N\to\infty} \mathbb{E} \sup_{t\in[0,T]} |\xi_g(t) - \xi_{g_N}(t)|^2_{r^2} = 0.$$

Making use of (1.33) and (1.34) we pass to the limit (as $N\to\infty$) in (1.32) and obtain (1.27).

1.3. In this item we shall at last obtain the desired form of integration by parts formula and deduce from it an equality of the type of (1.19).

Let E_1 and E_2 be finite-dimensional Euclidean spaces. Denote by $C^n_D(E_1,E_2)$ the space of n times differentiable functions from E_1 into E_2 with the property: for every multi-index α, where $0 \le |\alpha| \le n$, there exist $N\in\mathbf{R}_+$ and $k\in\mathbf{N}$ such that $\|D^\alpha f(x)\|_{E_2} \le N(1 + \|x\|^k_{E_1})$.

Also let $M^{r\otimes r}$ be a linear space of $r\times r$-dimensional matrices with the usual operations of summation and multiplication on real numbers. Define a norm on $M^{r\otimes r}$ by the formula

$$\|u\|^2 = \sum_{i,j=1}^{r} |u^{ij}|^2.$$

We need two additional standard diffusion processes $\eta(t)$ and $\overline{\eta}(t)$ in $\mathbb{R}^n$ and $\mathbb{R}^{\overline{n}}$ respectively. Let us construct the process $\eta_g(t)$ corresponding to the processes $\eta(t)$ and let $\xi(t) := (\eta(t), \eta_g(t), \overline{\eta}(t))$.

We also assume that the following additional hypothesis is fulfilled (it will be of crucial importance in the proof):

(*L*) *There exists a matrix* $\{g^{il}\}$, $i=1,2,\dots,n$, $l=1,2,\dots,d_0$, *with elements from* $\bigcap_{p>0} \mathbb{L}_p([0,T]; \mathcal{P}(\mathfrak{F}); \mathbb{R}^1)$ *such that the process* $\eta_g(t)$ *is invertible and*

(1.35) $$\eta_g^{-1}(t) \in \bigcap_{p>0} \mathbb{L}_p(\Omega; M^{n\otimes n}) \text{ for all } t\in]0,T].$$

Warning 3. *Hypothesis* (*L*) *is assumed to hold throughout this section.* □

Let us fix a vector v from $\mathbf{R}^n$ and denote $D_v\psi(x) := \psi_{;i}(x)v^i$.

Theorem 3. Suppose that functions $\psi \in C^1_D(\mathbf{R}^n, \mathbf{R}^1)$ *and* $\varphi \in C^\infty_D(M^{n\otimes n} \times \mathbf{R}^{\overline{n}}, \mathbf{R}^1)$ *are given.*

There exist a standard diffusion process ${}_1\overline{\eta}(t)$ *taking values in* $\mathbf{R}^{\overline{n}_1}$ *and a function* ${}_1\varphi \in C^\infty_D(M^{n\otimes n} \times \mathbf{R}^{\overline{n}_1}; \mathbf{R}^1)$ *such that for every* $t \in [0,T]$ *the following equality*

$$\mathsf{E}[D_v\psi(\eta(t))\ \varphi(\eta^{-1}(t), \overline{\eta}(t))\ e(t)] =$$

$$= \mathsf{E}[\psi(\eta(t))_1\ \varphi(\eta^{-1}(t), {}_1\overline{\eta}(t))\ e(t)] \tag{1.36}$$

*holds.** □

Proof. Let S be an element of $\mathbf{C}^\infty_b(M^{n\otimes n} M^{n\otimes n})$ such that $S(y) = y^{-1}$ if $\|y^{-1}\| \leq 1$ and $S(y) = 0$ if $det(y) = 0$ or $\|y^{-1}\| \geq 2$. Given $R \in \mathbf{R}_+$, denote $S_R(y) = RS(Ry)$.

Evidently

$$\lim_{R\to\infty} S_R(y) = y^{-1} \tag{1.37}$$

Define $\varphi_R(y, \overline{y}) := S_R(y)\ \varphi(S_R(y), \overline{y})$ and $\varphi_\infty(y, \overline{y}) = y^{-1}\ \varphi(y^{-1}, \overline{y})$.

Letting $f(\xi(t)) := \psi(\eta(t))\ (\varphi_R(\eta_g(t), \overline{\eta}(t))v)^k$, where k is the same coordinate as in formula (1.27), we apply Lemma 3 to the process $\xi(t)$.

Denote by $\eta_{gg}(t)$ and $\overline{\eta}_g(t)$ the processes corresponding to the processes $\eta_g(t)$ and η(t), respectively, by a transformation similar to one connecting $\eta(t)$ and $\eta_g(t)$. Write $\hat{\eta}_g(t) := (\eta_{gg}(t), \overline{\eta}_g(t))$. Then by formula (1.27) we get

$$\begin{aligned} \mathcal{T} &:= E[\psi_{;i}(\eta(t))\ (\varphi_R(\eta_g(t), \overline{\eta}(t))v)^k e(t)\ \eta_g^{ik}(t)] = \\ &- \mathsf{E}[\psi(\eta(t))\varphi_R(\eta_g(t), \overline{\eta}(t)v)^k_{;i}\ e(t)\ \hat{\eta}_g^{ik}(t)] \\ &- \mathsf{E}[\psi(\eta(t))\ (\varphi_R(\eta_g(t), \overline{\eta}(t)))v)^k\ (e^k(t) \\ &- H^k(t))e(t)] = - \mathcal{T}_1 - \mathcal{T}_2. \end{aligned} \tag{1.38}$$

---

*For our purpose it suffices to assume that $a(t, \xi(t), \omega)$ depends only on the first n coordinates of ξ (i.e. on η).

Making use of the structure of S_R we can rewrite the left-hand side of the last equality as follows:

$$\mathcal{T} = \mathbf{E}[\psi_i(\eta(t))\, \varphi_\infty^{kj}(\eta_g(t), \overline{\eta}(t))v^j e(t) \times$$

$$\times\, 1_{\{\|\eta_g^{-1}(t)\| \leq R\}}\, \eta_g^{ik}(t) +$$

(1.39)

$$+\, \mathbf{E}[\psi_i(\eta(t))\varphi_R^{kj}(\eta_g(t), \overline{\eta}(t))v^j e(t) \times$$

$$\times\, 1_{\{R < \|\eta_g^{-1}(t)\|\}}\, \eta_g^{ik}(t)] = U_1 + U_2.$$

Since

$$|\varphi_R^{kj}(S_R(\eta_g(t)), \overline{\eta}(t))| \leq RC(1 + R^\beta + |\overline{\eta}(t)|_{\overline{n}}^{\beta},$$

we have

$$U_2 \leq N\, \mathbf{E}|\sum_{k=1}^{n} \psi_i(\eta(t))\eta_g^{ik}(t)R(1 + |\overline{\eta}(t)|_{\overline{n}}^{\beta})) \times$$

$$\times\, 1_{\{\|\eta_g^{-1}(t)\| > R\}}\, | + N\mathbf{E}|\, (\sum_{k=1}^{n} \psi_i(\eta(t))\eta_g^{ik}(t)R^{\beta+1}) \times$$

$$\times\, 1_{\{\|\eta_g^{-1}(t)\| > R\}} := U_{21} + U_{22}.$$

By the Chebyshev's and Schwarz's inequalities we get that

$$U_{21} \leq N[\mathbf{E}(\sum_{k=1}^{n} \psi_i(\eta(t))\eta_g^{ik}(t)\, (1 + |\overline{\eta}(t)|_{\overline{n}}^{\beta}))^2]^{\frac{1}{2}} \times$$

$$\times\, R\mathbf{P}(\|\eta_g^{-1}(t)\| > R) \leq N_1\mathbf{E}\|\eta_g^{-1}(t)\|^p / R^{p-1}$$

and

$$U_{22} \leq N[\mathsf{E}(\sum_{i=1}^{n} \psi_i(\eta(t))\eta_g^{ik}(t))^2]^{\frac{1}{2}} R^{\beta+1} \times$$

$$\times \mathsf{P}(\|\eta_g^{-1}(t)\| > R) \leq N_2 \mathsf{E}\|\eta_g^{-1}(t)\|^{\beta+p}/R^{p-1}.$$

Thus in view of hypothesis (L) we obtain

$$\lim_{R\to\infty} U_2 = 0 \tag{1.40}$$

By the same arguments

$$\lim_{R\to\infty} U_1 = \mathsf{E}[\psi_i(\eta(t))\ (\varphi(\eta_g^{-1}(t), \overline{\eta}(t))v)^k e(t)] \tag{1.41}$$

and

$$\lim_{R\to\infty} \mathcal{T}_2 = \mathsf{E}\ \psi(\eta(t))\ (\eta_g^{-1}(t)\varphi(\eta_g^{-1}(t), \overline{\eta}(t))v)^k \times \tag{1.42}$$

$$\times\ (e^k(t) - H^k(t))\ e(t)].$$

Consider next $\mathcal{T}_1$. Since $\eta_g(t)$ is a matrix-valued process it is convenient to rewrite $\mathcal{T}_1$ making use of double indices for the first n^2 coordinates, namely in the form

$$\mathcal{T}_1 = \mathsf{E}[\psi(\eta(t))\ \frac{\partial \varphi_R^{ki}}{\partial y^{jl}}\ (y, \overline{\eta}(t))v^i|_{y=\eta_g(t)} \times$$

$$\times\ e(t)\eta_{gg}^{jlk}(t)] + \mathsf{E}[\psi(\eta(t))\ \frac{\partial \varphi_R^{kj}}{\partial z^i}\ (\eta_g(t), z)v^j|_{z=\overline{\eta}(t)} \times \tag{1.43}$$

$$\times\ e(t)\overline{\eta}_g^{ik}(t)] := \mathcal{T}_{11} + \mathcal{T}_{12}.$$

Then taking into account the equality $\frac{\partial}{\partial y^{jl}} (y^{-1})^{mr} = -(y^{-1})^{mj} (y^{-1})^{lr}$ we get

$$\mathcal{T}_{11} = -\mathbf{E}[\sum_{m,r} \psi(\eta(t)) \frac{\partial}{\partial y^{mr}} (y\varphi(y, \overline{\eta}(t))^{ki}|_{y=\eta_g^{-1}(t)} \times$$

(1.44)
$$\times\, v^i (\eta_g^{-1}(t))^{mj} (\eta_g^{-1}(t))^{lr}\, e(t)\, \eta_{gg}^{jlk}(t) \times$$

$$\times\, 1_{\{\|\eta_g^{-1}(t)\| \le R\}}] - \mathbf{E}[\psi(\eta(t)) \frac{\partial \varphi_R^{ki}}{\partial y^{jl}} (y, \overline{\eta}(t))|_{y=\eta_g(t)} \times$$

$$\times\, v^i e(t) \eta_{gg}^j(t)\, 1_{\{R < \|\eta_g^{-1}(t)\|\}}].$$

From (1.43) and (1.44) with the help of the Schwarz's and Chebyshev's inequalities in the same way as above, we obtain

(1.45)
$$\lim_{R\to\infty} \mathcal{T}_1 = \mathbf{E}[\psi(\eta(t))\, (\varphi_\infty(\eta_g(t), \overline{\eta}(t))v)_i^k\, e(t) \hat{\eta}_g^{ik}(t)].$$

Combining (1.38) to (1.42) and (1.45), and letting $G^k(t) := e^k(t) - H^k(t)$, we find that

$$\mathbf{E}[\psi_i(\eta(t))(\eta_g^{-1}(t)\varphi(\eta_g^{-1}(t), \overline{\eta}(t))v)^k\, e(t) \eta_g^{ik}(t)] =$$

(1.46)
$$= \mathbf{E}[\psi(\eta(t))\, e(t)\, \{\sum_{m,r} \frac{\partial}{\partial y^{mr}} (y\varphi(y, \overline{\eta}(t)))^{ki}|_{y=\eta_g^{-1}(t)}\, v^i \times$$

$$\times\, (\eta_g^{-1}(t))^{mj} (\eta_g^{-1}(t))^{lr}\, \eta_{gg}^{jlk} + \frac{\partial \varphi}{\partial z^i} (\eta_g^{-1}(t), z)|_{z=\overline{\eta}(t)} \times$$

$$\times\, v^k \overline{\eta}_g^{ik}(t) - \eta_g^{-1}(t)\, (\varphi(\eta_g^{-1}(t), \overline{\eta}(t))v)^k G^k(t))\}].$$

Evidently, the left-hand side in equality (1.46) is the same as that in (1.36). On the other hand, since the processes $\eta(t)$ and $\overline{\eta}(t)$ are by definition standard diffusion process, the same is true for the process $(\overline{\eta}(t), \overline{\eta}_g(t), G(t), \eta_{gg}(t))$. Thus equality (1.46) gives us formula (1.36) with ${}_1\overline{\eta}(t) := (\overline{\eta}(t), \overline{\eta}_g(t), G(t), \eta_{gg}(t))$.

Hence the theorem is proved. □

Corollary 3. *Assume that* $f \in C_D^k(\mathbf{R}^n, \mathbf{R}^1)$ *and* α *is a multi-index such that* $|\alpha| = k$. *Then there exist a standard diffusions process* ${}_k\overline{\eta}(t)$ *taking values in* $\mathbf{R}^{\overline{n}_k}$ *and a function* ${}_\alpha\varphi \in C_D^\infty(M^{n\otimes n} \times \mathbf{R}^{\overline{n}_k}; \mathbf{R}^1)$ *such that for every* $t \in [0,T]$ *the following equality holds*

$$
\begin{aligned}
&\mathsf{E}[D^\alpha f(\eta(t))e(t)] = \\
&= \mathsf{E}[f(\eta(t))\ {}_\alpha\varphi(\eta_g^{-1}(t), {}_\alpha\overline{\eta}(t))e(t)]. \quad \square
\end{aligned}
\tag{1.47}
$$

Proof. We apply Theorem 3 to the functions $\psi = D^\alpha f$ and $\varphi \equiv 1$, and get

$$
\begin{aligned}
&\mathsf{E}[D^\alpha f(\eta(t))e(t)] = \\
&= \mathsf{E}[D^{\alpha_1} f(\eta(t))_{\alpha_1}\ \varphi(\eta_g^{-1}(t), {}_{\alpha_1}\overline{\eta}(t))e(t)]
\end{aligned}
\tag{1.48}
$$

where $|\alpha_1| = |\alpha| - 1$.

Applying this lemma once more to the right-hand side of the formula (1.48) yields

$$
\begin{aligned}
&\mathsf{E}[D^\alpha f(\eta(t))e(t)] = \\
&= \mathsf{E}[D^{\alpha_2} f(\eta(t))_{\alpha_2}\ \varphi(\eta_g^{-1}(t), {}_{\alpha_2}\overline{\eta}(t))e(t)],
\end{aligned}
$$

where $|\alpha_2| = |\alpha| - 2$.

After K sequential applications of Theorem 3 we obtain formula (1.47). □

1.4. In this paragraph we prove Theorem 1. As mentioned before, to prove this theorem it suffices to verify equality (1.19). To prove equality (1.19) we apply Lemma 3 taking $\mathfrak{X}(t)$ for $\eta(t)$ and $\rho(t)$ for $e(t)$. However we have to choose a function $g(t)$ such that the process $\mathfrak{X}_g(t)$ satifies hypothesis (L) of the previous paragraph.

Under the assumptions of the theorem the process $\mathfrak{X}(t)$ has a verions which is $\mathbf{C}^{0,\infty}$-diffeomorphism of $\mathbf{R}^d$ to $\mathbf{R}^d$ (see §5.2.1), and in the future we consider just this version. Hence the inverse matrix $(\mathfrak{X}'(t)^{-1}$, where $\mathfrak{X}'(t) = (\partial \mathfrak{X}^i(t)/\partial x_0^j)$ is well defined. The elements of the matrix $(\mathfrak{X}'(t))^{-1}$ will be denoted $D\mathfrak{X}^{ij}(t)$. This matrix is a solution of the following system

$$dD\mathfrak{X}^{jm}(t) = D\mathfrak{X}^{jk}(t)\ B^k_m(Z(t))\,dt\ -$$

$$- D\mathfrak{X}^{jk}(t)\ \Sigma^{kl}_m\ (Z(t))\,d\nu^l(t) + \tag{1.48}$$

$$+ D\mathfrak{X}^{jk}(t)\ \Sigma^{kl}_i\ \Sigma^{il}_m\ (Z(t))\,dt,\ t\in]0,T],$$

$$D\mathfrak{X}^{jm}(0) = \delta^{jm},\ i,m=1,2,\ldots,d \tag{1.49}$$

where δ^{jm} is the Kronecker symbol.

To prove (1.48), (1.49) we differentiate using Ito's formula the products $(D\mathfrak{X}^{jm}(t))\ \mathfrak{X}^m{}_i(t)$ and $\mathfrak{X}^j_m(t)\ D\mathfrak{X}^{mi}(t)$.

Differentiation of the first expression gives zero and of the second one gives a linear stochastic differential equation whose only solution is the identity matrix.

Applying Theorem 1.4.3 to (1.48) we obtain inequality

$$\sum_{j,k=1}^{d} \mathsf{E} \sup_{t\in[0,T]} |D\mathfrak{X}^{jk}(t)|^p<\infty, \qquad \forall p>0. \tag{1.50}$$

Define the function g in the following way

$$g(t):= ((\mathfrak{X}'(t))^{-1}\ \tilde{\Sigma}\ (Z(t)))^*$$

where $\tilde{\Sigma}$ is the matrix consisting of the first d_0 columns of Σ and $*$ is the symbol of conjugating.

Making use of this function, we define a process $\mathfrak{X}_g(t)$ corresponding to $\mathfrak{X}(t)$ in the same way as the process $\xi_g(t)$ corresponding to $\xi(t)$ was defined in §1.3.

Note that $\mathfrak{X}_g(t)$ can be represented in the form

(1.51) $$\mathfrak{X}_g(t) = \mathfrak{X}'(t) \int_{[0,t]} (\mathfrak{X}'(s))^{-1}\, \tilde{\Sigma}(Z(s))g(s)\,ds.$$

The reader will easily verify this equality by differentiating its right-hand side. Write

$$C_t = \int_{[0,t]} (\mathfrak{X}'(s))^{-1}\, \tilde{\Sigma}(Z(s))((\mathfrak{X}'(s))^{-1}\, \tilde{\Sigma}(Z(s))^*\,ds.$$

Consider next the matrix $Z'(t) = (Z^i_j(t))$. Recall that $Z = (\mathfrak{X},\mathfrak{Y})$. The process $\mathfrak{Y}(t)$ is differentiable in y_0, which can be proved in the same way as that of $\mathfrak{X}(t)$ in x_0. We note also that the matrix $Z'(t)$ has the following form

$$\begin{pmatrix} \mathfrak{X}' & \mathfrak{X}'_y \\ 0 & \mathfrak{Y}' \end{pmatrix},$$

where

$$\mathfrak{Y}'(t) := \left\{\frac{\partial \mathfrak{Y}^i(t)}{\partial y^j_0}\right\},\ \mathfrak{X}'_y(t) := \left\{\frac{\partial \mathfrak{X}^i(t)}{\partial y^j_0}\right\}$$

and 0 is a $d_1 \times d$-dimensional zero matrix. The same arguments used earlier for $\mathfrak{X}'(t)$ show the existence of the inverse of $\mathfrak{Y}'(t)$. Thus

$$(Z'(t)^{-1} = \begin{pmatrix} (\mathfrak{X}'(t))^{-1} & U(t) \\ 0 & (\mathfrak{Y}'(t))^{-1} \end{pmatrix}.$$

where

$$U(t) := -(\mathfrak{X}'(t))^{-1}\, \mathfrak{X}'_y(t)\, (\mathfrak{Y}'(t))^{-1}$$

Note that only the first d rows of the matrix V are non-zero. These rows coincide with the corresponding rows of $\tilde{\Sigma}$. Hence

$$(Z^l(t))^{-1} V^l(Z(t)) := \begin{pmatrix} (\mathfrak{S}'(t))^{-1} \tilde{\Sigma}^l (Z(t)) \\ 0 \end{pmatrix}.$$

Theorem 4. Given any $t \in]0,T]$ *for all* $\xi \in \mathbf{R}^d$ $\mathbf{P}(C_t^{ij}\xi^i\xi^j > 0) = 1$. □

Proof. By definition, for $t>0$, $C_t^{ij}\xi^i\xi^j \geq 0$ ($\mathbf{P}$-*a.s.*). Denote by U_s the linear space generated by the vectors $(\mathfrak{S}'(s))^{-1} \Sigma^l(Z(s))$, $l=1.2,\dots,d$, and let $\mathfrak{U}_t$ be a linear space spanned by $\bigcup_{s\leq t} U_s$, and let $\mathfrak{U}_t^+ := \bigcap_{s>t} \mathfrak{U}_s$. Take η to be a non-random vector from $\mathbf{R}^d$. It is clear that the event $\{\eta \in \mathfrak{U}_0^+)$ is measurable with respect to $\mathfrak{F}_0^0 = \bigcap_{t>0} \mathfrak{F}_t^0$ and consequently non-random. Suppose that the matrix C_t is degenerate with a positive probability. Then dim $\mathfrak{U}_0^+ < d$. Indeed, assume to the contrary that dim $\mathfrak{U}_0^+ = d$ and denote $\tau = inf(t\colon \mathfrak{U}_t \neq \mathfrak{U}_0^+)$. Because of the right-continuity of $\mathfrak{U}_t$, $\mathbf{P}(\tau>0)=1$. By our assumption, there exists a vector $\xi\colon \Omega \to \mathbf{R}^d$ such that with a positive probability

$$\xi^* C_t \xi = 0.$$

However

$$\xi^* C_t \xi = \sum_{l=1}^{d_0} \int_{[0,t]} |\xi^*(\mathfrak{S}'(s))^{-1} \tilde{\Sigma}^l(Z(s))|^2 \, ds,$$

and consequently with a positive probability the vector ξ is orthogonal to all the vectors $(\mathfrak{S}'(s))^{-1} \tilde{\Sigma}^l (Z(s))$, $s \leq t$.

This contradicts the assumption that dim $\mathfrak{U}_0^+ = d$. Let θ^d be a non-zero vector from $\mathbf{R}^d$ orthogonal to $\mathfrak{U}_0^+$. Add to it any other d_1 arbitrary coordinates and denote the resulting $(d+d_1)$-vector by θ.

Clearly

$$\theta^*(Z^l(t))^{-1} V^l(Z(t)) = 0 \qquad (\mathbf{P}\text{-}a.s.) \tag{1.52}$$

for $t \leq \tau$, $l=1,2,\dots,d_0$.

The same arguments as that used in the proof of (1.48), (1.49) yield that $(Z'(t))^{-1}$ satisfies the following equation

$$d(Z'(t))^{-1} = - (Z'(t))^{-1} \frac{\partial}{\partial z} V^0 (Z(t))dt$$

$$- (Z'(t))^{-1} \frac{\partial}{\partial z} \hat{V}^l(Z(t))d\nu^l(t) +$$

$$+ (Z'(t))^{-1} \frac{\partial}{\partial z} \hat{V}^l (Z(t)) (\frac{\partial}{\partial z} \hat{V}^l(Z(t)))^* dt, \ t\in]0,T]$$

$$(Z'(0))^{-1} = \mathbf{I}$$

where $\mathbf{I}$ is the identity matrix.

Applying Ito's formula, it is easy to verify that for every $f\in \mathbf{C}_b^2(\mathbf{R}^{d+d_1}, \mathbf{R}^{d+d_1})$ and the vector $\theta\in\mathbf{R}^{d+d_1}$ which was defined above, we get

$$(1.53) \qquad d(\theta^*(Z'(t))^{-1} f(Z(T))) = \theta^*(Z'(t))^{-1} \times$$

$$\times [\tilde{V}^0, f] (Z(t))dt + \theta^*(Z'(t))^{-1} [\hat{V}^l, f] (Z(t))d\nu^l(t),$$

where $\tilde{V}^0(z) = \hat{V}^0(z) + \frac{1}{2} \sum_{l=1}^{d_0+d_1} [\hat{V}^l[V^l, f]](z)$.

Assume that $f(z) = V^m(z)$, $m=1,2,\dots,d_0$. Then from (1.52), (1.53), and from the well-known properties of a stochastic integral (see, e.g. [90]) it follows that

$$(1.54) \qquad \theta^*(Z'(t))^{-1} [\hat{V}^l, V^m] (Z(t)) = 0, \qquad (\mathbf{P}\text{-}a.s.),$$

for $t\leq\tau$, $l=1,2,\dots,d_0+d_1$ and $m=1,2,\dots,d_0$ relation

$$(1.55) \qquad \theta^*(Z'(t))^{-1} [\hat{V}^l[\hat{V}^l, V^m]] (Z(t)) = 0, \qquad (\mathbf{P}\text{-}a.s.),$$

for $t \leq \tau$ and $m=1,2,\dots d_0$.

From (1.52) to (1.55) it follows that

$$\theta^*(Z'(t))^{-1}\,[\hat{V}^0, V^m]\,(Z(t)) = 0, \qquad (\mathbf{P}\text{-}a.s.) \tag{1.56}$$

for $t \leq \tau$ and $m=1,2,\dots,d_0$.

Thus it is proved that for every $l=0,1,\dots,d_0+d_1$, $m=1,2,\dots,d_0$, the vector θ is orthogonal to all the vectors $(Z'(t))^{-1}\ [\hat{V}^l, V^m]\ (Z(t))$ for $t \leq \tau$ ($\mathbf{P}$-$a.s.$). Now by induction it is easy to prove that for every $l_i=0,1,\dots,d_0+d_1$, $m=1,2,\dots,d_0$, the vector θ is orthogonal to all the vectors of the form

$$(Z'(t))^{-1}\ [\hat{V}^{l_1},[\hat{V}^{l_2},\dots,[\hat{V}^{l_n}, V^m]\dots]\ (Z(t))$$

for $t \leq \tau$ and $\mathbf{P}$-a.a. ω.

Thus, since $(Z'(0))^{-1} = \mathbf{I}$, it is follows that the vector θ orthogonal to $V^1(z_0),\dots, V^{d_0}(z_0)$ and to all the Lie brackets $[\hat{V}^{l_1},[\hat{V}^{l_2},\dots,[\hat{V}^{l_n}, V^m]\dots]\ (z_0)$, where $l_i=0,1,\dots,d_0+d_1$, and $m=1,2,\dots,d_0$.

This together with the anti-symmetry of the Lie brackets and Jacoby's identity implies that the vector θ is orthogonal to all the Lie brackets of the form

$$[\hat{V}^{l_1}[\hat{V}^{l_2},\dots,[\hat{V}^{l_n}, V^m],\ \hat{V}^{l_{n+1}}],\dots, \hat{V}^{l_k}]\ (z_0)$$

Clearly the last d_1 coordinates of every Lie bracket of such a form are equal to 0. Therefore the non-zero d-dimensional vector θ^d is orthogonal to the projections of $V^1(z_0),\dots, V^m(z_0)$ and $[\hat{V}^{l_1}[\hat{V}^{l_2},\dots, \hat{V}^{l_n}, V^m],\ \hat{V}^{l_{n+1}}],\dots, \hat{V}^{l_k}](z_0)$on $\mathbf{R}^d$. By the same arguments, the dimension of the linear space generated by these vectors is the same as that of the linear space generated by their projections on $\mathbf{R}^d$. Thus we get a contradiction to the generalized Hörmander's hypothesis (H_1).

From (1.51) and the theorem it follows that for every $t\in[0,T]$, $\mathfrak{S}_g(t)$ is invertible ($\mathbf{P}$-a.s.) and $\mathfrak{S}_g^{-1}(t) = C_t^{-1}(\mathfrak{S}'(t))^{-1}$.

1.5. In view of (1.50), the following statement completes the proof that the process $\mathfrak{S}(t)$ satisfies the hypothesis (L).

Lemma 5. (*Kusuoka, Stroock*). *For all* $p>0$ *and* $\varepsilon\in]0,T]$ $\sup_{t\in[\varepsilon,T]}\|C_t^{-1}\| \in \mathbf{L}_p(\Omega;\mathbf{R}^1)$. □

For the proof of the lemma see e.g. [136].

7.2. The Filtering Transition Density and a Fundamental Solution of the Filtering Equation in Hypoelliptic and Superparabolic Cases.

2.0. In Chapter 6 we have considered the existence of the density problem for the filtering measure of a diffusion process $X(t)$ given the path of the "observable" diffusion process $Y(t)$. We also investigated the structure of this density and equations for it (the so-called filtering equations). Throughout Chapter 6 it was supposed and used frequently that the conditional distribution of $X(0)$ with respect to $Y(0)$ is absolutely continuous with respect to Lebesgue measure and the density belongs to $\mathbb{L}_p(\Omega;\mathbb{P};\mathbb{H}^1)$.

In this section we consider the same problems but without the above assumption. In other words, in terms of the theory of Markov proceses we consider in this Chapter the conditional transition density instead of the conditional density investigated in Chapter 6. In terms of theory of differential equations that would mean that we consider the fundamental solution of the filtering equation.

Warning 0. *In what follows the assumptions and notation of Chapter* 6 *except the assumption discussed above in this paragraph are still in force.* □

As in Chapter 6 we consider the $(d+d_1)$-dimensional diffusion process $Z(t)$ governed by the system of Ito's equations

$$Z^i(t) = z_0^i + \int_{[0,t]} b^i(s,Z(s))ds + \int_{[0,t]} \Sigma^{il}(s,Z(s))d\nu^l(s),$$

$$i,\ l=1,2,\dots,d+d_1,\ t\in[0,T].$$

However, contrary to Chapter 6 it is supposed that $z_0 := (z_0^1,\dots,z_0^{d+d_1})$ is a non-random vector.

We denote by $X(t)$ (x_0) the first d coordinates and by $Y(t)$ (y_0) last d_1 ones of $Z(t)$ (z_0, respectively).

We consider $X(t)$ as a "signal" process and $Y(t)$ as "observations".

We are interested in the existence and analytical properties of the transition filtering density

$$\pi(0,x_0,T_1,x) := \mathbb{P}(X(T_1)\in dx|\ \mathcal{Y}^0_{T_1})/dx$$

as well as the transition interpolation and extrapolation densities.

2.1. In this item we deal with the existence problem for the fundamental solution of the direct linear filtering equation in the hypoelliptic case. To be more precise, the degeneration of matrix ${}_X\Sigma$ is considered to be possible but some additional assumptiones, namely the generalized Hörmander's condition (see §7.1) and the infinite differentiability of the coefficients b and Σ are made.

We need some extra notation.

Denote ${}_XB^i := {}_Xb^i - \sigma^{il}h^l$ and put

$$V^{io}(z) := \begin{cases} {}_XB^i(z) & for\ i=1,2,\ldots,d \\ 0 & for\ i=d+1,\ldots,d+d_1 \\ 1 & for\ i=d+d_1+1 \end{cases}$$

$$\hat{V}^{il}(z) = \begin{cases} \Sigma^{il}(z) & for\ i=1,2,\ldots,d+d_1 \\ 0 & for\ i=d,\ldots,d_1+1 \end{cases}$$

$$\hat{V}^{io}(z) := V^{io}(z) - \frac{1}{2} \sum_{l=1}^{d+d_1+1} \hat{V}_j^{il}(z)\, \hat{V}^{jl}(z)$$

for $i=1,2,\ldots,d+d_1+1$,

$$V^{il}(z) = \hat{V}^{il}(z),\ i=1,2,..,d+d_1,\quad l=1,2,\ldots,d.$$

Throught this section the following additional conditions are assumed to be fulfilled:

(A) *For $i=1,2,\ldots,d+d_1+1$, $l=1,2,\ldots,d+d_1$ the coefficients $\hat{V}^{il} \in \mathbf{C}_b^\infty(\mathbf{R}^{d+d_1+1})$.*

(H_3) (*Hörmander's type condition*). *At z_0 the dimension of the linear space generated by the vector fields $V^1,\ldots,V^d$ and the Lie brackets of the vector fields $\hat{V}^0,\ldots,\hat{V}^{d+d_1+1}$ among which there is at least one vector field V^i, $i=1,2,\ldots,d$, is equal to d.*

Just as in Chapter 6 we write

$$(\rho_t^s)^{-1} := exp\Big\{- \int_{[s,t]} h^l(r,Z(r))dw^l(r) - \frac{1}{2} \int_{[s,t]} h^l h^l(r,Z(r))dr\Big\},$$

where $h^l := {}_Y b^i \, {}_0\sigma^{-il}$ (see §6.0.2) and

$$\tilde{\mathbf{P}}(\Gamma) := \int_\Gamma (\rho_t^0)^{-1} d\mathbf{P}(\omega) \qquad \forall \Gamma \in \mathcal{F}.$$

We recall that on the standard probability space $\tilde{\mathbf{F}} := (\Omega, \mathcal{F}, \{\mathcal{F}_t\}_{t\in[0,T]}, \tilde{\mathbf{P}})$ there exists a standard Wiener process $\tilde{\nu}(t)$ such that the process $Z(t)$ considered on this space satisfies the following system of Ito's equations

$$(2.1) \qquad X^i(t) = x_0^i + \int_{[0,t]} {}_X B^i(s, Z(s))\,ds + \int_{[0,t]} {}_X \Sigma^{il}(s, Z(s))\,d\tilde{\nu}^l(s), \quad i=1,2,\dots d,$$

$$(2.2) \qquad Y^i(t) = y_0^i + \int_{[0,t]} {}_0\sigma^{il}(s, Y(s))\,d\tilde{w}^l(s), \quad i=1,2,\dots,d_1$$

where $\tilde{w}^l = \tilde{\nu}^{d+l}$ for $l=1,2,\dots,d_1$ (see §6.1.1).

Let us define the operators $\mathcal{L}^*$ and $\mathcal{M}^{l*}$ by the formulas

$$\mathcal{L}^*(t,x,Y(t))u := (a^{ij}(t,x,Y(t))u)_{ij} - ({}_X f^i(t,x,Y(t))u)_i,$$

and

$$\mathcal{M}^{l*}(t,x,Y(t))u := -(\sigma^{il}(t,x,Y(t))u)_i + h^l(t,x,Y(t))u.$$

In view of Lemma 6.1.1, the σ-algebra $\mathcal{Y}_t^0$ (see §6.0.1) coincides with the σ-algebra $\mathcal{F}_t^0$ obtained as a completion with respect $\mathbf{P}$ of the σ-algebra generated by the Wiener process $\tilde{w}(s)$ for $s \leq t$. Denote by $\tilde{\mathbf{F}}(\mathcal{Y})$ the probability space $(\Omega; \mathcal{F}; \mathcal{Y}_t^0; \mathbf{P})$ and by $\mathcal{P}(\mathcal{Y})$ the σ-algebra of predictable sets on $\mathbf{F}(\mathcal{Y})$.

On the probability space $\tilde{\mathbf{F}}(\mathcal{Y})$ consider the direct linear filtering equation

$$(2.3) \qquad du(t,x) = \mathcal{L}^*(t,x,Y(t))u(t,x)\,dt + \mathcal{M}^{l*}(t,x,Y(t))u(t,x)\,d\tilde{w}^l(t),$$

$$(t,x,\omega)\in]0,T]\times\mathbf{R}^d\times\Omega.$$

Definition 1. The function $u(x_0,t,x,\omega)$: $]0,T]\times\mathbf{R}^d\times\Omega \to \mathbf{R}^1$ *is called the fundamental solution of equation* (2.3) *if it is* $\overline{\mathcal{B}(]0,T]\times\mathbf{R}^d)\otimes\mathcal{F}}$*-measurable and, for every* $x\in\mathbf{R}^d$, $\mathcal{P}(\mathcal{Y})$*-measurable, belongs to* $\mathbf{C}^2(\mathbf{R}^d)$ *for every* $t\in]0,T]$ *on a set of probability* 1, *and for* s, $t\in]0,T]$ *such that* $s<t$, *satisfies the integral equation*

$$(2.4)\qquad \begin{aligned} u(x_0,t,x) &= u(x_0,s,x) + \int_{[s,t]} \mathcal{L}^*(r,x,Y(r))u(x_0,r,x)\,dr + \\ &+ \int_{[s,t]} \mathcal{M}^{l*}(r,x,Y(r))u(x_0,r,x)\,d\tilde{w}^l(r), \end{aligned}$$

and for every $f\in\mathbf{C}_b^0(\mathbf{R}^d)$ *the following relation holds:*

$$(2.5)\qquad \lim_{t\downarrow 0} \int_{\mathbf{R}^d} u(x_0,t,x)f(x)\,dx = f(x_0). \quad \square$$

Theorem 1. If the conditions (*A*) *and* (H_3) *are fulfilled, then the direct linear filtering equation* (2.3) *has a fundamenal solution* $u(x_0,t,x)$.

This fundamental solution is non-negative and belongs to $\mathbf{C}^{0,\infty}(]0,T]\times\mathbf{R}^d)$ *for* $\mathbf{P}$*-a.a.* ω, *and for every* $\varepsilon\in]0,T]$, $p\geq 0$ *satisfies the inequality*

$$\mathbf{E} \sup_{t\in]\varepsilon,T]} \sup_{x\in\mathbf{R}^d} |u(x_0,t,x)|^p < \infty. \quad \square$$

Proof. By Theorem 1.0 equation (2.3) has a measure-valued solution $\mu_t(\omega,\cdot)$ corresponding to the initial condition

$$(2.6)\qquad \mu_0(\omega,\Gamma) = \begin{cases} 1, & x\in\Gamma \\ 0, & x\notin\Gamma \end{cases}$$

and for an arbitrary $\psi\in\mathbf{L}_\infty$ the following equality holds

$$\mu_t[\psi] = \tilde{\mathsf{E}}[\psi(X(t))\rho(t,0)|\ \mathcal{Y}_t^0], \qquad (\mathbf{P}\text{-}a.s.),$$

where $\rho(t,0) = exp\{ \int_{[0,t]} h^l(s,Z(s))d\tilde{w}^l(s) -$

$$- \frac{1}{2} \int_{[0,t]} h^l h^l(s,Z(s))ds\}.$$

Letting $\mathcal{X}(t) = X(t)$, $\mathcal{Y}(t) = (Y(t),t)$ and making use of the hypothesis (H_3) we apply Theorem 7.1 to the measure defined above. By this theorem for $t\in]0,T]$, there exists a density $u(x_0,t,x)$ of the measure-valued solution of equation (2.3) and

(2.7) $$u(x_0,t,x)\in \mathbf{C}^{0,\infty}(]0,T]\times\mathbf{R}^d), \qquad (\mathbf{P}\text{-}a.s.)$$

Moreover, by Remark 1.1 we have

(2.8) $$\mathsf{E} \sup_{t\in[\varepsilon,T]} \sup_{x\in\mathbf{R}^d} |u(x_0,t,x)|^p < \infty \qquad \forall \varepsilon\in]0,T],\ p\geq 0.$$

From (2.8) in view of the definition of a measure-valued solution equality (2.4) follows.

Equality (2.5) follows from the continuity of μ_t in t and the obvious equality $\mu_0[f] = f(x_0)$. Thus the theorem is proved. □

2.2. In this paragraph we prove the existence of a fundamental solution of the direct linear filtering equation in the superparabolic case.

More precisely, in this item instead of assumptions (A) and (H_3) made in §.1 we suppose that the following conditions are fulfilled:

(*B*) *There exists a number* $\delta>0$ *such that for all* $(t,x,y)\in]0,T]\times\mathbf{R}^d\times\mathbf{R}^{d_1}$,

$${}_1\sigma^{ij}(t,x,y)\xi^i\xi^j \geq \delta|\xi|_d^2, \qquad \forall\xi\in\mathbf{R}^d \ ^*$$

---

*We recall that ${}_1\sigma$ is a matrix consisting of the first d rows and columns of matrix Σ.

(*C*) *Given* $m\in\mathbf{N}$ the functions $a^{ij}(t,x,y)$, $a_k^{ij}(t,x,y)$, $b^i(t,x,y)$, $\sigma^{il}(t,x,y)$, $\sigma_k^{il}(t,x,y)$, *and* $h^l(t,x,y)\in\mathbf{C}_b^{0,m}(]0,T]\times\mathbf{R}^d)$ *for every* $y\in\mathbf{R}^{d_1}$, i, j, $k=1,2,\dots,d$, $l=1,2,\dots,d_1$.

Note that condition (*B*) is a superparabolic condition for the matrix a^{ij}, and it implies the condition (H_3). Condition (*C*) is clearly weaker than condition (*A*). In particular the first one does not require the smoothness of the coefficients with respect to t and y.

Let d^0 be the least integer greater than $d/2$. Also we shall use $(\cdot,\cdot)_r$ to denote the scalar product in $\mathbf{H}^r$.

Definition 2. *The function* $u(x_0,t,x,\omega)$ *is called a generalized fundamental solution of the direct linear filtering equation* (2.3), *if for every* $\varepsilon\in]0,T]$, *it is an element of* $L_2^\omega([\varepsilon,T];\ \mathcal{P}(\mathcal{Y});\ \mathbf{H}^1)\cap\mathbf{L}_1([0,T]\times\Omega;\ \mathbf{C}_\omega\mathbf{L}_1)$, *for all* $(t,\omega)\in[0,T]\times\Omega$ *belongs to the cone of non-negative functions from* $\mathbf{L}_1$, *and for all* $\psi\in\mathbf{C}_0^\infty(\mathbf{R}^d)$ *and* $t\in[0,T]$ *satisfies the following equality*

$$(u(x_0,t),\psi)_0 = \psi(x_0) + \int_{[0,t]} [-(a^{ij}u_i(x_0,s),\psi_j)_0 +$$

$$+((b^i - a_j^{ij})u_i(x_0,s),\ \psi)_0]ds \qquad (\mathbf{P}\text{-}a.s.) \tag{2.9}$$

$$+\int_{[0,t]} (\mathcal{M}^l u(x_0,s),\ \psi)_0\ d\tilde{w}^l(s),$$

and also for every function $f\in\mathbf{C}_b^0(\mathbf{R}^d)$

$$\lim_{t\downarrow 0}\int_{\mathbf{R}^d} u(x_0,t,x)f(x)dx = f(x_0) \qquad (\mathbf{P}\text{-}a.s.)\ \square \tag{2.10}$$

Theorem 2. *If* $m\geq 2d^0+2$, *then there exists a generalized fundamental solution* $u(x_0,t,x,\omega)$ *of the direct linear filtering equation* (2.3) *belonging to* $\mathbf{L}_2([\varepsilon,T];\ \mathcal{P}(\mathcal{Y});$ $\mathbf{H}^{m-2d^0-1})\cap\mathbf{L}_2(\Omega;\ \mathbf{C}([\varepsilon,T];\ ^{m-2d^0-2}))$, $\forall\varepsilon\in]0,T]$. *For every* $t\in[0,T]$ *and* $\psi\in\mathbf{L}_\infty$, *the following formula holds*

$$(u(x_0,t),\ \psi) = \tilde{\mathbf{E}}[\psi(X(t))\rho(t,0)|\ \mathcal{Y}_t^0] \qquad (\mathbf{P}\text{-}a.s.)$$

If both $u_1(x_0,t,x)$ and $u_2(x_0,t)$ are generalized fundamental solutions of the direct linear filtering equation, then

$$\mathbb{E}\int_{[0,T]}\|u_1(x_0,t)-u_2(x_0,t)\|_{1,2}^2dt = 0. \quad \square$$

Before proceeding to the proof we give a simple corollary of the theorem.

Corollary 2. If the assumptions of the theorem are fulfilled for some $n\in\mathbb{N}$ such that $m>n+2(d^0+1)+d/2$, then the generalized solution $u(x_0,t,x)$ of the direct linear filtering equation (2.3) has a version, which is $\mathcal{P}(\mathcal{Y})\otimes\mathcal{B}(\mathbf{R}^d)$-measurable, belongs to $\mathbf{C}^{0,n}_d(]0,T]\times\mathbf{R}^d)$ for all $\omega\in\Omega$, and for $n\geq 2$, satisfies equation (2.3), in a classical sense, for $t\in]0,T]$. $\square$

The proof of the corollary based on Proposition 4.1.3 and arguments used repeatedly above (see, e.g. Theorem 4.1.3). The details we leave to the reader.

Proof of the theorem. In view of Theorem 0, equation (2.3) with the initial condition (2.6) has a measure-valued solution μ_t determined by the equality

$$\mu_t[\psi] = \tilde{\mathbb{E}}[\psi(X(t))\rho(t,0)|\,\mathcal{Y}_t^0]. \qquad (\mathbf{P}\text{-}a.s.) \tag{2.11}$$

Let $\eta(t)$ be a smooth function such that $\eta(0)=0$, $\eta(t)>0$ for $t\in]0,T]$, and for every $\varepsilon\in]0,T]$

$$\sup_{t\in[\varepsilon,T]} |{}^r\eta(t)|^{-1} < \infty$$

where ${}^rg(t):=\int_{[0,t]}\int_{[0,t_{r-1}]}\dots\int_{[0,t_1]} g(s)\,ds\,dt_1,\dots,dt_{r-1}$.

Differentiating the product $\mu_t[\psi]\eta(t)$, we obtain that for all $\psi\in\mathbf{C}_0^\infty(\mathbf{R}^d)$, $t\in[0,T]$ and ω from a set of probability 1 the following equation holds

$$\mu_t[\psi]\eta(t) = \int_{[0,t]}\mu_s[\mathcal{L}(s)\psi]\eta(s)\,ds + \tag{2.12}$$

$$+ \int_{[0,t]} \mu_s[\mathcal{M}^l(s)\psi]\eta(s)dw^l(s) + \int_{[0,t]} \mu_s[\psi]\eta'(s)ds.$$

Proposition 5.4.3 implies that there exists a function $v \in \mathbf{L}_2([0,T];\ \mathcal{P}(\mathcal{Y});\ \mathbf{H}^{d^0})$ such that for all $(t,\omega) \in [0,T] \times \Omega$,

$$\mu_t[\psi] = (\psi,\ v(t))_{d^0}, \qquad \forall \psi \in \mathbf{C}_0^\infty(\mathbf{R}^d). \tag{2.13}$$

From this and (2.12) it follows that the function $u(t) = v(t)\eta(t) \in \mathbf{L}_2([0,T];$ $\mathcal{P}(\mathcal{Y});\ \mathbf{H}^{d^0})$ and for every $t \in [0,T]$ and all $\psi \in \mathbf{C}_0^\infty(\mathbf{R}^d)$ satisfies the equality

$$(u(t)),\psi)_{d^0} = \int_{[0,t]} (u(s),\mathcal{L}\psi)_{d^0} ds + \int_{[0,t]} (u(s),\mathcal{M}^l\psi)_{d^0} d\tilde{w}^l(s) + \qquad (\mathbf{P}\text{-}a.s.)$$

(2.14)

$$+ \int_{[0,t]} (v(s)\eta'(s),\psi)_{d^0} ds$$

Write ${}^0u(t) = \Lambda^{-1}u(s)$, where as before $\Lambda = \mathbf{I}\text{-}\Delta$ (see §3.4.2). Thus, in view of Proposition 3.4.2, $\Lambda^{-1}\tilde{v}(s)\eta'(s) \in \mathbf{L}_2([0,T];\ \mathcal{P}(\mathcal{Y});\ \mathbf{H}^{d^0+2})$ and for every $t \in [0,T]$ and $\psi \in \mathbf{C}_0^\infty(\mathbf{R}^d)$, ${}^0u(t)$ satisfies the following equation

$$({}^0u(t),\psi)_{d^0+1} = \int_{[0,t]} ({}^0u(s),\mathcal{L}\psi)_{d^0+1} ds + \int_{[0,t]} ({}^0u(s),\mathcal{M}^l\psi)_{d^0+1} d\tilde{w}^l(s) + \qquad (\mathbf{P}\text{-}a.s.)$$

(2.15)

$$+ \int_{[0,t]} (\Lambda^{-1}v(s)\eta'(s),\psi)_{d^0+1} ds.$$

Suppose that $n \in \mathbf{N} \cup \{0\}$, $\varphi_0 \in \mathbf{L}_2(\Omega;\ \mathcal{F}_0;\ \mathbf{H}^n)$ and $f \in \mathbf{L}_2([0,T];\ \mathcal{P}(\mathcal{Y});\ \mathbf{H}^{n-1})$ are given.

Consider the equation

$$(\varphi(t),\psi)_n = (\varphi_0,\psi)_n + \int_{[0,t]} (\varphi(s),\mathcal{L}\psi)_n ds +$$

$$+ \int_{[0,t]} [f(s),\psi]_n ds + \int_{[0,t]} (\varphi(s),\mathcal{M}^l(s)\psi)_n d\tilde{w}^l(s),$$

where $[\cdot,\cdot]_n$ is a CBF of $(\mathbf{H}^{n+1}, \mathbf{H}^n, \mathbf{H}^{n-1})$.

A function $\rho\in\mathbf{L}_2^{\omega}([0,T];\ \mathcal{P}(\mathcal{Y});\ \mathbf{H}^{n+1})$ satisfying equation (2.16) ($\mathbf{P}$-a.s.) on the interval $[0,T]$ for all $\psi\in\mathbf{C}_0^{\infty}(\mathbf{R}^d)$ will be called a generalized solution of this equation.

Lemma 2. Given $k\in\mathbf{N}\cup\{0\}$ *such that* $n+k\leq m$, *suppose that* $\varphi_0\in\mathbf{L}_2(\Omega;\ \mathcal{F}_0;\ \mathbf{H}^{n+k})$ *and* $f\in\mathbf{L}_2([0,T];\ \mathcal{P}(\mathcal{Y});\ \mathbf{H}^{n+k-1})$. *Then there exists a unique generalized solution* $\varphi(t)$ *of equation* (2.16) *and* $\varphi(t)\in\mathbf{L}_2([0,T];\ \mathcal{P}(\mathcal{Y});\ \mathbf{H}^{n+k+1})\ \cap\ \mathbf{L}_2(\Omega;\ \mathbf{C}([0,T];\ \mathbf{H}^{n+k}))$. □

Proof. By the same arguments as in Proposition 5.3.4 it is easy to show that equation (2.16) is equivalent to

$$u(t) = u_0 + \int_{[0,t]} [A(s)u(s)+f(s)]ds + \int_{[0,t]} B(s)u(s)d\tilde{w}(s) \tag{2.17}$$

in the normal triple $(\mathbf{H}^{n+1},\ \mathbf{H}^n,\ \mathbf{H}^{n-1})$, where the operators $A(s)$ and $B(\ s)$ are determined by the relations

$$[A(s)\varphi,\ \psi]_n:=(\varphi,\ \mathcal{L}(s)\psi)_n$$

and

$$(B(s)\varphi,\ \psi)_n^l:=(\varphi,\ \mathcal{M}^l(s)\psi)_n.$$

By arguments similar to that used in the proof of Proposition 5.3.4 (see (5.3.20) and (5.3.24) to (5.3.30)) we obtain easily that the LSES (2.17) satisfies the assumptions of Theorem 3.3.1 and 3.1.8, where $\mathbf{V}:=\mathbf{H}^{n+k+1}$, $\mathbf{U}:=\mathbf{H}^{n+k}$ and $\mathbf{V}':=\mathbf{H}^{n+k-1}$. Application of these theorems to (2.17) provides the proof of the lemma. □

From the lemma, ${}^0u(t)$ is the unique generalized solution of equation (2.15), and moreover ${}^0u\in\mathbf{L}_2([0,T];\ \mathcal{P}(\mathcal{Y});\ \mathbf{H}^{d^0+4})$. Hence $\Lambda^{-1}v(s)\eta'(s) =$

$u^0\eta^{-1}(s)\eta'(s)\in\mathbf{L}_2([0,T];\ \mathcal{P}(\mathcal{Y});\ \mathbf{H}^{d^0+4})$.

Define

$$^{r}u(t):=\Lambda^{-1}\, v(t)\int_{[0,t]}\int_{[0,t_{r-1}]}\cdots\int_{[0,t_1]}\eta(s)ds\, dt_1\ldots dt_{r-1}.$$

Clearly, $^{r}u(t)$ is a generalized solution of equation (2.16) for $n=d^0+1$, $f(t)={}^{r-1}u(t)$, and $\varphi_0=0$. Hence applying the lemma repeatedly we obtain that for

$2r\le m\text{-}3\text{-}d^0$ *, $^{r}u\in\mathbf{L}_2([0,T];\ \mathcal{P}(\mathcal{Y});\ \mathbf{H}^{d^0+4+2r})\cap\mathbf{L}_2(\Omega;\ \mathbf{C}[0,T];\ \mathbf{H}^{d^0+4+2r-1}))$

which in its turn gives $v(t)\in\mathbf{L}_2([\varepsilon,T];\ \mathcal{P}(\mathcal{Y});\ \mathbf{H}^{d^0+2(r+1)})\cap\mathbf{L}_2(\Omega;\ \mathbf{C}([\varepsilon,T];$ $\mathbf{H}^{d^0+2r+1}))$ for every $\varepsilon\in]0,T]$. From this and (2.13) it follows that for $t\in[0,T]$ and all $\varepsilon\in]0,T]$ and $\psi\in C_0^\infty(\mathbf{R}^d)$

$$\mu_t[\psi]=(\psi,\ u(x_0,t))_0,\qquad (\mathbf{P}\text{-}a.s.),\tag{2.18}$$

where $u(x_0,t)=\Lambda^{d^0}v(t)=\Lambda^{d^0+1}\,{}^{r}u(t)/{}^{r}\eta(t)\in\mathbf{L}_2([\varepsilon,T];\ \mathcal{P}(\mathcal{Y});\ \mathbf{H}^{2r+2-d^0}$

$\cap\ \mathbf{L}_2(\Omega;\ \mathbf{C}[\varepsilon,T];\ \mathbf{H}^{2r+1-d^0}))$. Letting $2r=m\text{-}3\text{-}d^0$, we obtain $u(x_0,t)\in\mathbf{L}_2([\varepsilon,T];$ $\mathcal{P}(\mathcal{Y});\ \mathbf{H}^{m-2d^0-1})\cap\mathbf{L}_2(\Omega;\ \mathbf{C}[\varepsilon,T];\ \mathbf{H}^{m-2d^0-2}))$, $\forall\varepsilon\in[0,T]$.

By the same arguments as those in Theorem 5.3.1 and Corollary 5.3.1 it is easy to verify that equality (2.18) is valid for every $\psi\in\mathbf{L}_\infty$ and the generalized fundamental solution has a version in $\mathbf{L}_1([0,T]\times\Omega;\ \mathbf{C}_w\mathbf{L}_1)$ belonging to the cone of non-negative functions in $\mathbf{L}_1$.

By the definition of a measure-valued solution, $u(x_0,t)$ satisfies equality (2.9) for all $t\in[0,T]$ and $\psi\in C_0^\infty(\mathbf{R}^d)$ on a ω-set of probability 1.

From (2.18) by a weak continuity argument, we obtain as well, that $u(x_0,t)$ satisfies equality (2.10).

Thus it is proved that $u(x_0,t)$ is a generalized solution of the direct linear filtering equation.

---

*The assumption $n+k\le m$ of the lemma implies this inequality.

It both u_1 and u_2 are generalized fundamental solutions of this equation, then their difference $\overline{u}$ is also a generalized solution of equation (2.3) with zero initial condition (in the sense of Definition 4.2.1). By Theorem 4.2.1

$$\mathbf{E} \int_{[0,t]} \|\overline{u}(s)\|_{1,2}^2 ds = 0. \quad \square$$

2.3. Here and in the next two paragraphs we investigate the filtering, interpolation and extrapolation transition probabilities and densities and their analytical properties in the superparabolic and hypoelliptic cases.

Throughout what follows condition (C) from the previous paragraph is required to hold. As in §6.0.1 the numbers T_0 and T_1 are supposed to be fixed and it is assumed that T_0, $T_1 \in]0,T]$ and $T_0 \leq T_1$.

Arguing exactly as in the proof of Theorem 6.2.1 we obtain from Theorem 2 the following statement.

Theorem 3. *Suppose that the assumptions of Theorem* 2 *are fulfilled, then there exists a function* $P^t_{\mathcal{Y}}(x_0)$: $\Omega \rightarrow \mathcal{M}(\mathbf{R}^d)$ with these properties:

a) for every $t \in]0,T]$, $P^t_{\mathcal{Y}}(x_0,\cdot,\omega)$ is a regular conditional distribution of $X(t)$ with respect to $\mathcal{Y}^0_t$;

b) for all $(t,\omega) \in]0,T] \times \Omega'$, where $P(\Omega')=1$, $P^t_{\mathcal{Y}}(x_0)$ is absolutely continuous with respect to Lebesgue measure on $\mathbf{R}^d$ and the Radon-Nikodym derivative is defined by the equality (in $\mathbf{L}_1$, for every $(t,\omega) \in]0,T] \times \Omega$)

$$\pi(x_0,t,x,\omega) = u(x_0,t,x,\omega) \Big(\int_{\mathbf{R}^d} u(x_0,t,x,\omega)\,dx \Big)^{-1}, \tag{2.19}$$

where $u(x_0,t,x,\omega)$ is generalized fundamental solution of equation (2.3).

c) for every $\Gamma \in \mathcal{B}(\mathbf{R}^d)$, $P^t_{\mathcal{Y}}(x_0,\Gamma,\omega)$ is a $\mathcal{P}(\mathcal{Y})$-measurable, continuous ($\mathbf{P}$-a.s.) stochastic process. $\square$

2.4. The following theorem can be proved in the same way.

Theorem 4. *If conditions* (A) *and* (H_3) *(see* §.2) *are fulfilled, then the statement of Theorem* 3 *is valid, and equality* (2.19), *where* $u(x_0,t,x,\omega)$ *stands for the fundamental solution of equation* (2.3) *in the sense of Definition* 2, *holds for all* $x \in \mathbf{R}^d$. $\square$

2.5. The functions $\pi(x_0,t,x,\omega)$ and $u(x_0,t,x,\omega)$ will be called the filtering transition density and the non-normalized filtering transition density, respectively.

Theorem 5 below can be proved exactly in the same way as Theorem 6.2.3.

Theorem 5. Under the assumptions of Theorem 2 the filtering transition density

$\pi(x_0,\cdot)\in \mathbf{L}_2([\varepsilon,T];\ \mathcal{P}(\mathcal{Y});\ \mathbf{H}^{m-2d^0-1}) \cap \mathbf{L}_2(\Omega;\ \mathbf{C}([\varepsilon,T];\ \mathbf{H}^{m-2d^0-2})$ *for every* $\varepsilon\in]0,T]$, *and it is also* $\mathcal{P}(\mathcal{Y})$*-measurable, as a mapping of* $[0,T]\times\Omega$ *to* $\mathbf{L}_1$, *a weakly continuous* $\mathbf{L}_1$*-process, and for all* $t\in[0,T]$, $\psi\in C_0^\infty(\mathbf{R}^d)$ *and* ω *from the same set of probability one satisfies the equality*

$$(\pi(x_0,t),\ \psi)_0 = \psi(x_0) + \int_{[0,t]} (\pi(x_0,s),\ \mathcal{L}(s)\psi)_0 ds +$$

$$+ \int_{[0,t]} [(\pi(x_0,s),\ \mathcal{M}^l(s)\psi)_0$$

$$- (\pi(x_0,s),\ h^l(s))_0(\pi(x_0,s),\psi)_0]d\bar{w}^l(s), \tag{2.20}$$

where $d\bar{w}(s) := d\tilde{w}^l(s) - (\pi(s),\ h^l(s))_0 ds$

If both $\pi^1(x_0,\cdot)$ *and* $\pi^2(x_0,\cdot)$ *are elements of* $\mathbf{L}_2^\omega([0,T];\ \mathcal{P}(\mathcal{Y});\ \mathbf{L}_1) \cap \mathbf{L}_2^\omega([\varepsilon,T];\ \mathcal{P}(\mathcal{Y});\ \mathbf{H}^1) \cap \mathbf{C}([\varepsilon,T];\ \mathcal{P}(\mathcal{Y});\ \mathbf{L}_2)$, $\varepsilon\in]0,T]$, *satisfying equation* (2.20) *for every* $t\in[0T]$, $\psi\in C_0^\infty(\mathbf{R}^d)$ ($\mathbf{P}$*-a.s.*), *then*

$$\mathbf{E}\int_{[0,T]} \|\pi^1(t) - \pi^2(t)\|_{1,2} dt = 0. \quad \square$$

In other words, this theorem states that the filteirng transition density is the unique (in a natural class) generalized fundamental solution of the direct filtering equation.

Corollary 4. Suppose that conditions (B) and (C) for $m=\infty$ *(from* §.2) *or conditions (A) and* (H_3) *(from* §.1) *are fulfilled, then there is a version of* $\pi(x_0,t,x,\omega)$ *which belongs to* $\mathbf{C}^{0,\infty}(]0,T]\times\mathbf{R}^d)$ *for all* $(x_0,\omega)\in\mathbf{R}^d\times\Omega$ *and is a fundamental solution (in* $\mathbf{C}^2$*) of the direct filtering equation* (6.2.5).

2.6. Here and in the next paragraph the assumptions of §6.3.1 are assuemd to be fulfilled.

Denote by $v_1(s,z)$ a continuous version of the r-generalized solution of the backward filtering equation (6.3.3), (6.3.5).

Theorem 6. If the conditions (A), (H_3) *of* §.1 *or the conditions* (B), (C) *for* $m \geq 2d^0+2$ *of* §.2 *are fulfilled, then there exists a function* $P_{\mathcal{Y}}^{T_1,T_0}(x_0)\colon \Omega \to \mathcal{M}(\mathbb{R}^d)$ *with these properties:*

(*a*) $P_{\mathcal{Y}}^{T_1,T_0}$ *is a regular conditional distribution of* $X(T_0)$ *relative to* $\mathcal{Y}^0_{T_1}$

(*b*) $P_{\mathcal{Y}}^{T_1,T_0}(x_0)$ *is absolutely continuous with respect to Lebesgue measure on* $\mathbb{R}^d$ *and the Radon-Nikodym derivative is defined by the equality*

$$\pi^{T_1,T_0}(x_0,x) = v_1(T_0,x,Y(t_0))u(x_0,T_0,x) \times \tag{2.21}$$
$$\times \Big(\int_{\mathbb{R}^d} v_1(T_0,x,Y(T_0))u(x_0,T_0,x)\,dx\Big)^{-1}. \qquad (\mathbf{P}\text{-}a.s.) \quad \square$$

Remark 6. If conditions (B), (C) *are fulfilled, the equality* (2.21) *is valid in* $\mathbf{L}_1$ *and* $u(x_0,t,x)$ *should be considered as a generalized solution of* (2.3). *On the other hand, conditions* (A) *and* (H_3) *provide the justification of this equality for all* $x \in \mathbb{R}^d$. *In this case,* $u(x_0,t,x)$ *has to be considered as a fundamental solution of equation* (2.3) *and* $v_1(s,z)$ *as a classic solution of problem* (6.3.3), (6.3.5). $\square$

Thus $\pi^{T_1,T_0}(x_0,x)$ is the interpolation transition density. Its analytical properties are determined by those of $v_1(s,z)$ and $u(x_0,t,x)$.

2.7. Evidently the formula

$$\mathbf{E}[f(X(T_1)) \mid \mathcal{Y}^0_{T_0}] =$$

$$= \int_{\mathbb{R}^d} v(T_0,(x,Y(T_0)))\pi(T_0,x)\,dx$$

established in §6.3.4 will also holds in this case when z_0 is a constant if we substitute the filtering transition density $\pi(x_0,T_0,x)$ for $\pi(T_0,x)$.

Thus we get

Theorem 2.7. If the condition (A), (H_3) of §.1 or the conditions (B), (C) for $m \geq 2d^0+2$ of §2 are fulfilled, then there exists a function $P_{\mathcal{Y}}^{T_1,T_0}(x_0): \Omega \to \mathcal{M}(\mathbf{R}^d)$ *with these properties:*

(a) $P_{\mathcal{Y}}^{T_1,T_0}$ *is a regular conditional distribution of $X(T_1)$ relative to* $\mathcal{Y}_{T_0}^0$.

(b) $P_{\mathcal{Y}}^{T_1,T_0}$ *is absolutely continuous with respect to Lebesgue measure on* $\mathbf{R}^d$

and the Radon-Nikodym derivative is defined by the equality

$$\pi^{T_1,T_0}(x_0,x) = \int_{\mathbf{R}^{d_1}} \int_{\mathbf{R}^d} p((x,y);\ T_1;\ (x',Y(T_0));T_0)\ \pi(x_0,T_0,x')\,dx'\,dy, \tag{2.22}$$

where $p(z,T_1,z',T_0)$ is the unconditional transition density of the process $Z(t)$. □

Remark 7. Formula (2.22) must be considered in the spirit of Remark 6. □

Notes

Chapter 1

The proof of Theorem 1.4.9 (Ito-Ventcel's formula) follows Rozovskiĭ [113]. Lemma 1.4.11 is due to Ventcel [137]. The Ito-Ventcel's formula was rediscovered and generalized by Bismut [12] and Kunita [75].

Chapter 2

Stochastic integrals with respect to square integrable martingales taking values in a Hilbert space was first systematically investigated in Kunita [74]. Afterwards the stochastic integration theory based on martingales taking values in infinite dimensional spaces was developed mainly by Métivier and has pupils (see e.g. [97], [98] and further references there). There is considerable overlap between the results presented in sections 2.1 to 2.3 with those of [98]. However the construction of a stochastic integral in this chapter differs from those developed in [74] and [98]. It is based on the idea outlined in an article by Krylov, Rozovskiĭ [70].

The notion of a normal triple given in seciton 2.4 is a version of Gelfand's triple. Theorem 2.4.2 (the statement and the proof) is almost identical to Theorem 3.1 from [70]. In this connection see also Gyoňgy, Krylov [38] and Grigelionis, Mikulevicius [37].

Chapter 3

The existence and uniqueness theorem for LSES (3.0.1) in the coercive case follows Pardoux [105], [108]. The results of section 2 appears to be new. Similar results were announced in Rozovskiǐ [116]. Theorem 3.1.1 is new. Exposition of the results concerning the approximation of LSES (3.0.1), the Markov property of its solution (Sect. 3), and the solvability of the first boundary problem (Sect. 4) follows Krylov, Rozovskiǐ [70], where corresponding results were developed for non-linear systems. We note that all the results of Sect. 1,3, and 4 could be carried over to the case of monotone coercive non-linear systems (see Krylov, Rozovskiǐ [70], Pardoux [105].

Chapter 4

Sections 1 and 2 are based on Krylov, Rozovskiǐ [67], [72]. The superparabolic condition for a second-order stochastic partial differential equations (SPDE) was introduced independently by Pardoux [105] and Krylov, Rozovskiǐ [66]. The results of Sect. 3 are due to the author.

The proofs of the existence theorems for parabolic and superparabolic Ito's equations given in this chapter are based on the Galerkin method. This method is still applicable in a much more general situations. In some particular cases, e.g. if the operator $\mathcal{L}$ is not random, results could be obtained by semi-grouped methods. For details see [17], [19], [96], [115], [122], [123].

Chapter 5

Averaging over characteristic (AOC) formulas (5.1.2) and (5.1.3), and Corollary 5.1.5 (maximum principle) are due to Krylov, Rozovskiĭ [71], [72]. Formula (5.1.3) in the case of uniformly non-degenerate matrix $(A^{ij}) := (2a^{ij} - \sum \sigma^{il}\sigma^{jl})$ was proved by Pardoux [108]. This additional assumption appears to be a quite restrictive one. For example, for both Liouville's equations for diffusion processes, $(A^{ij}) \equiv 0$. Subsequently, formulas similar to (5.1.2) and (5.1.3) for classical solutions of the corresponding problems were obtained by Kunita [79].

Theorem 5.1.2 for r=0 was proved by Krylov [64]. A statement very close to that in Lemma 5.1.4 can be found in Hida [44].

The forward and backward Liouville's equations, and in particular, the foward equation of inverse diffusion and the backward diffusion equation were derived by Krylov, Rozovskiĭ [69], [71], and [72], and Rozovskiĭ [118]. Independently the backward diffusion equation was obtained under different assumptions and by a different method in Kunita [78].

Thereom 5.2.1 was first published in Krylov, Rozovskiĭ [73]. An equation equivalent to the forward equation of inverse diffusion and a formula similar to (5.2.6) were derived in Kunita [75] under additional assumptions that the coefficients are non-random and possess some extra derivatives.

The derivation of the backward diffusion equation is taken from Krylov, Rozovskii [71]. Another development can be found in Malliavin [93]. See also the book of Ikeda and Watanabe [49].

That the mapping $X(t,\cdot)$: $x \rightarrow X(t,x)$ is a diffeomorphism has been known to many authors (e.g. [12], [49], [71], [73], [75], [80]).

Theorem 5.2.2 is due to the author. The idea to reduce a second-orderparabolic Ito's equation to a second-order parabolic deterministic equations (although to one with random coefficients) goes back to Ventcel [137]. Subsequently this idea was systematically used by Rozovskiǐ [112], [113] in the study of the filtering equations. Later an analogous idea was used by Kunita [77].

Averaging over characteristic formula (5.3.3) and its corollaries are due to the author. Note that the methods used in the proof had been used earlier in Krylov, Rozovskiǐ [68], and Rozovskiǐ [117] in the study of absolute continuity of the filtering measure with respect to the Lebesgue measure.

Theorem 5.3.2 is well known (see e.g. Lipster, Shiryayev [90], Rozovskiǐ, Shiryayev [114], and also Krylov, Rozovskiǐ [68]).

Chapter 6

Different versions of Bayes' formula were traditionally used in the development of filtering theory. The references are e.g. Kallianpur [53], Lipster, Shiryayev [90]. Note that in these books the reader can find the general theory of filtering, interpolation and extrapolation for semimartingales. Lemma 6.1.1 is taken from Loéve [91]. Theorem 6.1.2 is new, but certainly has is predecessors (see e.g. [53], [90] cited above).

Section 2 is based on Rozovskiĭ [117]. Proposition 6.2.2 is in Lipster, Shiryayev [90]. Similar problems for discontinuous processes were considered by Grigelionis, Mikulevicius [37].

A forward linear filtering equation (for non-normalized filtering density) was first derived by Zakai [148] in a particular case. The equivalence of the foward linear filtering equation and non-linear filtering equations in quite a general situation was proved in Rozovskiĭ, Shiryayev [114] (see also Krylov, Rozovskiĭ [68] and Lipster, Shiryayev [90]).

Theorems 6.3.1-6.3.4 are due to the author. The first results about the backward filtering equation were obtained in Kushner [84] and Pardoux [108].

Results related to those of the present chapter were published earlier by many authors. For example, filtering in bounded domains were considered in Margulis [96], Pardoux [107].

Chapter 7

In 1967 Hörmander published his famous results on the hypoellipticity of second-order degenerate parabolic equations. Since then, these results have been elaborated on by many authors. Malliavin [93], [94] developed a probability approach to the proof of Hörmander's theorem. Later Bismut [13], [14] presented a more graphic probability approach to the problem, which was more or less equivalent to that of Malliavin.

In this chapter we treat the hypoelliptic property for second-order parabolic Ito's equation on the basis of Bismut's version of Malliavin calculus.

Theorem 7.0.1. overlaps partly with the result of Kunita [76]. The prototype of Theorem 7.1.1 was developed for filtering equation in Bismut, Michel [14] and Kusuoka, Stroock [85]. The idea to prove the hypoellipticity property of deterministic second-order parabolic-elliptic equations via the application of Proposition 7.1.1 belongs to Malliavin. The formula of stochastic integration by parts in the form close to ours was first derived in Haussmann [43], where it was used in the investigation of the structure of square integrable martingales. Bismut [13] showed that it is an indespensable tool in hypoellipticity. The general scheme of the proof of Theorem 7.1.1 runs along the lines of Veretennikov [138]. Lemma 7.1.5. was first published in Stroock [136].

The proof of the existence and uniqueness of a generalized fundamental solution of the forward linear filtering equation in the superparabolic case mainly follows

Rozovskiǐ, Shimizu [119].

The results of Section 7.3 concerning the existence of conditional transition densities and their analytical properties are similar to those of 6.3.1-6.3.4.

In the hypoelliptic case similar results were obtained in Bismut, Michel [14], Kusuoka, Stroock [85], and Michel [101].

Further references are Chaleyat-Maurel, Michel [16], Ichihara, Kunita [48], Kunita [77], Shikegawa [124], Stroock [132]-[135], Veretennikov [139], and Zakai [149].

REFERENCES

1. Arnold, V.I., Ordinary differential equations , Cambridge: MIT Press, 1973.

2. Arnold, L., In: Dynamics of Synergetic Systems. Proc. of the Intern. Symp. on Synergetics, Bielefeld, 1979. Berlin etc.: Springer Verlag, 1980, 107-118.

3. Bachelier, L., Ann. Sci. Ecole Norm. Supér., 1900, 17, no. 3, 21-86.

4. Baklan, V.V., Dopovidi AN URSR, 1963, no. 10, 1299-1303 (in Ukrainian).

5. Balakrishnan, A.V., Introduction to optimization theory in a Hilbert space. New York etc.: Springer-Verlag, 1971.

6. Bellman, R., Stability theory of differential equations. New York etc.: McGraw-Hill, 1953.

7. Bilopolska, Ya. I., Dalečkii, Yu. L., Trudy Moskov. Mat. Obščhestva, 1978, 37, 107-141..

8. Bensoussan, A., Filtrage optimale des systemes lineaires. Paris: Dunod, 1971.

9. Bensoussan, A., Lions J.-L. Applications des inéquations variationelles an contrôle stochastique. Paris: Dunod, 1978.

10. Bernstein, I.N., Proceedings of Steklov institute of physics and mathematics, 5, 95-124, Moscow, 1934 (in Russian).

11. Bismut, J.-M. Mécanique aléatoire, Lecture Notes in Math., 866, Berlin etc.: Springer-Verlag, 1981.

12. Bismut, J.-M., Z. Wahrsch., 1981, 55, 331-350.

13. Bismut, J.-M., Z. Wahrsch., 1981, 56, 469-505.

14. Bismut, J.-M., Michel, D., J. Functional Anal., 1981, 44, 174-211; II. Ibid 1982, 45, 274-292.

15. Blagoveshchenskii, Yu.N., Freidlin, M.I., Soviet Math. Dokl., 1961, 138, 633-636.

16. Chaleyat-Maurel, M., Michel, D., Z. Wahrsch., 1984, 65, 573-597.

17. Curtain, R.F., Pritchard, A.J., Infinite dimensional linear systems theory Lecture Notes Contr. and Inf. Sci., 8, Berlin etc.: Springer Verlag, 1978.

18. Dalečkii, Yu.L., Russian Math. Surveys, 1967, 22 no. 4, 1-53.

19. Dalečkii, Yu.L., Fomin, S.V., Measures and differential equations in infinite-dimensional spaces, Moscow: Nauka, 1983 (in Russian).

20. Davis, M.H.A., In: Stochastic Systems: The mathematics of filtering and identification. Proc. of the NATO Adv. Study Inst., Les Aros, 1980/Ed. M. Hazewinkel, J.C. Willems. Dordrecht etc.: D. Reidel Publ. Co., 1981, 505-528.

21. Dawson, D.A., J. Multivar. Anal., 1975, 5, no. 1, 1-52.

22. Dellacherie, C. Capacités et processus stochastiques. Berlin etc.: Springer-Verlag, 1972.

23. Dellacheire, C., Meyer, P.A., Probabilités et potentials. Theorie des martingales. Paris: Herman, 1980.

24. Doob, J.L., Stochastic processes. New York: John Wiley, London: Chapman and Hall, 1953.

25. Dunford, N., Schwarz, J.T., Linear operators. Part I: General theory. New York etc.: Interscience Publishers, 1958.

26. Einstein, A., The collected papers of Albert Einstein, Princeton, N.J.: Princeton University Press, 1987.

27. Ershov, M.P., Teor. Prob. Appl., 1970, 15, no. 4, 705-717.

28. Gelafand, I.M., Vilenkin, N.Ya., Generalized functions. V. 4. Applications of harmonic analysis. New York: Academic Press, 1969.

29. Gihman, I.I., Ukrain. Matem. Z., 1950, 2, no. 3, 45-69 (in Russian).

30. Gihman, I.I., I. Ukrain. Matem. Z., 1950, 2, no. 4, 37-63, II. ibid, 1951, 3, no. 3, 317-339 (in Russian).

31. Gihman, I.I., Ukrain. Matem. Z., 1979, 3, no. 5, 483-489. (in Russian)

32. Gihman, I.I., Theory of random processes, Kiev: Naukova Dumka, 1980, 8, 20-31 (in Russian).

33. Gihman, I.I., In: Qualitative methods of investigations of non-linear equations and non-linear oscillation, Kiev.: Institut matematiki AN USSR, 1981, 25-59 (in Russian).

34. Gihman, I.I., Skorokhod, A.V., Stochastic differential equations, Berlin etc.: Springer-Verlag, 1972.

35. Gihman, I.I., Skorokhod, A.V., The theory of stochastic processes, Berlin etc.: Springer-Verlag, 1974.

36. Gihman, I.I., Skorokhod, A.V., Stochastic differential equations and its applications, Kiev: Naukova Dumka, 1982 (in Russian).

37. Grigelionis, B., Mikulevicius, R., Lecture Notes in Contr. Inf. Sci., 49, Berlin etc.: Springer-Verlag, 1983, 49-88.

38. Gyöngy, I., Krylov, N.V., Stochastics, 1981/82, 6, no. 3-4, 153-173.

39. Fleming, W.H., Lect. Notes Econ. and Math. Syst., 1975, 107, 179-191.

40. Fleming, W.H., Rishel, R.W., Deterministic and stochastic optimal control. Berlin etc.: Springer-Verlag, 1975.

41. Friedman, A., Partial differential equations of parabolic type, Englewood Cliffs: Prentice-Hall, Inc., 1964.

42. Functional analysis. Ed. by S.G. Krein. Groningen.: Wolters-Noordhoff, 1972.

43. Haussmann, U., Stochastics, 1979, v.3, 17-27.

44. Hida, T., Stationary stochastic processes, Princeton: Princeton Univ. Press, 1970.

45. Hida, T., Streit, L., Nagoya Math. J., 1977, 68, no. 12, 21-34.

46. Hörmander, L., Acta Mathematica, 1967, 119, 147-171.

47. Jacod, J., Calcul stochastique et processes de martingales, Lect. Notes Math., 714, Berlin etc.: Springer-Verlag, 1979.

48. Ichihara, K., Kunita H., Z. Wahrsch., 1974, 30, 235-254.

49. Ikeda, N., Watanabe, S., Stochastic differential equations and diffusion processes. Amsterdam; Tokyo: North-Holland, Kodansha, 1981.

50. Ito, K., Proc. Jap. Acad., 1946, 22, 32-35.

51. Ito, K., On stochastic differential equations, New York: American Mathematical Society, 1951.

52. Ito, K., McKean H.P., Diffusion processes and their sample paths, Berlin etc.: Springer-Verlag, 1965.

53. Kallianpur, G., Stochastic filtering theory, New York etc.: Springer-Verlag, 1980.

54. Kalman, R.E., Bucy, R.S., Trans. ASME D, 1961, 83, 95-108.

55. Khinchin, A.Ya., Asymptotic laws in probability theory, Moscow: ONTI, 1936 (in Russian).

56. Kleptsina, M.L., Veretennikov, A.Yu., In: Statistics and control of stochastic processes. Steklov seminar, 1984 (Ed. by N.V. Krylov, et al.) New York: Optimization soft., 1975, 179-195.

57. Klyackin ,V.I., Statistical description of dynamical systems with fluctuating parameters, Moscow: Nauka, 1975 (in Russian).

58. Klyackin, V.I., Stochastic equations and vawes in random heterogeneous medium, Moscow: Nauka, 1980 (in Russian).

59. Kolmogorov, A.N., Math. Ann., 1931, 104, 415-458.

60. Kolmogorov, A.N., Izv. Akad. Nauk SSSR. Ser. mat. 1941, 5, no. 1, 3-14 (in Russian).

61. Kolmogorov, A.N., Fomin, S.V., Introductory real analysis, New Jersey: Prentice-Hall, 1970.

62. Krein, S.G., Petunin, Yu.U., Semenov, E.M., Interpolation of linear operators, Providence, R.I.: American Mathematical Society, 1982.

63. Krylov, N.V., Controlled diffusion processes Berlin etc.: Springer-Verlag, 1980.

64. Krylov, N.V., Math. USSR Sbornik, 1980, 37, no. 1, 133-149.

65. Krylov, N.V., Teor. Prob. Appl., 1983, 28, no. 1, 151-155.

66. Krylov, N.V., Rozovskii, B.L., In: Proc. III Soviet - Japan Symp. Prob. Math. Stat., Tashkent, 1974, 171-173.

67. Krylov, N.V., Rozovskii, B.L., Math. USSR, Izvestija, 1977, 41, no. 6, 1267-1284.

68. Krylov, N.V., Rozovskii, B.L., Math. USSR, Izvestija, 1978, 12, no. 2, 336-356.

69. Krylov, N.V., Rozovskii, B.L., Uspekhi matematicheskikh nauk, 1980, 35, no. 4, 147 (in Russian).

70. Krylov, N.V., Rozovskii, B.L., J. Soviet Math., 1981, 16, 1233-1276.

71. Krylov, N.V., Rozovskii, B.L., Lect. Notes Contr. and Inf. Sci. Stochastic Differential Systems. Proc. of the 3rd IFIP-WG 7/1 Working Conf., Visegard, Hungary, 1980. B. etc.: Springer-Verlag, 1981, 36, 117-125.

72. Krylov, N.V., Rozovskii, B.L., J. Soviet Math., 1982, 32, 336-348.

73. Krylov, N.V., Rozovski, B.L., Russian Math. Surveys, 1982, 37, no. 6, 75-95.

74. Kunita, H., Nagoya Math. J., 1970, 38, no. 1, 41-52.

75. Kunita, H., In: Proc. Durham Conf. Stoch. Integrals. Lect. Notes in Math., Berlin etc.: Springer-Verlag, 1981, no. 851, 213-255.

76. Kunita, H., Syst. and Contr. Let., 1981, 1, no. 1, 37-41.

77. Kunita, H., Syst. and Contr. Let., 1981, 1, no. 2, 100-104.

78. Kunita, H., Stochastics, 1982, 6, 293-313.

79. Kunita, H., Lecture Notes in Math., 972, Berlin etc.: Springer-Verlag, 1981, 100-168.

80. Kunita, H., Lecture Notes in Math., 1097, Berlin etc.: Springer-Verlag, 1984, 149-303.

81. Kuo, H.-H., Gaussian measures in Banach spaces, Lecture Notes in Math., Berlin etc.: Springer-Verlag, 1975.

82. Kuratovski, K., Topology. v.I. New York etc.: Academic Press, 1966.

83. Kurzhauskii, A.B., Control and observation under uncertainity conditions, Moscow: Nauka, 1977 (in Russian).

84. Kushner, H.J., Probability methods for approximations in stochastic control and for elliptic equations. New York etc.: Academic press, 1977.

85. Kusuoka, S., Strook, D.W., Stochastics, 1984, 12, 83-142.

86. Kwakernaak H., Sivan, R., Linear optimal control systems. New York etc.: Wiley-Interscience, 1972.

87. Ladyzhenskaya, O.A., Solonnikov, V.A., Uraltseva, N.N., Linear and quasi-linear equations of parabolic type, Providence, R.I.: 1968.

88. Lévy, P., Processus stochastiques et mouvement Brownien, Paris: Gauthier-Villars, 1965.

89. Lions, J.-L., Magenes, E., Problemes aux limites non homogénes et applications, v.1., Paris: Dunod, 1968.

90. Liptser, R.S., Siryayev, A.N., Statistics of random processes., New York etc.: Springer-Verlag, I -1977, II - 1978.

91. Loéve, M., Probability theory., Princeton etc.: D. Van Nostrand Company Inc., 1960.

92. Mahno, S.Ya., In: Intern. Symp. on Stochastic Differential Equations: Abstr. of Commun., Vilnius, 1978, 73-77.

93. Malliavin, P., In: Proc. Intern. Symp. on Stochastic Differential Equations Kyoto, 1976, Tokyo (ed. by K. Ito), Tokio: Kimokuniya, 1978, 195-265.

94. Malliavin, P., In: Stochastic Analysis (Ed. by A. Friedman and M. Pinsky) New York etc.: Academic Press, 1978, 199-214.

95. Margulis, L.G., Rozovskii, B.L., Uspekhi matematicheskikh nauk, 1978, 33, no. 2, 197, (in Russian).

96. Margulis, L.G., In: Markovian random processes and applications, Saratov: Saratov State Univ. Publ., 1980, 1, 50-63 (in Russian).

97. Métivier, M., Semimartingales, a course on stochastic processes. Berlin etc.: Walter de Gruyter, 1982.

98. Métivier, M., Pellaumoil S. Stochastic integration. New York etc.: Academic Press, 1980.

99. Meyer, P.A., In: Séminaire de Prob. XI., Lect. Notes Math., 581, Berlin etc.: Springer-Verlag,1977, 446-463.

100. Meyer, P.A., Probability and potentials, Waltham Mass.: Blaisdel Publ. Co., 1966.

101. Michel, D., J. Funct. Anal., 1981, 40, no. 1, 8-36.

102. Michel, D., In: Proceedings of Taniguchi Intern. Symp. on Stochastic Analysis. Katata and Kioto 1982 (Ed. by K. Ito), Amsterdam etc.: North-Holland, 1984, p. 387-408.

103. Nikolskii, S.M., Approximation of functions of several variables and imbedding theorems, M.: Nauka, Berlin etc.: Springer-Verlag, 1975.

104. Oleinik, O.A., Radkevich, E.V., Second order equations with nonnegative characteristic form, Providence, R.I.: American Mathematical Society, 1973.

105. Pardoux, E., Equations aux dérivees partielles stochastiques non linéaries monotones. Etude de solutions fortes de type Ito: Thes. P., 1975.

106. Pardoux, E., Integrales stochastiques hilbertiennes, Univ. Paris-Dauphine, Cahiers de math. de la decis. 1976, no. 7617.

107. Pardoux, E., Lecture Notes Math., 636, Berlin etc.: Springer-Verlag, 1977, 163-188.

108. Pardoux, E., Stochastics, 1979, 3, 127-167.

109. Pardoux, E., Stochastic, 1982, 6, 193-231.

110. Riesz, F., Sz., Nagy, B., Leçons d'analyse fonctionelle, Budapest: Akadémiai Kiado, 1972.

111. Rozanov, Iu.A., Markov random fields, New York etc.: Springer-Verlag, 1982.

112. Rozovskii, B.L., Ph.D. These, Moscow State Univ., 1972 (in Russian).

113. Rozovskii, B.L., Moscow University Mathematics Bulletin, 1973, 28, no. 1, 22-26.

114. Rozovskii, B.L., Shirjaev, A.N., In: Suppl. to Prepr. of IFAC Symp. on Stochastic control. Budapest: 1974, 59-61.

115. Rozovskii, B.L., Math. USSR Sbornik, 1975, 25, 295-322.

116. Rozovskii, B.L., Proceedings of the Second International Vilnius Conference in Probability and Mathematical Statistics, Vilnius: 1977, 3, 196-197.

117. Rozovskii, B.L., Teor. Prob. Appl., 1980, 25, no. 1, 147-151.

118. Rozovskii, B.L., In: XIV All Union School in Probability and Statistics Bakuriani, 1980: Proceedings. Tbilisi: Mecnieraba, 1980, 26-28 (in Russian).

119. Rozovskii, B.L., Shimizu, A., Nagoya Math. J., 1981, v. 84, 195-208.

120. Rozovskii, B.L., Backward diffusion. In: Proceedings of the Third International Vilnius Conference in Probability and Mathematical Statistics, Vilnius: 1981, v.3, 291-292.

121. Rytov, S.M., Kravtsov, Yu.A., Tatarskij, V.I., Introduction to statistical radiophysics. Random fields. Part II, Moscow, 1978 (in Russian).

122. Shimizu, A., Nagoya Math. J., 1977, 66, no. 10, 23-36.

123. Shimizu, A., Nagoya Math. J., 1979, 68, no. 1, 37-41.

124. Shigekawa, I., J. Math. Kyoto Univ., 1980, 20, 263-289.

125. Shilov, G.E., Generalized functions and partial differential equations, New York: Gordon and Breach, 1968.

126. Shubin, M.A., Pseudodifferential operators and spectral theory, Berlin etc.: Springer-Verlag, 1987.

127. Simon, B., The $P(\varphi)_2$ Euclidian (quantum) Field Theory, Princeton N.J.: Princeton Univ. Press, 1974.

128. Skorohod, A.V., Studies in the theory of random processes, Reading, Mass.: Addison-Wesley Pub. Co., 1965.

129. Skorohod, A.V., Random linear operators, Dordrecht etc.: D. Reidel Pub. Co., 1984.

130. Skorohod, A.V., Russian Math. Surveys, 1982, 37:6, 177-204.

131. Sobolev, S.L., Applications of functional analysis in mathematical physics, Providence, R.I.: American Mathematical Society, 1963.

132. Stroock, D.W, Varadhan S.R.S., Multidimensional diffusion processes. New York etc.: Springer-Verlag, 1979.

133. Stroock, D.W., Lecture Notes in Math., Berlin etc.: Springer-Verlag, 1981, 851, 394-432.

134. Stroock, D.W., Math. System Theory, 1981, 14, 25-65, 1981, v. 14, 141-171.

135. Stroock, D.W., J. Functional Anal., 1981, 44, 212-257.

136. Stroock, D.W., Lecture Notes in Math., 976, Berlin etc.: Springer-Verlag 1983, 261-282.

137. Ventcel, A.D., Teor. Prob. Appl. 1965, 10, no. 2, 357-361.

138. Verentennikov, A.Yu., Russian Math. Surveys, 1983, 38, 127-140.

139. Veretennikov, A.Yu., Math. USSR Izvestiya, 1985, 25, no. 3, 455-473.

140. Veršik, A.M., Ladyzenskaja, O.A., Soviet Math. Dokl. 1976, 17, no. 1, 23-25.

141. Viõt, M., Solutions faibles d'équations aux dérivées partielles stochastiques non linéaires: Thes. doct. sci. Univ. Pierre et Marie Curie. P., 1976.

142. Višik, M.I., Komeč, A.I., Soviet Math. Dokl., 1981, 23, no. 2, 444-447.

143. Višik, M.I., Komeč, A.I., In: Proceedings of Petrovski seminar, Moscow: Moscow State Univ. Pub., 1982, 8, 86-119 (in Russian).

144. Vishik, M.I., Fursikov, A.V., Mathematical problems of statistical hydromechanics, Dordrecht etc.: Kluwer Academic Publ., 1988.

145. Wiener, N., J. Math. Phys., 1923, 2, 131-174.

146. Wiener, N., Extrapolation, interpolation and smoothing of stationary time series. New York: J. Wiley and Sons, 1949.

147. Yosida, K., Functional Analysis, Berlin etc.: Springer-Verlag, 1965.

148. Zakai, M., Z. Wahrsch., 1969, 11, 230-243.

149. Zakai, M., Acta Applicandae Math., 1985, 3, 175-207.

INDEX

9 780792 300373